T0309447

Random Matrices
and Random Partitions

Normal Convergence

World Scientific Series on Probability Theory and Its Applications

Series Editors: Zenghu Li *(Beijing Normal University, China)*
Yimin Xiao *(Michigan State University, USA)*

Vol. 1   Random Matrices and Random Partitions: Normal Convergence
by Zhonggen Su (Zhejiang University, China)

World Scientific Series on
**Probability Theory and
Its Applications**

Volume 1

Random Matrices and Random Partitions

Normal Convergence

Zhonggen Su

Zhejiang University, China

World Scientific

NEW JERSEY · LONDON · SINGAPORE · BEIJING · SHANGHAI · HONG KONG · TAIPEI · CHENNAI

Published by

World Scientific Publishing Co. Pte. Ltd.
5 Toh Tuck Link, Singapore 596224
USA office: 27 Warren Street, Suite 401-402, Hackensack, NJ 07601
UK office: 57 Shelton Street, Covent Garden, London WC2H 9HE

Library of Congress Cataloging-in-Publication Data
Su, Zhonggen.
   Random matrices and random partitions normal convergence / by Zhonggen Su (Zhejiang University, China).
       pages cm. -- (World scientific series on probability theory and its applications ; volume 1)
   Includes bibliographical references and index.
   ISBN 978-9814612227 (hardcover : alk. paper)
   1. Random matrices. 2. Probabilities. I. Title.
   QA273.43.S89 2015
   519.2'3--dc23
                      2015004842

British Library Cataloguing-in-Publication Data
A catalogue record for this book is available from the British Library.

Printed in Singapore

To Yanping and Wanning

Preface

This book is intended to provide an introduction to remarkable probability limit theorems in random matrices and random partitions, which look rather different at a glance but have many surprising similarities from a probabilistic viewpoint.

Both random matrices and random partitions play a ubiquitous role in mathematics and its applications. There have been a great deal of research activities around them, and an enormous exciting advancement had been seen in the last three decades. A couple of excellent and big books have come out in recent years. However, the work on these two objects are so rich and colourful in theoretic results, practical applications and research techniques. No one book is able to cover all existing materials. Needless to say, these are rapidly developing and ever-green research fields. Only recently, a number of new interesting works emerged in literature. For instance, based on Johansson's work on deformed Gaussian unitary ensembles, two groups led respectively by Erdös-Yau and Tao-Vu successfully solved, around 2010, the long-standing conjecture of Dyson-Gaudin-Mehta-Wigner's bulk universality in random matrices by developing new techniques like the comparison principles and rigidity properties. Another example is that with the help of concepts of determinantal point processes coined by Borodin and Olshanski, around 2000, in the study of symmetric groups and random partitions, a big breakthrough has been made in understanding universality properties of random growth processes. Each of them is worthy of a new book.

This book is mainly concerned with normal convergence, namely central limit theorems, of various statistics from random matrices and random partitions as the model size tends to infinity. For the sake of writing and learning, we shall only focus on the simplest models among which are circular unitary ensemble, Gaussian unitary ensemble, random uniform partitions and random Plancherel partitions. As a matter of fact, many of the results addressed in this book are found valid for more general models. This book consists of three parts as follows.

We shall first give a brief survey on normal convergence in Chapter 1. It includes the well-known laws of large numbers and central limit theorems for independent identically distributed random variables and a few methods widely used in dealing with normal convergence. In fact, the central limit theorems are arguably regarded as one of the most important universality principles in describing laws of random phenomena. Most of the materials can be found in any standard probability theory at graduate level. Because neither the eigenvalues of a random matrix with all entries independent nor the parts of random partitions are independent of each other, we need new tools to treat statistics of dependent random variables. Taking this into account, we shall simply review the central limit theorems for martingale difference sequences and Markov chains. Besides, we shall review some basic concepts and properties of convergence of random processes. The statistic of interest is sometimes a functional of certain random process in the study of random matrices and random partitions. We will be able to make use of functional central limit theorems if the random processes under consideration is weakly convergent. Even under the stochastic equicontinuity condition, a slightly weaker condition than uniform tightness, the Gikhmann-Skorohod theorem can be used to guarantee convergence in distribution for a wide class of integral functionals.

In Chapters 2 and 3 we shall treat circular unitary ensemble and Gaussian unitary ensemble respectively. A common feature is that there exists an explicit joint probability density function for eigenvalues of each matrix model. This is a classic result due to Weyl as early as the 1930s. Such an explicit formula is our starting point and this makes delicate analysis possible. Our focus is upon the second-order fluctuation, namely asymptotic distribution of a certain class of linear functional statistics of eigenvalues. Under some smooth conditions, a linear eigenvalue statistic satisfies the central limit theorem without normalizing constant $\sqrt{n}$, which appears in classic Lévy-Feller central limit theorem for independent identically distributed random variables. On the other hand, either indicator function or logarithm function does not satisfy the so-called smooth condition. It turns out that the number of eigenvalues in an interval and the logarithm of characteristic polynomials do still satisfy the central limit theorem after suitably normalized by $\sqrt{\log n}$. The $\log n$-phenomena is worthy of more attention since it will also appear in the study of other similar models. In addition to circular and Gaussian unitary ensembles, we shall consider their extensions like circular β matrices and Hermite β matrices where $\beta > 0$ is a model parameter. These models were introduced and studied at length by Dyson in the early 1960s to investigate energy level behaviors in complex dynamic systems. A remarkable contribution at this direction is that there is a five (resp. three) diagonal sparse matrix model representing circular β ensemble (resp. Hermite β ensemble).

In Chapters 4 and 5 we shall deal with random uniform partitions and random Plancherel partitions. The study of integer partitions dates back to Euler as early as the 1750s, who laid the foundation of partition theory by determining the number of all distinct partitions of a natural number. We will naturally produce a probability space by assigning a probability to each partition of a natural number. Uniform measure and Plancherel measure are two best-studied objects. Young diagram and Young tableau are effective geometric representation in analyzing algebraic, combinatorial and probabilistic properties of a partition. Particularly interesting, there exists a nonrandom limit shape (curve) for suitably scaled Young diagrams under both uniform and Plancherel measure. This is a kind of weak law of large numbers from the probabilistic viewpoint. To proceed, we shall further investigate the second-order fluctuation of a random Young diagram around its limit shape. We need to treat separately three different cases: at the edge, in the bulk and integrated. It is remarkable that Gumbel law, normal law and Tracy-Widom law can be simultaneously found in the study of random integer partitions. A basic strategy of analysis is to construct a larger probability space (grand ensemble) and to use the conditioning argument. Through enlarging probability space, we luckily produce a family of independent geometric random variables and a family of determinantal point processes respectively. Then a lot of well-known techniques and results are applicable.

Random matrices and random partitions are at the interface of many science branches and they are fast-growing research fields. It is a formidable and confusing task for a new learner to access the research literature, to acquaint with terminologies, to understand theorems and techniques. Throughout the book, I try to state and prove each theorem using language and ways of reasoning from standard probability theory. I hope it will be found suitable for graduate students in mathematics or related sciences who master probability theory at graduate level and those with interest in these fields. The choice of results and references is to a large extent subjective and determined by my personal point of view and taste of research. The references at the end of the book are far from exhaustive and in fact are rather limited. There is no claim for completeness.

This book started as a lecture note used in seminars on random matrices and random partitions for graduate students in the Zhejiang University over these years. I would like to thank all participants for their attendance and comments. This book is a by-product of my research project. I am grateful to the National Science Foundation of China and Zhejiang Province for their generous support in the past ten years. I also take this opportunity to express a particular gratitude to my teachers, past and present, for introducing me to the joy of mathematics. Last, but not least, I wish to thank deeply my family for their kindness and love which is indispensable in completing this project.

I apologize for all the omissions and errors, and invite the readers to report any remarks, mistakes and misprints.

Zhonggen Su
Hangzhou
December 2014

Contents

Chapter 1

Normal Convergence

1.1 Classical central limit theorems

Throughout the book, unless otherwise specified, we assume that $(\Omega, \mathcal{A}, P)$ is a large enough probability space to support all random variables of study. E will denote mathematical expectation with respect to P.

Let us begin with Bernoulli's law, which is widely recognized as the first mathematical theorem in the history of probability theory. In modern terminology, the Bernoulli law reads as follows. Assume that $\xi_n, n \geq 1$ is a sequence of independent and identically distributed (i.i.d.) random variables, $P(\xi_n = 1) = p$ and $P(\xi_n = 0) = 1 - p$, where $0 < p < 1$. Denote $S_n = \sum_{k=1}^{n} \xi_k$. Then we have

$$\frac{S_n}{n} \xrightarrow{P} p, \quad n \to \infty. \tag{1.1}$$

In other words, for any $\varepsilon > 0$,

$$P\left(\left|\frac{S_n}{n} - p\right| > \varepsilon\right) \to 0, \quad n \to \infty.$$

It is this law that first provide a mathematically rigorous interpretation about the meaning of probability p that an event A occurs in a random experiment. To get a feeling of the true value p (unknown), what we need to do is to repeat independently a trial n times (n large enough) and to count the number of A occurring. According to the law, the larger n is, the higher the precision is.

Having the Bernoulli law, it is natural to ask how accurate the frequency S_n/n can approximate the probability p, how many times one should repeat the trial to attain the specified precision, that is, how big n should be.

With this problem in mind, De Moivre considered the case $p = 1/2$ and

1

proved the following statement:

$$P\left(a \le \frac{S_n - \frac{n}{2}}{\frac{1}{2}\sqrt{n}} \le b\right) \approx \frac{1}{\sqrt{2\pi}} \int_a^b e^{-x^2/2} dx. \qquad (1.2)$$

Later on, Laplace further extended the work of De Moivre to the case $p \ne 1/2$ to obtain

$$P\left(a \le \frac{S_n - np}{\sqrt{np(1-p)}} \le b\right) \approx \frac{1}{\sqrt{2\pi}} \int_a^b e^{-x^2/2} dx. \qquad (1.3)$$

Formulas (1.2) and (1.3) are now known as De Moivre-Laplace central limit theorem (CLT).

Note $ES_n = np$, $Var(S_n) = np(1-p)$. So $(S_n - np)/\sqrt{np(1-p)}$ is a normalized random variable with mean zero and variance one. Denote $\phi(x) = e^{-x^2/2}/\sqrt{2\pi}$, $x \in \mathbb{R}$. This is a very nice function from the viewpoint of function analysis. It is sometimes called bell curve since its graph looks like a bell, as shown in Figure 1.1.

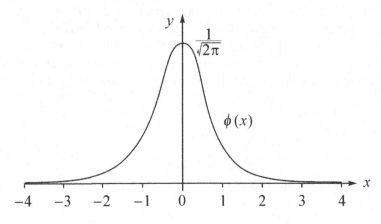

Fig. 1.1   Bell curve

The Bernoulli law and De Moivre-Laplace CLT have become an indispensable part of our modern daily life. See Billingsley (1999a, b), Chow (2003), Chung (2000), Durrett (2010) and Fischer (2011) for a history of the central limit theorem and the link to modern probability theory. But what is the proof? Any trick? Let us turn to De Moivre's original proof of (1.2). To control the left hand side of (1.2), De Moivre used the binomial formula

$$P(S_n = k) = \binom{n}{k} \frac{1}{2^n}$$

and invented together with Stirling the well-known Stirling formula (it actually should be called De Moivre-Stirling formula)

$$n! = n^n e^{-n} \sqrt{2\pi n}(1 + o(1)).$$

Setting $k = n/2 + \sqrt{n}x_k/2$, where $a \leq x_k \leq b$, we have

$$P\left(\frac{S_n - \frac{n}{2}}{\frac{1}{2}\sqrt{n}} = x_k\right) = \frac{1}{2^n} \cdot \frac{n^n e^{-n}\sqrt{2\pi n}(1 + o(1))}{k^k e^{-k}(n-k)^{n-k}e^{-(n-k)}\sqrt{2\pi k}\sqrt{2\pi(n-k)}}$$

$$= \frac{1}{\sqrt{2\pi n}}e^{-x_k^2/2}(1 + o(1)).$$

Taking sum over k yields the integral of the right hand side of (1.2).

Given a random variable X, denote its distribution function $F_X(x)$ under P. Let X, X_n, $n \geq 1$ be a sequence of random variables. If for each continuity point x of F_X,

$$F_{X_n}(x) \to F_X(x), \quad n \to \infty,$$

then we say X_n converges in distribution to X, and simply write $X_n \overset{d}{\longrightarrow} X$. In this terminology, (1.3) is written as

$$\frac{S_n - np}{\sqrt{np(1-p)}} \overset{d}{\longrightarrow} N(0,1), \quad n \to \infty,$$

where $N(0,1)$ stands for a standard normal random variable.

As the reader may notice, the Bernoulli law only deals with frequency and probability, i.e., Bernoulli random variables. However, in practice people are faced with a lot of general random variables. For instance, measure length of a metal rod. Its length, μ, is intrinsic and unknown. How do we get to know the value of μ? Each measurement is only a realization of μ. Suppose that we measure repeatedly the metal rod n times and record the observed values ξ_1, ξ_2, $\cdots$, ξ_n. It is believed that $\sum_{k=1}^n \xi_k/n$ give us a good feeling of how long the rod is. It turns out that a claim similar to the Bernoulli law is also valid for general cases. Precisely speaking, assume that ξ is a random variable with mean μ. ξ_n, $n \geq 1$ is a sequence of i.i.d. copy of ξ. Let $S_n = \sum_{k=1}^n \xi_k$. Then

$$\frac{S_n}{n} \overset{P}{\longrightarrow} \mu, \quad n \to \infty. \tag{1.4}$$

This is called the Khinchine law of large numbers. It is as important as the Bernoulli law. As a matter of fact, it provides a solid theoretic support for a great deal of activity in daily life and scientific research.

The proof of (1.4) is completely different from that of (1.1) since we do not know the exact distribution of ξ_k. To prove (1.4), we need to invoke the following Chebyshev inequality. If X is a random variable with finite mean μ and variance σ^2, then for any positive $x > 0$

$$P(|X - \mu| > x) \leq \frac{\sigma^2}{x^2}.$$

In general, we have

$$P(X > x) \leq \frac{Ef(X)}{f(x)},$$

where $f : \mathbb{R} \mapsto \mathbb{R}$ is a nonnegative nondecreasing function. We remark that the Chebyshev inequalities have played a fundamental role in proving limit theorems like the law of large numbers.

Having (1.4), we next naturally wonder what the second order fluctuation is of S_n/n around μ? In other words, is there a normalizing constant $a_n \to \infty$ such that $a_n(S_n - n\mu)/n$ converges in distribution to a certain random variable? What is the distribution of the limit variable? To attack these problems, we need to develop new tools and techniques since the De Moivre argument using binomial distribution is no longer applicable.

Given a random variable X with distribution function F_X, define for every $t \in \mathbb{R}$,

$$\psi_X(t) = Ee^{\mathrm{i}tX}$$
$$= \int_{\mathbb{R}} e^{\mathrm{i}tx} dF_X(x).$$

Call $\psi_X(t)$ the characteristic function of X. This is a Fourier transform of $F_X(x)$. In particular, if X has a probability density function $p_X(x)$, then

$$\psi_X(t) = \int_{\mathbb{R}} e^{\mathrm{i}tx} p_X(x) dx;$$

while if X takes only finitely or countably many values, $P(X = x_k) = p_k$, $k \geq 1$, then

$$\psi_X(t) = \sum_{k=1}^{\infty} e^{\mathrm{i}tx_k} p_k.$$

Note the characteristic function of any random variable is always well defined no matter whether its expectation exists.

Example 1.1. (i) If X is a normal random variable with mean μ and variance σ^2, then $\psi_X(t) = e^{\mathrm{i}\mu t - \sigma^2 t^2/2}$;
(ii) If X is a Poisson random variable with parameter λ, then $\psi_X(t) = e^{\lambda(e^{\mathrm{i}t}-1)}$;
(iii) If X is a standard Cauchy random variable, then $\psi_X(t) = e^{-|t|}$.

Some basic properties are listed below.

(i) $\psi_X(0) = 1$, $|\psi_X(t)| \leq 1$ for any $t \in \mathbb{R}$.

(ii) $\psi_X(t)$ is uniformly continuous in any finite closed interval on $\mathbb{R}$.

(iii) $\psi_X(t)$ is nonnegative definite.

According to Bochner's theorem, if any function satisfying (i), (ii) and (iii) above must be a characteristic function of a random variable.

(iv) $\overline{\psi_X(t)} = \psi_X(-t)$ for any $t \in \mathbb{R}$.

(v) If $E|X|^k < \infty$, then $\psi_X(t)$ is k times differentiable, and

$$\psi_X^{(k)}(0) = \mathbf{i}^k EX^k. \tag{1.5}$$

Hence we have the Taylor expansion at zero: for any $k \geq 1$

$$\psi_X(t) = \sum_{l=0}^{k} \frac{\mathbf{i}^l m_l}{l!} t^l + o(|t|^k) \tag{1.6}$$

and

$$\psi_X(t) = \sum_{l=0}^{k-1} \frac{\mathbf{i}^l m_l}{l!} t^l + \beta_k \theta_k \frac{|t|^k}{k!},$$

where $m_l := m_l(X) = EX^l$, $\beta_k = E|X|^k$, $|\theta_k| \leq 1$.

(vi) If X and Y are independent random variables, then

$$\psi_{X+Y}(t) = \psi_X(t)\psi_Y(t).$$

Obviously, this product formula can be extended to any finitely many independent random variables.

(vii) The distribution function of a random variable is uniquely determined by its characteristic function. Specifically speaking, we have the following inversion formula: for any F_X-continuity points x_1 and x_2

$$F_X(x_2) - F_X(x_1) = \lim_{T \to \infty} \frac{1}{2\pi} \int_{-T}^{T} \frac{e^{-\mathbf{i}tx_1} - e^{-\mathbf{i}tx_2}}{\mathbf{i}t} \psi_X(t) dt.$$

In particular, if $\psi_X(t)$ is absolutely integrable, then X has density function

$$p_X(x) = \frac{1}{2\pi} \int_{-\infty}^{\infty} e^{-\mathbf{i}tx} \psi_X(t) dt.$$

On the other hand, if $\psi_X(t) = \sum_{k=1}^{\infty} a_k e^{\mathbf{i}tx_k}$ with $a_k > 0$ and $\sum_{k=1}^{\infty} a_k = 1$, then X is a discrete random variable, and

$$P(X = x_k) = a_k, \quad k = 1, 2, \cdots.$$

In addition to basic properties above, we have the following Lévy continuity theorem, which will play an important role in the study of convergence in distribution.

Theorem 1.1. *(i)* $X_n \xrightarrow{d} X$ *if and only if* $\psi_{X_n}(t) \to \psi_X(t)$ *for any* $t \in \mathbb{R}$.
(ii) If ψ_{X_n} *converges pointwise to a function* ψ, *and* ψ *is continuous at* $t = 0$, *then* $\psi(t)$ *is a characteristic function of some random variable, say* X, *and so* $X_n \xrightarrow{d} X$.

Having the preceding preparation, we can easily obtain the following CLT for sums of independent random variables.

Theorem 1.2. *Let* $\xi, \xi_n, n \geq 1$ *be a sequence of i.i.d. random variables with mean* μ *and variance* σ^2. *Let* $S_n = \sum_{k=1}^{n} \xi_k$. *Then*

$$\frac{S_n - n\mu}{\sigma\sqrt{n}} \xrightarrow{d} N(0,1), \quad n \to \infty. \tag{1.7}$$

This is often referred to as Feller-Lévy CLT. Its proof is purely analytic. For sake of comparison, we quickly review the proof.

Proof. Without loss of generality, we may and do assume $\mu = 0$, $\sigma^2 = 1$. By hypothesis it follows

$$\psi_{S_n/\sqrt{n}}(t) = \left(\psi_\xi\left(\frac{t}{\sqrt{n}}\right)\right)^n.$$

Also, using (1.6) yields

$$\psi_\xi\left(\frac{t}{\sqrt{n}}\right) = 1 - \frac{t^2}{2n} + O\left(\frac{1}{n}\right)$$

for each t. Hence we have

$$\psi_{S_n/\sqrt{n}}(t) = \left(1 - \frac{t^2}{2n} + O\left(\frac{1}{n}\right)\right)^n$$
$$\to e^{-t^2/2}.$$

By Theorem 1.1 and (i) of Example 1.1, we conclude the desired (1.7). $\square$

Remark 1.1. Under the assumption that $\xi_n, n \geq 1$ are i.i.d. random variables, the condition $\sigma^2 < \infty$ is also necessary for (1.7) to hold. See Chapter 10 of Ledoux and Talagrand (2011) for a proof.

In many applications, it is restrictive to require that $\xi_n, n \geq 1$ are i.i.d. random variables. Therefore we need to extend the Feller-Lévy CLT to the non-i.i.d. cases. In fact, there have been a great deal of work toward this direction. For the sake of reference, we will review below some most commonly used CLT, including the Lindeberg-Feller CLT for independent not necessarily identically distributed random variables, the martingale CLT, the CLT for ergodic stationary Markov chains.

Assume that ξ_n, $n \geq 1$ is a sequence of independent random variables, $E\xi_n = \mu_n$, $Var(\xi_n) = \sigma_n^2 < \infty$. Let $S_n = \sum_{k=1}^n \xi_k$, $B_n = \sum_{k=1}^n \sigma_k^2$. Assume further that $B_n \to \infty$. Introduce the following two conditions. Feller condition:

$$\frac{1}{B_n} \max_{1 \leq k \leq n} \sigma_k^2 \to 0.$$

Lindeberg condition: for any $\varepsilon > 0$

$$\frac{1}{B_n} \sum_{k=1}^n E(\xi_k - \mu_k)^2 \mathbf{1}_{(|\xi_k - \mu_k| \geq \varepsilon \sqrt{B_n})} \to 0.$$

Obviously, Feller condition is a consequence of Lindeberg condition. Moreover, we have the following Lindeberg-Feller CLT.

Theorem 1.3. *Under Feller condition, the ξ_n satisfies the CLT, that is*

$$\frac{S_n - \sum_{k=1}^n \mu_k}{\sqrt{B_n}} \xrightarrow{d} N(0,1), \quad n \to \infty$$

if and only if Lindeberg condition holds.

It is easy to see that if there is a $\delta > 0$ such that

$$\frac{1}{B_n^{1+\delta/2}} \sum_{k=1}^n E|\xi_k|^{2+\delta} \to 0, \tag{1.8}$$

then Lindeberg condition is satisfied. The condition (1.8) is sometimes called Lyapunov condition.

Corollary 1.1. *Assume that ξ_n, $n \geq 1$ is a sequence of independent Bernoulli random variables, $P(\xi_n = 1) = p_n$, $P(\xi_n = 0) = 1 - p_n$. If $\sum_{n=1}^\infty p_n(1 - p_n) = \infty$, then*

$$\frac{\sum_{k=1}^n (\xi_k - p_k)}{\sqrt{\sum_{k=1}^n p_k(1 - p_k)}} \xrightarrow{d} N(0,1), \quad n \to \infty. \tag{1.9}$$

The Lindeberg-Feller theorem has a wide range of applications. In particular, it implies that the normal law exists universally in nature and human society. For instance, the error in measurement might be caused by a large number of independent factors. Each factor contributes only a very small part, but none plays a significant role. Then the total error obeys approximately a normal law.

Next turn to the martingale CLT. First, recall some notions and basic properties of martingales. Assume that $\mathcal{A}_n$, $n \geq 0$ is a sequence of non-decreasing sub-sigma fields of $\mathcal{A}$. Let X_n, $n \geq 0$ be a sequence of random variable with $X_n \in \mathcal{A}_n$ and $E|X_n| < \infty$. If for each $n \geq 1$

$$E(X_n|\mathcal{A}_{n-1}) = X_{n-1} \quad \text{a.e.}$$

we call $\{X_n, \mathcal{A}_n, n \geq 0\}$ a martingale. If $\mathcal{A}_n = \sigma\{X_0, X_1, \cdots, X_n\}$, we simply write $\{X_n, n \geq 0\}$ a martingale. If $\{X_n, \mathcal{A}_n, n \geq 0\}$ is a martingale, setting $d_n := X_n - X_{n-1}$, then $\{d_n, \mathcal{A}_n, n \geq 0\}$ forms a martingale difference sequence, namely

$$E(d_n|\mathcal{A}_{n-1}) = 0 \quad \text{a.e.}$$

Conversely, given a martingale difference sequence $\{d_n, \mathcal{A}_n, n \geq 0\}$, we can form a martingale $\{X_n, \mathcal{A}_n, n \geq 0\}$ by

$$X_n = \sum_{k=1}^{n} d_k.$$

Example 1.2. (i) Assume that $\xi_n, n \geq 1$ is a sequence of independent random variables with mean zero. Let $S_0 = 0$, $S_n = \sum_{k=1}^{n} \xi_k$, $\mathcal{A}_0 = \{\emptyset, \Omega\}$, $\mathcal{A}_n = \sigma\{\xi_1, \cdots, \xi_n\}$, $n \geq 1$. Then $\{S_n, \mathcal{A}_n, n \geq 0\}$ is a martingale.
(ii) Assume that X is a random variable with finite expectation. Let $\mathcal{A}_n$, $n \geq 0$ is a sequence of non-decreasing sub-sigma fields of $\mathcal{A}$. Let

$$X_n = E(X|\mathcal{A}_n) \quad \text{a.e.}$$

Then $\{X_n, \mathcal{A}_n, n \geq 0\}$ is a martingale.

We now state a martingale CLT due to Brown (1971).

Theorem 1.4. *Assume that* $\{d_n, \mathcal{A}_n, n \geq 0\}$ *is a martingale difference sequence, set* $S_n = \sum_{k=1}^{n} d_k$, $B_n = \sum_{k=1}^{n} E d_k^2$. *If the following three conditions*
(i)

$$\frac{1}{B_n} \max_{1 \leq k \leq n} E\big(d_k^2|\mathcal{A}_{k-1}\big) \xrightarrow{P} 0, \quad n \to \infty$$

(ii)

$$\frac{1}{B_n} \sum_{k=1}^{n} E\big(d_k^2|\mathcal{A}_{k-1}\big) \xrightarrow{P} 1, \quad n \to \infty$$

(iii) for any $\varepsilon > 0$

$$\frac{1}{B_n} \sum_{k=1}^{n} E\big(d_k^2 \mathbf{1}_{(|d_k| \geq \varepsilon \sqrt{B_n})}|\mathcal{A}_{k-1}\big) \xrightarrow{P} 0, \quad n \to \infty$$

are satisfied, then

$$\frac{S_n}{\sqrt{B_n}} \xrightarrow{d} N(0,1), \quad n \to \infty.$$

(1974) presented an improved version under the following slightly weaker conditions:

(i') there is a constant $M > 0$ such that

$$\frac{1}{B_n} E \max_{1 \le k \le n} d_k^2 \le M, \quad \forall n \ge 1;$$

(ii')

$$\frac{1}{B_n} \max_{1 \le k \le n} d_k^2 \xrightarrow{P} 0, \quad n \to \infty;$$

(iii')

$$\frac{1}{B_n} \sum_{k=1}^{n} d_k^2 \xrightarrow{P} 1, \quad n \to \infty.$$

The interested reader is referred to Hall and Heyde (1980) for many other limit theorems related to martingales.

Let $\mathcal{E}$ be a set of at most countable points. Assume that X_n, $n \ge 0$ is a random sequence with state space $\mathcal{E}$. If for any states i and j, any time $n \ge 0$,

$$\begin{aligned}
P(X_{n+1} = j | X_n = i, X_{n-1} &= i_{n-1}, \cdots, X_0 = i_0) \\
&= P(X_{n+1} = j | X_n = i) \\
&= P(X_1 = j | X_0 = i), \quad (1.10)
\end{aligned}$$

then we call X_n, $n \ge 0$ a time-homogenous Markov chain. Condition (1.10), called the Markov property, implies that the future is independent of the past given its present state.

Denote p_{ij} the transition probability:

$$p_{ij} = P(X_{n+1} = j | X_n = i).$$

The matrix $\mathbf{P} := (p_{ij})$ is called the transition matrix. It turns out that both the X_0 and the transition matrix $\mathbf{P}$ will completely determine the law of a Markov chain. Denote the n-step transition matrix $\mathbf{P}^{(n)} = (p_{ij}^{(n)})$, where $p_{ij}^{(n)} = P(X_n = j | X_0 = i)$. Then a simple chain rule manipulation shows

$$\mathbf{P}^{(n+m)} = \mathbf{P}^{(n)} \mathbf{P}^{(m)}.$$

This is the well-known Chapman-Kolmogorov equation. Moreover, $\mathbf{P}^{(n)} = \mathbf{P}^n$.

State j is accessible from state i, denoted by $i \to j$, if $p_{ij}^{(n)} > 0$ for some $n \geq 1$. State i and j communicate with each other, denoted by $i \leftrightarrow j$, if $i \to j$ and $j \to i$. A Markov chain is irreducible if any two states communicate each other.

The period d_i of a state i is the greatest common divisor of all n that satisfy $p_{ii}^{(n)} > 0$. State i is aperiodic if $d_i = 1$, and otherwise it is periodic.

Denote by τ_i the hitting time

$$\tau_i = \min\{n \geq 1; X_n = i\}.$$

A state i is transient if $P(\tau_i = \infty | X_0 = i) > 0$, and i is recurrent if $P(\tau_i < \infty | X_0 = i) = 1$. A recurrent state i is positive recurrent if

$$E(\tau_i | X_0 = i) < \infty.$$

An irreducible aperiodic positive recurrent Markov chain is called ergodic.

If a probability distribution π on $\mathcal{E}$ satisfies the following equation:

$$\pi = \pi \mathbf{P},$$

then we call π a stationary distribution. If we choose π to be an initial distribution, then the X_n is a stationary Markov chain. In addition, if for any i, j

$$\pi_i p_{ij} = \pi_j p_{ji},$$

then the X_n is reversible. In particular,

$$(X_0, X_1) \overset{d}{=} (X_1, X_0).$$

An irreducible aperiodic Markov chain is ergodic if and only if it has a stationary distribution.

Theorem 1.5. *Assume that $X_n, n \geq 0$ is an ergodic Markov chain with stationary distribution π. Assume that $f : \mathcal{E} \mapsto \mathbb{R}$ is such that $\sum_{i \in \mathcal{E}} f(i)\pi_i$ is absolutely convergent. Then*

$$\lim_{n \to \infty} \frac{1}{n} \sum_{i=0}^{n-1} f(X_i) = \sum_{i \in \mathcal{E}} f(i)\pi_i \quad a.e.$$

This is a type of law of large numbers for Markov chains. See Serfozo (2009) for a proof.

Let L_2^0 be the subspace of $L_2(\pi)$ consisting of functions $f : \mathcal{E} \mapsto \mathbb{R}$ with $E_\pi f := \sum_{i \in \mathcal{E}} f(i)\pi_i = 0$. We shall give a sufficient condition under which the linear sum $S_n(f) := \sum_{i=0}^{n-1} f(X_i)$ satisfies the CLT. To this end, we introduce the transition operator T

$$Tg(i) = \sum_{j \in \mathcal{E}} g(j)p_{ij}.$$

Trivially, $Tg(i) = E\big(g(X_1)\big|X_0 = i\big)$. Assume that there exists a function g such that

$$f = g - Tg. \tag{1.11}$$

Then it easily follows from the martingale CLT

$$\frac{S_n(f)}{\sqrt{n}} \xrightarrow{d} N\big(0, \sigma_f^2\big)$$

with limit variance

$$\sigma_f^2 = \|g\|_2^2 - \|Tg\|_2^2,$$

where $\| \cdot \|_2$ denotes the L_2-norm.

If f is such that

$$\sum_{n=0}^{\infty} T^n f \tag{1.12}$$

is convergent in $L_2(\pi)$, then the solution of the equation (1.11) do exist. Indeed, $g = \sum_{n=0}^{\infty} T^n f$ just solves the equation.

It is too restrictive to require that the series in (1.12) is L_2-convergent. An improved version is

Theorem 1.6. *Let $X_n, n \geq 0$ is an ergodic Markov chain with stationary distribution π. Assume $f \in L_2^0$ satisfies the following two conditions:*
(i)

$$T^n f \xrightarrow{L_2} 0, \quad n \to \infty;$$

(ii)

$$\sum_{n=0}^{\infty} \big(\|T^n f\|_2^2 - \|T^{n+1} f\|_2^2\big)^{1/2} < \infty.$$

Then we have

$$\frac{S_n(f)}{\sqrt{n}} \xrightarrow{d} N\big(0, \sigma_f^2\big)$$

with limit variance

$$\sigma_f^2 = \lim_{n \to \infty} \Big(\big\| \sum_{k=0}^{n} T^k f \big\|_2^2 - \big\| \sum_{k=0}^{n+1} T^k f \big\|_2^2\Big).$$

In the preceding paragraphs, we have seen that the characteristic functions is a powerful tool in proving convergence in distribution and identifying the limit distribution. It is particularly successful in the study of partial sums of independent or asymptotically independent random variables. However, it is sometimes not an easy task to compute the characteristic function of a random variable of interest. In the rest of this section and next sections we will briefly introduce other methods and techniques, among which are the moment method, the replacement trick, the Stein method and the Stieltjes transform method.

The moment method is closely related to an interesting old problem. Is the distribution of X uniquely determined by its moments? If not, what extra conditions do we require? Suppose X has finite moments of all orders. Then according to (1.5), $\psi_X^{(k)}(0) = i^k m_k$ where $m_k = EX^k$, $k \geq 0$. However, it does not necessarily follow

$$\psi_X(t) = \sum_{k=0}^{\infty} \frac{i^k m_k}{k!} t^k. \tag{1.13}$$

Example 1.3. Consider two random variables X and Y, whose probability density functions are as follows

$$p_X(x) = \begin{cases} \frac{1}{\sqrt{2\pi} x} e^{-(\log x)^2/2}, & x > 0 \\ 0, & x \leq 0 \end{cases}$$

and

$$p_Y(x) = \begin{cases} \frac{1}{\sqrt{2\pi} x} e^{-(\log x)^2/2}(1 + \sin(2\pi \log x)), & x > 0 \\ 0, & x \leq 0. \end{cases}$$

Then it is easy to check that X and Y have all moments finite and $EX^k = EY^k$ for any $k \geq 1$.

If the following Carleman condition

$$\sum_{k=1}^{\infty} m_{2k}^{-1/2k} = \infty \tag{1.14}$$

is satisfied, then (1.13) holds, and so the distribution of X is uniquely determined by its moments. A slightly stronger condition for (1.14) to hold is

$$\liminf_{k \to \infty} \frac{1}{k} m_{2k}^{1/2k} < \infty.$$

Example 1.4. (i) Assume $X \sim N(0, \sigma^2)$, then for $k \geq 1$

$$\begin{cases} m_{2k}(X) = \sigma^{2k}(2k-1)!!, \\ m_{2k+1}(X) = 0. \end{cases}$$

(ii) Assume X is a Poisson random variable with parameter λ, then for $k \geq 1$

$$EX(X-1)\cdots(X-k+1) = \lambda^k.$$

(iii) Assume X is a random variable with density function given by

$$\rho_{sc}(x) = \begin{cases} \frac{1}{2\pi}\sqrt{4-x^2}, & |x| \leq 2, \\ 0, & \text{otherwise}, \end{cases} \tag{1.15}$$

then for $k \geq 1$

$$\begin{cases} m_{2k}(X) = \frac{1}{k+1}\binom{2k}{k}, \\ m_{2k+1}(X) = 0. \end{cases}$$

(iv) Assume X is a random variable with density function given by

$$\rho_{MP}(x) = \begin{cases} \frac{1}{2\pi}\sqrt{\frac{4-x}{x}}, & 0 < x \leq 4, \\ 0, & \text{otherwise}, \end{cases} \tag{1.16}$$

then for $k \geq 1$

$$m_k(X) = \frac{1}{k+1}\binom{2k}{k}.$$

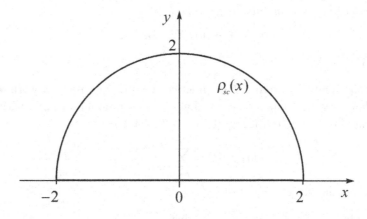

Fig. 1.2   Wigner semicircle law

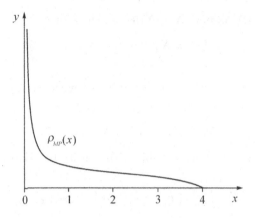

Fig. 1.3   Marchenko-Pastur law

We remark that ρ_{sc} and ρ_{MP} (see Figures 1.2 and 1.3) are often called Wigner semicircle law and Marchenko-Pastur law in random matrix literature. They are respectively the expected spectrum distributions of Wigner random matrices and sample covariance matrices in the large dimensions. It is now easy to verify that these moments satisfy the Carleman condition (1.14). Therefore normal distribution, Poisson distribution, Wigner semicircle law and Marchenko-Pastur law are all uniquely determined by their moments.

Theorem 1.7. *Let X_n, $n \geq 1$ be a sequence of random variables with all moments finite. Let X be a random variable whose law is uniquely determined by its moments. If for each $k \geq 1$*

$$m_k(X_n) \to m_k(X), \quad n \to \infty$$

then $X_n \xrightarrow{d} X$.

When applying Theorem 1.7 in practice, it is often easier to work with cumulants rather than moments. Let X be a random variable with all moments finite. Expand $\log \psi_X(t)$ at $t = 0$ as follows

$$\log f_X(t) = \sum_{k=1}^{\infty} \frac{i^k \tau_k}{k!} t^k.$$

We call τ_k the kth cumulant of X.

Example 1.5. (i) If $X \sim N(\mu, \sigma^2)$, then

$$\tau_1 = \mu, \quad \tau_2 = \sigma^2, \quad \tau_k = 0, \quad \forall k \geq 3.$$

(ii) If X is a Poisson random variable with parameter λ, then

$$\tau_k = \lambda, \quad \forall k \geq 1.$$

The cumulants possess the following nice properties. Fix a a constant.
(i) shift equivariance:

$$\tau_1(X + a) = \tau_1(X) + a;$$

(ii) shift invariance:

$$\tau_k(X + a) = \tau_k(X), \quad \forall k \geq 2;$$

(iii) homogeneity:

$$\tau_k(aX) = a^k \tau_k(X), \quad \forall k \geq 2;$$

(vi) additivity: if X and Y are independent random variables, then

$$\tau_k(X + Y) = \tau_k(X) + \tau_k(Y), \quad \forall k \geq 2;$$

(v) relations between cumulants and moments:

$$\tau_k = m_k - \sum_{l=1}^{k-1} \binom{k-1}{l-1} \tau_l m_{k-l}$$

$$= \sum_{\alpha_1,\cdots,\alpha_k} (-1)^{\alpha_1 + \cdots + \alpha_k - 1}(\alpha_1 + \cdots + \alpha_k - 1)! \prod_{l=1}^{k} \frac{k!}{\alpha_l! l^{\alpha_l}} m_l^{\alpha_l}, \quad (1.17)$$

where the summation is extended over all nonnegative integer solutions of the equation $\alpha_1 + 2\alpha_2 + \cdots + k\alpha_k = k$, and

$$m_k = \sum_{\lambda} \prod_{B \in \lambda} \tau_{|B|}$$

where λ runs through the set of all partitions of $\{1, 2, \cdots, n\}$, $B \in \lambda$ means one of the blocks into which the set if partitioned and $|B|$ is the size of the block.

Since knowledge of the moments of a random variable is interchangeable with knowledge of its cumulants, Theorem 1.7 can be reformulated as

Theorem 1.8. *Let X_n, $n \geq 1$ be a sequence of random variables with all moments finite. Let X be a random variable whose law is uniquely determined by its moments. If for each $k \geq 1$*

$$\tau_k(X_n) \to \tau_k(X), \quad n \to \infty$$

then $X_n \xrightarrow{d} X$.

This theorem is of particular value when proving asymptotic normality. Namely, if

$$\tau_1(X_n) \to 0, \quad \tau_2(X_n) \to 1$$

and for $k \geq 3$

$$\tau_k(X_n) \to 0,$$

then $X_n \xrightarrow{d} N(0,1)$.

To conclude this section, we will make a brief review about Lindeberg replacement strategy by reproving the Feller-Lévy CLT. To this end, we need an equivalent version of convergence in distribution, see Section 1.4 below.

Lemma 1.1. *Let X, $X_n, n \geq 1$ be a sequence of random variables. Then $X_n \xrightarrow{d} X$ if and only if for each bounded thrice continuously differentiable function f with $\|f^{(3)}\|_\infty < \infty$,*

$$Ef(X_n) \longrightarrow Ef(X), \quad n \to \infty.$$

Theorem 1.9. *Let ξ_n, $n \geq 1$ be a sequence of i.i.d. random variables with mean zero and variance 1 and $E|\xi_n|^3 < \infty$. Let $S_n = \sum_{k=1}^n \xi_k$. Then it follows*

$$\frac{S_n}{\sqrt{n}} \xrightarrow{d} N(0,1), \quad n \to \infty.$$

Proof. Let η_n, $n \geq 1$ be a sequence of i.i.d. normal random variables with mean zero and variance 1, and let $T_n = \sum_{k=1}^n \eta_k$. Trivially,

$$\frac{T_n}{\sqrt{n}} \sim N(0,1).$$

According to Lemma 1.1, it suffices to show that for any bounded thrice continuously differentiable function f with $\|f^{(3)}\|_\infty < \infty$,

$$Ef\left(\frac{S_n}{\sqrt{n}}\right) - Ef\left(\frac{T_n}{\sqrt{n}}\right) \longrightarrow 0 \quad n \to \infty.$$

To do this, set

$$R_{n,k} = \sum_{l=1}^{k-1} \xi_l + \sum_{l=k+1}^n \eta_l, \quad 1 \leq k \leq n.$$

Then

$$Ef\left(\frac{S_n}{\sqrt{n}}\right) - Ef\left(\frac{T_n}{\sqrt{n}}\right)$$

$$= \sum_{k=1}^{n}\left[Ef\left(\frac{1}{\sqrt{n}}(R_{n,k} + \xi_k)\right) - Ef\left(\frac{1}{\sqrt{n}}(R_{n,k} + \eta_k)\right)\right]$$

$$= \sum_{k=1}^{n}\left[Ef\left(\frac{1}{\sqrt{n}}(R_{n,k} + \xi_k)\right) - Ef\left(\frac{1}{\sqrt{n}}R_{n,k}\right)\right]$$

$$- \sum_{k=1}^{n}\left[Ef\left(\frac{1}{\sqrt{n}}(R_{n,k} + \eta_k)\right) - Ef\left(\frac{1}{\sqrt{n}}R_{n,k}\right)\right]. \tag{1.18}$$

Applying the Taylor expansion of f at $R_{n,k}/\sqrt{n}$, we have by hypothesis

$$Ef\left(\frac{1}{\sqrt{n}}(R_{n,k} + \xi_k)\right) - Ef\left(\frac{1}{\sqrt{n}}R_{n,k}\right)$$

$$= Ef'\left(\frac{1}{\sqrt{n}}R_{n,k}\right)\frac{\xi_k}{\sqrt{n}} + \frac{1}{2}Ef''\left(\frac{1}{\sqrt{n}}R_{n,k}\right)\frac{\xi_k^2}{n}$$

$$+ \frac{1}{6}Ef^{(3)}(R^*)\frac{\xi_k^3}{n^{3/2}}$$

$$= \frac{1}{2n}Ef''\left(\frac{1}{\sqrt{n}}R_{n,k}\right) + \frac{1}{6n^{3/2}}Ef^{(3)}(R^*)\xi_k^3, \tag{1.19}$$

where R^* is between $(R_{n,k} + \xi_k)/\sqrt{n}$ and $R_{n,k}/\sqrt{n}$.

Similarly, we also have

$$Ef\left(\frac{1}{\sqrt{n}}(R_{n,k} + \xi_k)\right) - Ef\left(\frac{1}{\sqrt{n}}R_{n,k}\right)$$

$$= \frac{1}{2n}Ef''\left(\frac{1}{\sqrt{n}}R_{n,k}\right) + \frac{1}{6n^{3/2}}Ef^{(3)}(R^{**})\eta_k^3, \tag{1.20}$$

where R^{**} is between $(R_{n,k} + \xi_k)/\sqrt{n}$ and $R_{n,k}/\sqrt{n}$.

Putting (1.19) and (1.20) back into (1.18) yields

$$Ef\left(\frac{S_n}{\sqrt{n}}\right) - Ef\left(\frac{T_n}{\sqrt{n}}\right)$$

$$= \sum_{k=1}^{n}\frac{1}{6n^{3/2}}Ef^{(3)}(R^*)\xi_k^3 + \frac{1}{6n^{3/2}}Ef^{(3)}(R^{**})\eta_k^3.$$

Noting $\|f^{(3)}\|_\infty < \infty$ and $E|\xi_k|^3 < \infty$ and $E|\eta_k|^3 < \infty$, we obtain

$$Ef\left(\frac{S_n}{\sqrt{n}}\right) - Ef\left(\frac{T_n}{\sqrt{n}}\right) = O(n^{-1/2}).$$

The assertion is now concluded. $\qquad\qquad\qquad\qquad\qquad\qquad\qquad\square$

Proof of Theorem 1.2 To apply Theorem 1.9, we need to use the truncation technique. For any constant $a > 0$, define

$$\bar{\xi}_k(a) = \xi_k \mathbf{1}_{(|\xi_k| \le a)}, \quad 1 \le k \le n.$$

Obviously, $\bar{\xi}_k(a)$, $1 \le k \le n$ are i.i.d. bounded random variables. Let $\mu_k(a) = E\bar{\xi}_k(a)$ and $\bar{\sigma}_k^2(a) = Var(\bar{\xi}_k(a))$ for $1 \le k \le n$. So according to Theorem 1.9, it follows

$$\frac{\sum_{k=1}^n (\bar{\xi}_k(a) - \mu_k(a))}{\bar{\sigma}_1(a)\sqrt{n}} \xrightarrow{d} N(0,1), \quad n \to \infty.$$

Since $a > 0$ is arbitrary, then by selection principle, there is a sequence of constants $a_n > 0$ such that $a_n \to \infty$ and

$$\frac{\sum_{k=1}^n (\bar{\xi}_k(a_n) - \mu_k(a_n))}{\bar{\sigma}_1(a_n)\sqrt{n}} \xrightarrow{d} N(0,1). \tag{1.21}$$

In addition, it is easy to see

$$\mu_k(a_n) \to 0, \quad \bar{\sigma}_n(a_n) \to 1.$$

Hence by (1.21),

$$\frac{\sum_{k=1}^n (\bar{\xi}_k(a_n) - \mu_k(a_n))}{\sqrt{n}} \xrightarrow{d} N(0,1). \tag{1.22}$$

Finally, it follows from the Chebyshev inequality

$$\frac{1}{\sqrt{n}}\left(S_n - \sum_{k=1}^n (\bar{\xi}_k(a_n) - \mu_k(a_n))\right) \xrightarrow{P} 0. \tag{1.23}$$

Combining (1.22) and (1.23) together, we conclude the assertion (1.7). $\square$

Remark 1.2. The Lindeberg replacement strategy makes clear the fact that the CLT is a local phenomenon. By this we mean that the structure of the CLT does not depend on the behavior of any fixed number of the increments. Only recently was it successfully used to establish the Four Moment Comparison theorem for eigenvalues of random matrices, which in turn solves certain long-standing conjectures related to universality of eigenvalues of random matrices. See Tao and Vu (2010).

1.2 The Stein method

The Stein method was initially conceived by Stein (1970, 1986) to provide errors in the approximation by the normal distribution of the distribution of the sum of dependent random variables of a certain structure. However, the ideas presented are sufficiently abstract and powerful to be able to work well beyond that purpose, applying to approximation of more general random variables by distributions other than normal. Besides, the Stein method is a highly original technique and useful in quantifying the error in the approximation of one distribution by another in a variety of metrics. This subsection serves as a basic introduction of the fundamentals of the Stein method. The interested reader is referred to nice books and surveys, say, Chen, Goldstein and Shao (2010), Ross (2011).

A basic starting point is the following Stein equation.

Lemma 1.2. *Assume that ξ is a random variable with mean zero and variance σ^2. Then ξ is normal if and only if for every bounded continuously differentiable function f ($\|f\|_\infty$, $\|f'\|_\infty < \infty$),*

$$E\xi f(\xi) = \sigma^2 E f'(\xi). \tag{1.24}$$

Proof. Without loss of generality, assume $\sigma^2 = 1$. If $\xi \sim N(0,1)$, then by the integration by part formula, we easily obtain (1.24).

Conversely, assume (1.24) holds. Fix $z \in \mathbb{R}$ and consider the following first order ordinary differential equation

$$f'(x) - x f(x) = \mathbf{1}_{(-\infty, z]}(x) - \Phi(z), \tag{1.25}$$

where $\Phi(\cdot)$ denotes the standard normal distribution function. Then a simple argument shows that there exists a unique solution:

$$f_z(x) = \begin{cases} e^{x^2/2} \int_x^\infty e^{-t^2/2}(\Phi(z) - \mathbf{1}_{(-\infty, z]}(t))dt, \\ e^{x^2/2} \int_x^\infty e^{-t^2/2}(\Phi(z) - \mathbf{1}_{(-\infty, z]}(t))dt. \end{cases}$$

Note

$$1 - \Phi(z) \sim \frac{1}{z} e^{-z^2/2}, \quad z \to \infty.$$

It is not hard to see $\|f_z\|_\infty < \sqrt{\pi/2}$ and $\|f_z'\|_\infty \leq 2$. By hypothesis, it follows

$$P(\xi \leq z) - \Phi(z) = E f_z'(\xi) - E\xi f_z(\xi) = 0.$$

We now conclude the proof. $\square$

As an immediate corollary to the proof, we can derive the following Stein continuity theorem.

Theorem 1.10. *Assume that ξ_n, $n \geq 1$ is a sequence of random variables with mean zero and variance 1. If for every bounded continuously differentiable function f,*

$$E\xi_n f(\xi_n) - Ef'(\xi_n) \to 0, \quad n \to \infty$$

then $\xi_n \overset{d}{\longrightarrow} N(0,1)$.

Remark 1.3. The above Stein equation can be extended to a non-normal random variable. Assume that ξ has the $(q+2)$th moment finite, f is $(q+1)$ times bounded continuously differentiable, then

$$E\xi f(\xi) = \sum_{k=0}^{q} \frac{\tau_{k+1}}{k!} Ef^{(k)}(\xi) + \varepsilon_q,$$

where τ_k is the kth culumant of ξ, the remainder term admits the bound

$$\varepsilon_q \leq c_q \|f^{(q+1)}\| E|\xi|^{q+2}, \quad c_q \leq \frac{1 + (3 + 2q)^{q+2}}{(q+1)!}.$$

As the reader may notice, if we replace the indicator function $\mathbf{1}_{(-\infty,z]}$ by a smooth function in the preceding differential equation (1.25), then its solution will have a nicer regularity property. Let $\mathcal{H}$ be a family of 1-Lipschitz functions, namely

$$\mathcal{H} = \big\{h : \mathbb{R} \mapsto \mathbb{R}, \ |h(x) - h(y)| \leq |x - y|\big\}.$$

Consider the following differential equation:

$$f'(x) - xf(x) = h(x) - Eh(\xi), \tag{1.26}$$

where $\xi \sim N(0,1)$.

Lemma 1.3. *Assume $h \in \mathcal{H}$. There exists a unique solution of (1.26):*

$$f_h(x) = \begin{cases} e^{x^2/2} \int_x^\infty e^{-t^2/2}(\Phi(z) - h(t))dt, \\ e^{x^2/2} \int_x^\infty e^{-t^2/2}(\Phi(z) - h(t))dt. \end{cases}$$

Moreover, f_h satisfies the following properties

$$\|f_h\|_\infty \leq 2, \quad \|f_h'\|_\infty \leq \sqrt{\frac{\pi}{2}}, \quad \|f_h''\|_\infty \leq 2.$$

We omit the proof, which can be found in Chen, Goldstein and Shao (2010). It is easily seen that for any random variable W of interest,

$$Eh(W) - Eh(\xi) = Ef'_h(W) - EWf_h(W).$$

Given two random variables X and Y, the Wasserstein distance is defined by

$$d_W(X, Y) = \sup_{h \in \mathcal{H}} |Eh(X) - Eh(Y)|.$$

Note the Wasserstein distance is widely used in describing the distributional approximations. In particular,

$$\sup_{x \in \mathbb{R}} |P(X \le x) - P(Y \le x)| \le \sqrt{\frac{\pi}{2}} \sqrt{d_W(X, Y)}.$$

Let $\mathcal{G}$ be the family of bounded continuously differentiable functions with bounded first and second derivatives, namely

$$\mathcal{G} = \left\{ f : \mathbb{R} \mapsto \mathbb{R}, \|f\|_\infty \le 2, \|f'\|_\infty \le \sqrt{\frac{\pi}{2}}, \|f''\|_\infty \le 2 \right\}.$$

Taking Lemma 1.3 into account, we immediately get

Theorem 1.11. *Let $\xi \sim N(0, 1)$, W a random variable. Then we have*

$$d_W(W, \xi) \le \sup_{f \in \mathcal{G}} |EWf(W) - Ef'(W)|.$$

To illustrate the use of the preceding Stein method, let us take a look at the normal approximation of sums of independent random variables below.

Example 1.6. Suppose that ξ_n, $n \ge 1$ is a sequence of independent random variables with mean zero, variance 1 and $E|\xi_n|^3 < \infty$. Let $S_n = \sum_{i=1}^n \xi_i$, then

$$d_W\left(\frac{S_n}{\sqrt{n}}, \xi\right) \le \frac{4}{n^{3/2}} \sum_{i=1}^n E|\xi_i|^3. \tag{1.27}$$

Proof. Writing $W_n = S_n/\sqrt{n}$, we need only to control the supremum of $EW_nf(W_n) - Ef'(W_n)$ over $\mathcal{G}$. Set $W_{n,i} = (S_n - \xi_i)/\sqrt{n}$. Then by independence and noting $E\xi_i = 0$

$$EW_nf(W_n) = \frac{1}{\sqrt{n}} \sum_{i=1}^n E\xi_i f(W_n)$$

$$= \frac{1}{\sqrt{n}} \sum_{i=1}^n E\xi_i \big(f(W_n) - f(W_{n,i})\big). \tag{1.28}$$

Using the Taylor expansion of f at $W_{n,i}$, we have

$$f(W_n) - f(W_{n,i}) = f'(W_{n,i})\frac{\xi_i}{\sqrt{n}} + \frac{1}{2}f''(W_{n,i}^*)\frac{\xi_i^2}{n}, \tag{1.29}$$

where $W_{n,i}^*$ is between W_n and $W_{n,i}$.

Inserting (1.29) into (1.28) and noting $E\xi_i^2 = 1$ yields

$$EW_n f(W_n) = \frac{1}{n}\sum_{i=1}^n Ef'(W_{n,i}) + \frac{1}{2n^{3/2}}\sum_{i=1}^n Ef''(W_{n,i}^*)\xi_i^3.$$

Subtracting $Ef'(W_n)$ in both sides gives

$$EW_n f(W_n) - Ef'(W_n) = \frac{1}{n}\sum_{i=1}^n E\big(f'(W_{n,i}) - f'(W_n)\big)$$

$$+ \frac{1}{2n^{3/2}}\sum_{i=1}^n Ef''(W_{n,i}^*)\xi_i^3. \tag{1.30}$$

It follows from the mean value theorem

$$f'(W_{n,i}) - f'(W_n) = f''(W_{n,i}^{\ddagger})\frac{\xi_i}{\sqrt{n}}, \tag{1.31}$$

where $W_{n,i}^{\ddagger}$ is between W_n and $W_{n,i}$.

Thus combining (1.30) and (1.31), and noting $\|f''\|_\infty \leq 2$ and $E|\xi_i| \leq 1 \leq E|\xi_i|^3$, we have (1.27) as desired. $\square$

Recall the ordered pair (W, W') of random variables is exchangeable if

$$(W, W') \overset{d}{=} (W', W).$$

Trivially, if (W, W') is an exchangeable pair, then $W \overset{d}{=} W'$. Also, assuming $g(x, y)$ is antisymmetric, namely $g(x, y) = -g(y, x)$, then

$$Eg(W, W') = 0$$

if the expectation exists.

Theorem 1.12. *Assume that (W, W') is an exchangeable pair, and assume there exists a constant $0 < \tau \leq 1$ such that*

$$E(W'|W) = (1 - \tau)W.$$

If $EW^2 = 1$, then

$$d_W(W, \xi) \leq \frac{1}{\sqrt{2\pi}}\sqrt{E\Big(1 - \frac{E(|W' - W|^2|W)}{2\tau}\Big)^2} + \frac{1}{3\tau}E|W' - W|^3.$$

Proof. Given $f \in \mathcal{G}$, define $F(x) = \int_0^x f(t)dt$. Then it obviously follows

$$EF(W') - EF(W) = 0. \tag{1.32}$$

On the other hand, using the Taylor expansion for F at W, we obtain

$$\begin{aligned}
F(W') - F(W) &= F'(W)(W' - W) + \frac{1}{2}F''(W)(W' - W)^2 \\
&\quad + \frac{1}{6}F'''(W^*)(W' - W)^3 \\
&= f(W)(W' - W) + \frac{1}{2}f'(W)(W' - W)^2 \\
&\quad + \frac{1}{6}f''(W^*)(W' - W)^3,
\end{aligned}$$

where W^* is between W and W'.

This together with (1.32) in turn implies

$$Ef(W)(W' - W) + \frac{1}{2}Ef'(W)(W' - W)^2 + \frac{1}{6}Ef''(W^*)(W' - W)^3 = 0.$$

Note by hypothesis $E(W'|W) = (1 - \tau)W$,

$$\begin{aligned}
Ef(W)(W' - W) &= E\big(f(W)(W' - W)|W\big) \\
&= Ef(W)E\big(W' - W|W\big) \\
&= -\tau EWf(W).
\end{aligned}$$

Hence we have

$$EWf(W) = \frac{1}{2\tau}Ef'(W)(W' - W)^2 + \frac{1}{6\tau}Ef''(W^*)(W' - W)^3.$$

Subtracting $Ef'(W)$ yields

$$\begin{aligned}
EWf(W) - Ef'(W) &= -Ef'(W)\Big(1 - \frac{1}{2\tau}E\big((W' - W)^2|W\big)\Big) \\
&\quad + \frac{1}{6\tau}Ef''(W^*)(W' - W)^3.
\end{aligned}$$

Thus it follows from the Cauchy-Schwarz inequality and the fact $\|f'\|_\infty \leq \sqrt{2/\pi}$ and $\|f''\|_\infty \leq 2$,

$$\begin{aligned}
|EWf(W) - Ef'(W)| &\leq \sqrt{\frac{2}{\pi}}\sqrt{E\Big(1 - \frac{1}{2\tau}E\big((W' - W)^2|W\big)\Big)^2} \\
&\quad + \frac{1}{3\tau}E|W' - W|^3.
\end{aligned}$$

The proof is complete. $\qquad\square$

As an application, we shall revisit below the normal approximation of sums of independent random variables. Use notations in Example 1.6. A key ingredient is to construct W_n' in such a way that (W_n, W_n') is an exchangeable pair and $E(W_n'|W_n) = (1 - \tau)W_n$ for some $0 < \tau \leq 1$.

Let $\{\xi_n', n \geq 1\}$ be an independent copy of $\{\xi_n, n \geq 1\}$. Let I be a uniform random variable taking values $1, 2, \cdots, n$, independent of all other random variables. Define

$$W_n' = \frac{1}{\sqrt{n}}\left(S_n - \xi_I + \xi_I'\right).$$

Let $\mathcal{A}_n = \sigma\{\xi_1, \cdots, \xi_n\}$. Trivially, $\mathcal{A}_n$ is a sequence of increasing σ-fields and $W_n \in \mathcal{A}_n$. Some simple manipulation shows

$$E\left(W_n' - W_n|\mathcal{A}_n\right) = \frac{1}{\sqrt{n}}E\left(-\xi_I + \xi_I'|\mathcal{A}_n\right)$$

$$= -\frac{1}{n}W_n,$$

which implies $\tau = 1/n$. In addition,

$$E\left((W_n' - W_n)^2|\mathcal{A}_n\right) = \frac{1}{n}E\left((\xi_I - \xi_I')^2|\mathcal{A}_n\right)$$

$$= \frac{1}{n} + \frac{1}{n^2}\sum_{i=1}^{n}\xi_i^2$$

and

$$E\left((W_n' - W_n)^2|W_n\right) = \frac{1}{n} + \frac{1}{n^2}E\left(\sum_{i=1}^{n}\xi_i^2|W_n\right).$$

Hence we have

$$E\left(1 - \frac{1}{2\tau}E\left((W_n' - W_n)^2|W_n\right)\right)^2$$

$$= E\left(1 - \frac{1}{2\tau}\left(\frac{1}{n} + \frac{1}{n^2}E\left(\sum_{i=1}^{n}\xi_i^2|W_n\right)\right)\right)^2$$

$$= \frac{1}{4n^2}E\left(\sum_{i=1}^{n}(\xi_i^2 - E\xi_i^2)|W_n\right)^2$$

$$\leq \frac{1}{4n^2}E\left(\sum_{i=1}^{n}(\xi_i^2 - E\xi_i^2)\right)^2$$

$$\leq \frac{1}{4n^2}\sum_{i=1}^{n}E(\xi_i^2 - E\xi_i^2)^2 \leq \frac{1}{n^2}\sum_{i=1}^{n}E\xi_i^4.$$

Finally, note

$$E\left|W_n' - W_n\right|^3 = \frac{1}{n^{3/2}}E\left|\xi_I - \xi_I'\right|^3$$

$$\leq \frac{8}{n^{3/2}}\frac{1}{n}\sum_{i=1}^n E|\xi_i|^3.$$

Applying Theorem 1.12, we immediately obtain

$$d_W\left(\frac{S_n}{\sqrt{n}},\xi\right) \leq \sqrt{\frac{1}{2\pi}\frac{1}{n}}\sqrt{\sum_{i=1}^n E\xi_i^4 + \frac{8}{3n^{3/2}}\frac{1}{n}\sum_{i=1}^n E|\xi_i|^3}.$$

In particular, when $\xi_n, n \geq 1$ are i.i.d. random variables with $E\xi_n^4 < \infty$, then

$$d_W\left(\frac{S_n}{\sqrt{n}},\xi\right) \leq \frac{A}{\sqrt{n}}$$

for some numerical constant A.

1.3 The Stieltjes transform method

Stieltjes transforms, also called Cauchy transforms, of functions of bounded variation are other important tools in the study of convergence of probability measures. It actually plays a particularly significant role in the asymptotic spectrum theory of random matrices. Given a probability measure μ on the real line $\mathbb{R}$, its Stieltjes transform is defined by

$$s_\mu(z) = \int_{\mathbb{R}} \frac{1}{x - z}d\mu(x)$$

for any z outside the support of μ. In particular, it is well defined for all complex numbers z in $\mathbb{C} \setminus \mathbb{R}$. Some elementary properties about $s_\mu(z)$ are listed below. Set $z = a + i\eta$.

(i)

$$\overline{s_\mu(z)} = s_\mu(\bar{z}).$$

So we may and do focus on the upper half plane, namely $\eta > 0$ in the sequel.

(ii)

$$|s_\mu(z)| \leq \frac{1}{|\eta|}.$$

(iii)
$$Im(s_\mu(z)) = Im(z) \int_\mathbb{R} \frac{1}{|x - z|^2} d\mu(x).$$
So $Im(s_\mu(z))$ has the same sign as $Im(z)$.
(iv)
$$s_\mu(z) = -\frac{1}{z}(1 + o(1)), \quad z \to \infty.$$
(v) If $m_k(\mu) := \int_\mathbb{R} x^k d\mu(x)$ exists and are finite for every $k \geq 0$, then it follows
$$s_\mu(z) = -\frac{1}{z} \sum_{k=0}^\infty \frac{m_k(\mu)}{z^k}.$$
So $s_\mu(z)$ is closely related to the moment generating function of μ.
(vi) $s_\mu(z)$ is holomorphic outside the support of μ.

Example 1.7. (i) Let $\mu := \mu_{sc}$ be a probability measure on $\mathbb{R}$ with density function ρ_{sc} given by (1.15). Then its Stieltjes transform is
$$s_{sc}(z) = -\frac{z}{2} + \frac{\sqrt{z^2 - 4}}{2}. \tag{1.33}$$
(ii) Let $\mu := \mu_{MP}$ be a probability measure on $\mathbb{R}$ with density function given by (1.16). Then its Stieltjes transform is
$$s_{MP}(z) = -\frac{1}{2} + \frac{1}{2z}\sqrt{z(z - 4)}.$$

Theorem 1.13. *Let μ be a probability measure with Stieltjes transform $s_\mu(z)$. Then for any μ-continuity points a, b $(a < b)$*
$$\mu(a, b) = \lim_{\eta \to 0} \frac{1}{\pi} \int_a^b \frac{s_\mu(\lambda + i\eta) - s_\mu(\lambda - i\eta)}{2i} d\lambda$$
$$= \lim_{\eta \to 0} \frac{1}{\pi} \int_a^b Im(s_\mu(\lambda + i\eta)) d\lambda.$$

Proof. Let X be a random variable whose law is μ. Let Y be a standard Cauchy random variable, namely Y has probability density function
$$p_Y(y) = \frac{1}{\pi(1 + y^2)}, \quad y \in \mathbb{R}.$$
Then the random variable $Z_\eta := X + \eta Y$ has probability density function
$$p_\eta(\lambda) = \int_\mathbb{R} \frac{1}{\eta} p_Y\left(\frac{\lambda - y}{\eta}\right) d\mu(y)$$
$$= \frac{1}{\pi} \int_\mathbb{R} \frac{\eta}{(\lambda - y)^2 + \eta^2} d\mu(y)$$
$$= \frac{1}{\pi} Im(s_\mu(\lambda + i\eta)).$$

Note that $Z_\eta \xrightarrow{d} X$ as $\eta \to 0$. This implies

$$\lim_{\eta \to 0} P(Z_\eta \in (a,b)) = P(X \in (a,b)).$$

We now conclude the proof. $\qquad\qquad\qquad\qquad\qquad\qquad\qquad\qquad$ $\square$

Compared with the Fourier transform, an important advantage of Stieltjes transform is that one can easily find the density function of a probability measure via the Stieltjes transform. In fact, let $F(x)$ be a distribution function induced by μ and $x_0 \in \mathbb{R}$. Suppose that $\lim_{z \to x_0} Im(s_\mu(z))$ exists, denoted by $Im(s_\mu(x_0))$. Then F is differentiable at x_0, and $F'(x_0) = Im(s_\mu(x_0))/\pi$.

The Stieltjes continuity theorem reads as follows.

Theorem 1.14. *(i) If μ, μ_n, $n \geq 1$ is a sequence of probability measures on $\mathbb{R}$ such that*

$$\mu_n \Rightarrow \mu, \quad n \to \infty$$

then for each $z \in \mathbb{C} \setminus \mathbb{R}$,

$$s_{\mu_n}(z) \to s_\mu(z), \quad n \to \infty.$$

(ii) Assume that μ_n, $n \geq 1$ is a sequence of probability measures on $\mathbb{R}$ such that as $n \to \infty$,

$$s_{\mu_n}(z) \to s(z), \quad \forall z \in \mathbb{C} \setminus \mathbb{R}$$

for some $s(z)$. Then there exists a sub-probability measure μ ($\mu(\mathbb{R}) \leq 1$) such that

$$s(z) = \int_{\mathbb{R}} \frac{1}{x - z} d\mu(x)$$

and for any continuous function f decaying to 0 at infinity,

$$\int_{\mathbb{R}} f(x) d\mu_n(x) \to \int_{\mathbb{R}} f(x) d\mu(x), \quad n \to \infty.$$

(iii) Assume μ is a deterministic probability measure, and μ_n, $n \geq 1$ is a sequence of random probability measures on $\mathbb{R}$. If for any $z \in \mathbb{C} \setminus \mathbb{R}$,

$$s_{\mu_n}(z) \xrightarrow{P} s_\mu(z), \quad n \to \infty$$

then μ_n weakly converges in probability to μ. Namely, for any bounded continuous function f,

$$\int_{\mathbb{R}} f(x) d\mu_n(x) \xrightarrow{P} \int_{\mathbb{R}} f(x) d\mu(x), \quad n \to \infty.$$

The reader is referred to Anderson, Guionnet and Zeitouni (2010), Bai and Silverstein (2010), Tao (2012) for its proof and more details.

To conclude, let us quickly review the Riesz transform for a continual diagram. Let

$$\Omega(u) = \begin{cases} \frac{2}{\pi}\left(u\arcsin\frac{u}{2} + \sqrt{4 - u^2}\right), & |u| \leq 2, \\ |u|, & |u| > 2 \end{cases} \tag{1.34}$$

its Riesz transform is defined by

$$R_\Omega(z) = \frac{1}{z}\exp\left(\int_{-2}^{2} \frac{(\Omega(u) - |u|)'}{u - z}du\right)$$

for each $z \notin [-2, 2]$.

It is easy to compute

$$R_\Omega(z) = -\frac{z}{2} + \frac{1}{2}\sqrt{z^2 - 4},$$

which implies by (1.33)

$$R_\Omega(z) = s_{sc}(z). \tag{1.35}$$

This is *not* a coincidence! As we will see in Chapter 5, $\Omega(u)$ (see Figure 1.4 below) turns out to be the limit shape of a typical Plancherel Young diagram. The equation (1.35) provides another evidence that there is a close link between random matrices and random Plancherel partitions.

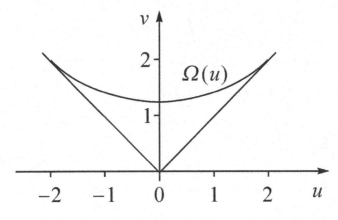

Fig. 1.4 Ω curve

1.4 Convergence of stochastic processes

Let (S, ρ) be a metric space, $\mathcal{S}$ a σ-field generated by its topology. A mapping $X : \Omega \mapsto S$ is said to be measurable if for any $A \in \mathcal{S}$

$$X^{-1}(A) = \{\omega : X(\omega) \in A\} \in \mathcal{A}.$$

We also call X a S-valued random element. The most commonly studied random elements include real (complex) random variable, random vector, random processes, Banach-valued random variable. Denote by P_X the law of X under P:

$$P_X(A) := P \circ X^{-1}(A) = P(\omega \in \Omega : X(\omega) \in A), \quad A \in \mathcal{S}.$$

By definition, a sequence of random variables X_n weakly converges to a random variable X if $P_{X_n} \Rightarrow P_X$, and write simply $X_n \Rightarrow X$. The following five statements are equivalent:
(i) for any bounded continuous function f,

$$Ef(X_n) \to Ef(X), \quad n \to \infty;$$

(ii) for any bounded uniformly continuous function f,

$$Ef(X_n) \to Ef(X), \quad n \to \infty;$$

(iii) for any closed set F,

$$\limsup_{n \to \infty} P(X_n \in F) \le P(X \in F);$$

(iv) for any open set G,

$$\liminf_{n \to \infty} P(X_n \in G) \ge P(X \in G);$$

(v) for any measurable X-continuity set A,

$$\lim_{n \to \infty} P(X_n \in A) = P(X \in A).$$

The reader is referred to Billingsley (1999a) for the proof and more details. In addition, (ii) can be replaced by
(ii') for any bounded infinitely differentiable function f,

$$Ef(X_n) \to Ef(X), \quad n \to \infty.$$

It can even be replaced by
(ii'') for any continuous function f with compact support,

$$Ef(X_n) \to Ef(X), \quad n \to \infty.$$

In the special cases $S = \mathbb{R}$ and $\mathbb{R}^k$, $X_n \Rightarrow X$ is equivalent to $X_n \xrightarrow{d} X$. In the case $S = \mathbb{R}^\infty$, $X_n \Rightarrow X$ if and only if for each $k \geq 1$

$$X_n \circ \pi_k^{-1} \Rightarrow X \circ \pi_k^{-1}, \quad n \to \infty$$

where π_k denotes the projection from $\mathbb{R}^\infty$ to $\mathbb{R}^k$.

The case of $C[0,1]$ is more interesting and challenging. Assume $X_n \Rightarrow X$. Then for any $k \geq 1$ and any k points $t_1, t_2, \cdots, t_k \in [0,1]$

$$X_n \circ \pi_{t_1,\cdots,t_k}^{-1} \Rightarrow X \circ \pi_{t_1,\cdots,t_k}^{-1}, \quad n \to \infty \tag{1.36}$$

where $\pi_{t_1,\cdots,t_k}$ is a projection from $C[0,1]$ to $\mathbb{R}^k$, i.e., $\pi_{t_1,\cdots,t_k}(x) = (x(t_1), \cdots, x(t_k))$.

However, the condition (1.36) is not a sufficient condition for X_n to weakly converge to X. We shall require additional conditions. The X_n is said to be weakly relatively compact if every subsequence has a further convergent subsequence in the sense of weak convergence. According to the subsequence convergence theorem, X_n is weakly convergent if all the limit variables are identical in law. Another closely related concept is uniform tightness. The X_n is uniformly tight if for any $\varepsilon > 0$, there is a compact subset K_ε in (S, ρ) such that

$$P(X_n \notin K_\varepsilon) < \varepsilon, \quad \text{for all } n \geq 1.$$

The celebrated Prohorov's theorem tells that the X_n must be weakly relatively compact if X_n is uniformly tight. The converse is also true in a separable complete metric space. A major point of this theorem is that the weak convergence of probability measures rely on how they concentrate in a compact subset in a metric space. In $C[0,1]$, the so-called Ascoli-Arzelà lemma completely characterizes a relatively compact subset: $K \subseteq C[0,1]$ is relatively compact if and only if

(i) uniform boundedness:

$$\sup_{t \in \mathbb{R}} \sup_{x \in K} |x(t)| < \infty,$$

(ii) equi-continuity: for any $\varepsilon > 0$ there is a $\delta > 0$ such that

$$\sup_{|s-t| \leq \delta} \sup_{x \in K} |x(s) - x(t)| < \varepsilon.$$

Note that under condition (ii), (i) can be replaced by the condition (i')

$$\sup_{x \in K} |x(0)| < \infty.$$

Combining the Ascoli-Arzelà lemma and Prohorov's theorem, we can readily give a criterion for X_n to weakly converge to X in $C[0,1]$. Assume that we are given a sequence of continuous random processes X and X_n, $n \geq 1$ in $C[0,1]$. Then $X_n \Rightarrow X$ if and only if
(i) finite dimensional distributions converge, namely (1.36) holds;
(ii) for any $\varepsilon > 0$, there is a finite positive constant M such that

$$P(|X_n(0)| > M) < \varepsilon, \quad \text{for all } n \geq 1; \tag{1.37}$$

(iii) for any $\varepsilon > 0$ and $\eta > 0$, there is a $\delta > 0$ such that

$$P\left(\sup_{|s-t|<\delta} |X_n(s) - X_n(t)| > \eta \right) < \varepsilon, \quad \text{for all } n \geq 1. \tag{1.38}$$

To illustrate how to use the above general framework, we shall state and prove the Donsker invariance principle.

Theorem 1.15. *Let $\xi_n, n \geq 1$ be a sequence of i.i.d. real random variables defined in a common probability space $(\Omega, \mathcal{A}, P)$, and $E\xi_1 = 0$ and $Var(\xi_1) = 1$. Define $S_n = \sum_{i=1}^{n} \xi_i$, $n \geq 1$, and define*

$$X_n(t) = \frac{1}{\sqrt{n}} S_{[nt]} + \frac{nt - [nt]}{\sqrt{n}} \xi_{[nt]+1}, \quad 0 \leq t \leq 1. \tag{1.39}$$

Then

$$X_n \Rightarrow B, \quad n \to \infty$$

where $B = (B(t), 0 \leq t \leq 1)$ is a standard Brownian motion.

Proof. We need to verify the conditions (1.36), (1.37) and (1.38). Indeed, (1.36) directly follows from the Feller-Lévy CLT. (1.37) is trivial since $X_n(0) = 0$, and (1.38) follows from the Lévy maximal inequality for sums of independent random variables. The detail is left to the reader. $\square$

We remark that the random process constructed in (1.39) is a polygon going through points $(k/n, S_k/\sqrt{n})$. It is often referred to as a partial sum process. The Donsker invariance principle has found a large number of applications in a wide range of fields. To apply it, we usually need the following mapping theorem. Let (S_1, ρ_1) and (S_2, ρ_2) be two metric spaces, $h : S_1 \mapsto S_2$ a measurable mapping. Assume that X, X_n, $n \geq 1$ is a sequence of S_1-valued random elements and $X_n \Rightarrow X$. It is natural to ask under what hypothesis the $h(X_n)$ still weakly converges to $h(X)$. Obviously, if h is continuous, then we have $h(X_n) \Rightarrow h(X)$. Moreover, the same still holds if h is a measurable mapping and $P_X(D_h) = 0$ where D_h is the set of all discontinuity points of h. As a simple example, we can compute the limiting distribution of $\max_{1 \leq k \leq n} S_k/\sqrt{n}$. Indeed, let $h(x) = \sup_{0 \leq t \leq 1} x(t)$ where $x \in C[0,1]$. Then h is continuous and

$$h(X_n) = \frac{1}{\sqrt{n}} \max_{1 \leq k \leq n} S_k, \quad h(B) = \sup_{0 \leq t \leq 1} B(t).$$

Hence it follows

$$\frac{1}{\sqrt{n}} \max_{1 \leq k \leq n} S_k \xrightarrow{d} \sup_{0 \leq t \leq 1} B(t), \quad n \to \infty.$$

Another example is to compute the limiting distribution of the weighted sum $\sum_{k=1}^{n} k\xi_k/n^{3/2}$. Let $h(x) = \int_0^1 x(t)dt$ where $x \in C[0,1]$. Then h is continuous and

$$h(X_n) = \frac{1}{n^{3/2}} \sum_{k=1}^{n} k\xi_k + o_p(1), \quad h(B) = \int_0^1 B(t)dt,$$

where $o_p(1)$ is negligible. Hence it follows

$$\frac{1}{n^{3/2}} \sum_{k=1}^{n} k\xi_k \xrightarrow{d} \int_0^1 B(t)dt.$$

More interesting examples can be found in Billingsley (1999a). In addition to $\mathbb{R}^\infty$ and $C[0,1]$, one can also consider weak convergence of random processes in $D[0,1]$, $C(0,\infty)$ and $D(0,\infty)$.

As the reader might notice, in proving the weak convergence of X_n in $C[0,1]$, the most difficult part is to verify the uniform tightness condition (1.38). A weaker version than (1.38) is stochastic equicontinuity: for every $\varepsilon > 0$ and $\eta > 0$, there is a $\delta > 0$ such that

$$\sup_{|s-t|<\delta} P(|X_n(s) - X_n(t)| > \eta) < \varepsilon, \quad \text{for all } n \geq 1. \qquad (1.40)$$

Although (1.40) does not guarantee that the process X_n converges weakly, we can formulate a limit theorem for comparatively narrow class of functionals of integral form.

Theorem 1.16. *Suppose that $X = (X(t), 0 \leq t \leq 1)$ and $X_n = (X_n(t), 0 \leq t \leq 1)$ is a sequence of random processes satisfying (1.36) and (1.40). Suppose that $g(t,x)$ is a continuous function and there is a nonnegative function $h(x)$ such that $h(x) \uparrow \infty$ as $|x| \to \infty$ and*

$$\lim_{a \to \infty} \sup_{0 \leq t \leq 1} \sup_{|x| > a} \frac{|g(t,x)|}{h(x)} = 0.$$

If $\sup_{n \geq 1} \sup_{0 \leq t \leq 1} E|h(X_n(t))| < \infty$, then as $n \to \infty$

$$\int_0^1 g(t, X_n(t))dt \xrightarrow{d} \int_0^1 g(t, X(t))dt.$$

This theorem is sometimes referred to as Gikhman-Skorohod theorem. Its proof and applications can be found in Chapter 9 of Gikhman and Skorohod (1996).

Chapter 2

Circular Unitary Ensemble

2.1 Introduction

For $n \in \mathbb{N}$, a complex $n \times n$ matrix U_n is said to be unitary if

$$U_n^* U_n = U_n U_n^* = I_n.$$

This is equivalent to saying U_n is nonsingular and $U_n^* = U_n^{-1}$. The set $\mathcal{U}_n$ of unitary matrices forms a remarkable and important set, a compact Lie group, which is generally referred to as the unitary group. This group has a unique regular probability measure μ_n that is invariant under both left and right multiplication by unitary matrices. Such a measure is called Haar measure. Thus we have induced a probability space $(\mathcal{U}_n, \mu_n)$, which is now known as Circular Unitary Ensemble (CUE).

By definition the columns of an $n \times n$ random unitary matrix are orthogonal vectors in the n dimensional complex space $\mathbb{C}^n$. This implies that the matrix elements are not independent and thus are statistically correlated. Before discussing statistical correlation properties, we shall have a quick look at how to generate a random unitary matrix.

Form an $n \times n$ random matrix $Z_n = (z_{ij})_{n \times n}$ with i.i.d. complex standard normal random variables. Recall $Z = X + iY$ is a complex standard normal random variable if X and Y are i.i.d. real normal random variable with mean 0 and variance $1/2$. The Z_n is almost sure of full rank, so apply Gram-Schmidt orthonormalization to its columns: normalize the first column to have norm one, take the first column out of the second column and normalize to have norm one, and so on. Let T_n be the map induced by Gram-Schmidt algorithm, then the resulting matrix $T_n(Z_n)$ is unitary and even is distributed with Haar measure. This is easy to prove and understand. Indeed it holds $T_n(U_n Z_n) = U_n T_n(Z_n)$ for any unitary matrix $U_n \in \mathcal{U}_n$. Since $U_n Z_n \stackrel{d}{=} Z_n$, so $U_n T_n(Z_n) \stackrel{d}{=} T_n(Z_n)$ as required.

Given a unitary matrix U_n, consider the equation

$$U_n \mathbf{v} = \lambda \mathbf{v},$$

where λ is a scalar and $\mathbf{v}$ is a vector in $\mathbb{C}^n$. If a scalar λ and a nonzero vector $\mathbf{v}$ happen to satisfy this equation, then λ is called an eigenvalue of U_n and $\mathbf{v}$ is called an eigenvector associated with λ. The eigenvalues of U_n are zeros of the characteristic polynomial $\det(zI_n - U_n)$. It turns out that all eigenvalues are on the unit circle $\mathbb{T} := \{z \in \mathbb{C}; |z| = 1\}$ and are almost surely distinct with respect to product Lebesgue measure. Call these $\{e^{i\theta_1}, \cdots, e^{i\theta_n}\}$ with $0 \leq \theta_1, \cdots, \theta_n < 2\pi$. Note that for any sequence of n points on $\mathbb{T}$ there are matrices in $\mathcal{U}_n$ with these points as eigenvalues. The collection of all matrices with the same set of eigenvalues constitutes a conjugacy class in $\mathcal{U}_n$.

The main question of interest in this chapter is: pick a $U_n \in \mathcal{U}_n$ according to Haar measure, how are $\{e^{i\theta_1}, \cdots, e^{i\theta_n}\}$ distributed? The most celebrated result is the following Weyl formula.

Theorem 2.1. *The joint probability density for the unordered eigenvalues of a Haar distributed random matrix in $\mathcal{U}_n$ is*

$$p_n\big(e^{i\theta_1}, \cdots, e^{i\theta_n}\big) = \frac{1}{(2\pi)^n n!} \prod_{1 \leq j < k \leq n} \big|e^{i\theta_j} - e^{i\theta_k}\big|^2, \qquad (2.1)$$

where the product is by convention 1 when $n = 1$.

The proof is omitted, the reader is referred to Chapter 11 of Mehta (2004). See also Chapter 2 of Forrester (2010).

Weyl's formula is the starting point of the following study of the CUE. In particular, it gives a simple way to perform averages on $\mathcal{U}_n$. For a class function f that are constant on conjugacy classes,

$$\int_{\mathcal{U}_n} f(U_n) d\mu_n = \int_{[0,2\pi]^n} f\big(e^{i\theta_1}, \cdots, e^{i\theta_n}\big) p_n\big(e^{i\theta_1}, \cdots, e^{i\theta_n}\big) d\theta_1 \cdots d\theta_n.$$

Obviously, U_1 has only one eigenvalue and is uniformly distributed on $\mathbb{T}$. U_2 has two eigenvalues whose joint probability density is

$$p_2\big(e^{i\theta_1}, e^{i\theta_2}\big) = \frac{1}{8\pi^2} \big|e^{i\theta_1} - e^{i\theta_2}\big|^2.$$

It is easy to compute the marginal density for each eigenvalue by integrating out the other argument. In particular, each eigenvalue is also uniformly distributed on $\mathbb{T}$. But these two eigenvalues are not independent of each other; indeed $p_2\big(e^{i\theta_1}, e^{i\theta_2}\big)$ tends to zero as θ_1 and θ_2 approach each other.

Interestingly, these properties hold for general n. To see this, let for any set of n-tuple complex numbers $(x_1, \cdots, x_n)$

$$\Delta(x_1, \cdots, x_n) = \det\left(x_k^{j-1}\right)_{1 \le j,k \le n}.$$

Then the Vandermonde identity shows

$$\Delta(x_1, \cdots, x_n) = \prod_{1 \le j < k \le n} (x_k - x_j).$$

Define

$$S_n(\theta) = e^{-i(n-1)\theta/2} \sum_{k=1}^{n} e^{ik\theta} = \frac{\sin n\theta/2}{\sin \theta/2}, \qquad (2.2)$$

where by convention $S_n(0) = n$. Thus we have by using the fact that the determinant of a matrix and its transpose are the same

$$\prod_{1 \le j < k \le n} \left|e^{i\theta_j} - e^{i\theta_k}\right|^2 = \left|\Delta(e^{i\theta_1}, \cdots, e^{i\theta_n})\right|^2$$

$$= \det\left(S_n(\theta_k - \theta_j)\right)_{1 \le j,k \le n}.$$

This formula is very useful to computing some eigenvalue statistics. In order to compute the m-dimensional marginal density, we also need a formula of Gaudin (see Conrey (2005)), which states

$$\int_0^{2\pi} S_n(\theta_j - \theta)S_n(\theta - \theta_k)d\theta = 2\pi S_n(\theta_j - \theta_k).$$

As a consequence, we have

$$\int_0^{2\pi} \det\left(S_n(\theta_j - \theta_k)\right)_{1 \le j,k \le n} d\theta_n = 2\pi \det\left(S_n(\theta_j - \theta_k)\right)_{1 \le j,k \le n-1}.$$

Repeating this yields easily

$$p_{n,m}\left(e^{i\theta_1}, \cdots, e^{i\theta_m}\right) := \frac{1}{n!(2\pi)^n} \int_{[0,2\pi]^{n-m}} p_n\left(e^{i\theta_1}, \cdots, e^{i\theta_n}\right) d\theta_{m+1} \cdots d\theta_n$$

$$= \frac{1}{n!(2\pi)^n} \int_{[0,2\pi]^{n-m}} \det\left(S_n(\theta_j - \theta_k)\right)_{n \times n} d\theta_{m+1} \cdots d\theta_n$$

$$= \frac{(n-m)!}{n!} \frac{1}{(2\pi)^m} \det\left(S_n(\theta_j - \theta_k)\right)_{1 \le j,k \le m}.$$

In particular, the first two marginal densities are

$$p_{n,1}\left(e^{i\theta}\right) = \frac{1}{2\pi}, \quad 0 \le \theta \le 2\pi \qquad (2.3)$$

and

$$p_{n,2}\left(e^{i\theta_1}, e^{i\theta_2}\right) = \frac{1}{n(n-1)} \frac{1}{(2\pi)^2} \left(1 - (S_n(\theta_1 - \theta_2))^2\right). \qquad (2.4)$$

Hence each eigenvalue is still uniformly distributed on $\mathbb{T}$.

The CUE and its eigenvalue distributions naturally appear in a variety of problems from particle physics to analytic number theory. It is indeed a very special example of three ensembles introduced and studied by Dyson in 1962 with a view to simplifying the study of energy level behavior in complex quantum systems, see Dyson (1962). More generally, consider n identically charged particles confined to move on $\mathbb{T}$ in the complex plane. Each interacts with the others through the usual Coulomb potential, $-\log\left|e^{i\theta_j} - e^{i\theta_k}\right|$, which gives rise to the Hamiltonian

$$H_n(\theta_1, \cdots, \theta_n) = \sum_{1 \le j < k \le n} -\log\left|e^{i\theta_j} - e^{i\theta_k}\right|.$$

This induces the Gibbs measure with parameter n and $\beta > 0$

$$p_{n,\beta}\left(e^{i\theta_1}, \cdots, e^{i\theta_n}\right) = \frac{1}{(2\pi)^n Z_{n,\beta}} \prod_{1 \le j < k \le n} \left|e^{i\theta_j} - e^{i\theta_k}\right|^\beta, \qquad (2.5)$$

where n is the number of particles, β stands for the inverse temperature, and $Z_{n,\beta}$ is given by

$$Z_{n,\beta} = \frac{\Gamma(\frac{1}{2}\beta n + 1)}{[\Gamma(\frac{1}{2}\beta + 1)]^n}.$$

The family of probability measures defined by (2.5) is called Circular β Ensemble (CβE). The CUE corresponds to $\beta = 2$. Viewed from the opposite perspective, one may say that the CUE provides a matrix model for the Coulomb gas at the inverse temperature $\beta = 2$. In Section 2.5 we shall see a matrix model for general β.

In this chapter we will be particularly interested in the asymptotic behaviours of various eigenvalue statistics as n tends to infinity. Start with the average spectral measures. Let $e^{i\theta_1}, \cdots, e^{i\theta_n}$ be eigenvalues of a Haar distributed unitary matrix U_n. Put them together as a probability measure on $\mathbb{T}$:

$$\nu_n = \frac{1}{n} \sum_{k=1}^{n} \delta_{e^{i\theta_k}}.$$

Theorem 2.2. *As $n \to \infty$,*

$$\nu_n \Rightarrow \mu \quad in \ P,$$

where μ is a uniform measure on $\mathbb{T}$.

The proof is basically along the line of Diaconis and Shahshahani (1994). We need a second-order moment estimate as follows.

Lemma 2.1. *For any integer $l \neq 0$,*

$$E\left|\sum_{k=1}^{n} e^{il\theta_k}\right|^2 = \min(|l|, n). \tag{2.6}$$

Proof. It is easy to see

$$E\left|\sum_{k=1}^{n} e^{il\theta_k}\right|^2 = n + n(n-1)Ee^{il(\theta_1-\theta_2)}. \tag{2.7}$$

In turn, in virtue of (2.2) and (2.4) it follows

$$n(n-1)Ee^{il(\theta_1-\theta_2)}$$

$$= \frac{1}{(2\pi)^2} \int_0^{2\pi}\int_0^{2\pi} e^{il(\theta_1-\theta_2)}\left(1 - (S_n(\theta_1-\theta_2))^2\right)d\theta_1 d\theta_2$$

$$= -\frac{1}{(2\pi)^2} \int_0^{2\pi}\int_0^{2\pi} e^{il(\theta_1-\theta_2)}\left|\sum_{k=1}^{n} e^{ik(\theta_1-\theta_2)}\right|^2 d\theta_1 d\theta_2$$

$$= -\frac{1}{(2\pi)^2} \int_0^{2\pi}\int_0^{2\pi} e^{il(\theta_1-\theta_2)}\left(n + \sum_{1\leq m\neq k\leq n} e^{i(m-k)(\theta_1-\theta_2)}\right)d\theta_1 d\theta_2$$

$$= -\sharp\{(m,k): m-k=l, \quad 1\leq m\neq k\leq n\}$$

$$= \begin{cases} |l|-n, & |l|\leq n, \\ 0, & |l|>n. \end{cases} \tag{2.8}$$

Substituting (2.8) into (2.7) immediately yields (2.6). $\qquad\square$

Proof of Theorem 2.2. It is enough to show that the Fourier transform of ν_n converges in probability to that of μ. Let for each integer $l \neq 0$

$$\hat{\nu}_n(l) = \frac{1}{2\pi}\int_0^{2\pi} e^{-il\theta}d\nu_n(\theta), \quad \hat{\mu}(l) = \frac{1}{2\pi}\int_0^{2\pi} e^{-il\theta}d\mu(\theta).$$

Trivially $\hat{\mu}(l) = 0$. We shall need only prove

$$E\hat{\nu}_n(l) = 0, \quad E|\hat{\nu}_n(l)|^2 \to 0. \tag{2.9}$$

Since each eigenvalue is uniform over $\mathbb{T}$, then

$$E\hat{\nu}_n(l) = E\frac{1}{n}\sum_{k=1}^{n} e^{-il\theta_k}$$

$$= Ee^{-il\theta_1} = 0.$$

Also, it follows from (2.6)

$$E|\hat{\nu}_n(l)|^2 = \frac{|l|}{n^2}$$

whenever $n \geq |l|$. Thus (2.9) holds as $n \to \infty$. The proof is concluded. $\square$

Theorem 2.2 means that for every bounded continuous function f

$$\frac{1}{n} \sum_{k=1}^{n} f(e^{i\theta_k}) \xrightarrow{P} \frac{1}{2\pi} \int_0^{2\pi} f(e^{i\theta}) d\theta.$$

This is a kind of law of large numbers, and is very similar to the Khinchine law of large numbers for sums of i.i.d. random variables in standard probability theory, see (1.4). One cannot see from such a first-order average the difference between eigenvalues and sample points chosen at random from the unit circle $\mathbb{T}$. However, a significant feature will appear in the second-order fluctuation, which is the main content of the following sections.

2.2   Symmetric groups and symmetric polynomials

We shall first introduce the irreducible characters of symmetric groups and state without proofs character relations of two kinds. Then we shall define four classes of symmetric polynomials and establish a Schur orthonormality formula of Schur polynomials and power polynomials with respect to Haar measure. Most of the materials can be found in Macdonald (1995) and Sagan (2000).

Throughout the section, n is a fixed natural number. Consider the symmetric group , $\mathcal{S}_n$, consisting of all permutations of $\{1, 2, \cdots, n\}$ using composition as the multiplication. Assume $\sigma \in \mathcal{S}_n$, denote by r_k the number of cycles of length k in σ. The cycle type, or simply the type, of σ is an expression of the form

$$\left(1^{r_1}, 2^{r_2}, \cdots, n^{r_n}\right).$$

For example, if $\sigma \in \mathcal{S}_5$ is given by

$$\sigma(1) = 2, \, \sigma(2) = 3, \, \sigma(3) = 1, \, \sigma(4) = 4, \, \sigma(5) = 5$$

then it has cycle type $\left(1^2, 2^0, 3^1, 4^0, 5^0\right)$. The cycle type of the identity permutation is (1^n).

In $\mathcal{S}_n$, permutations σ and τ are conjugates if

$$\sigma = \vartheta \tau \vartheta^{-1}$$

for some $\vartheta \in \mathcal{S}_n$. The set of all permutations conjugate to a given σ is called the conjugacy class of σ and is denoted by K_σ. Conjugacy is an equivalent relation, so the distinct conjugacy classes partition $\mathcal{S}_n$. It is not hard to see that two permutations are in the same conjugacy class if and only if they have the same cycle type, and the size of a conjugacy class with cycle type $(1^{r_1}, 2^{r_2}, \cdots, n^{r_n})$ is given by $n!/\prod_{i=1}^{n} i^{r_i} r_i!$.

Let $\mathcal{M}_d$ be the set of all $d \times d$ matrices with complex entries, and $\mathcal{L}_d$ the group of all invertible $d \times d$ matrices with multiplication. A matrix representation of $\mathcal{S}_n$ is a group homomorphism

$$X : \mathcal{S}_n \longrightarrow \mathcal{L}_d.$$

Equivalently, to each $\sigma \in \mathcal{S}_n$ is assigned $X(\sigma) \in \mathcal{L}_d$ such that
(i) $X(1^n) = I_d$ the identity matrix, and
(ii) $X(\sigma\tau) = X(\sigma)X(\tau)$ for all $\sigma, \tau \in \mathcal{S}_n$.

The parameter d is called the degree of the representation. Given a matrix representation X of degree d, let V be the vector space of all column vectors of length d. Then we can multiply $\mathbf{v} \in V$ by $\sigma \in \mathcal{S}_n$ using

$$\sigma\mathbf{v} = X(\sigma)\mathbf{v}.$$

This makes V into a $\mathcal{S}_n$-module of dimension d. If a subspace $W \subseteq V$ is closed under the action of $\mathcal{S}_n$, that is,

$$\mathbf{w} \in W \Rightarrow \sigma\mathbf{w} \in W \quad \text{for all } \sigma \in \mathcal{S}_n,$$

then we say W is a $\mathcal{S}_n$-submodule of V.

A non-zero matrix representation X of degree d is reducible if the $\mathcal{S}_n$-module V contains a nontrivial submodule W. Otherwise, X is said to be irreducible. Equivalently, X is reducible if V has a basis $\mathcal{B}$ in which every $\sigma \in \mathcal{S}_n$ is assigned a block matrix of the form

$$X(\sigma) = \begin{pmatrix} A(\sigma) & \mathbf{0} \\ \mathbf{0} & B(\sigma) \end{pmatrix},$$

where the $A(\sigma)$ are square matrices, all of the same size, and $\mathbf{0}$ is a nonempty matrix of zeros.

Two representations X and Y are equivalent if and only if there exists a fixed matrix T such that

$$Y(\sigma) = TX(\sigma)T^{-1}, \quad \text{for all } \sigma \in \mathcal{S}_n.$$

The number of inequivalent irreducible representations is equal to the number of conjugacy classes of $\mathcal{S}_n$.

A classical theorem of Maschke implies that every matrix representation of $\mathcal{S}_n$ having positive dimension is completely reducible. In particular, let X be a matrix representation of $\mathcal{S}_n$ of degree $d > 0$, then there is a fixed matrix T such that every matrix $X(\sigma)$, $\sigma \in \mathcal{S}_n$ has the form

$$TX(\sigma)T^{-1} = \begin{pmatrix} X^{(1)}(\sigma) & 0 & \cdots & 0 \\ 0 & X^{(2)}(\sigma) & \cdots & 0 \\ \vdots & \vdots & \ddots & \vdots \\ 0 & 0 & \cdots & X^{(k)}(\sigma) \end{pmatrix},$$

where each $X^{(i)}$ is an irreducible matrix representation of $\mathcal{S}_n$.

To every matrix representation X assigns one simple statistic, the character defined by

$$\chi_X(\sigma) = tr X(\sigma),$$

where tr denotes the trace of a matrix. Otherwise put, χ_X is the map

$$\mathcal{S}_n \xrightarrow{tr X} \mathbb{C}.$$

It turns out that the character contains much of the information about the matrix representation. Here are some elementary properties of characters.

Lemma 2.2. *Let X be a matrix representation of $\mathcal{S}_n$ of degree d with character χ_X, then*
(i) $\chi_X(1^n) = d$;
(ii) χ_X is a class function, that is, $\chi_X(\sigma) = \chi_X(\tau)$ if σ and τ are in the same conjugacy class;
(iii) if Y is a matrix representation equivalent to X, then their characters are identical: $\chi_X \equiv \chi_Y$.

Let χ and ψ be any two functions from $\mathcal{S}_n$ to $\mathbb{C}$. The inner product of χ and ψ is defined by

$$\langle \chi, \psi \rangle = \frac{1}{n!} \sum_{\sigma \in \mathcal{S}_n} \chi(\sigma)\overline{\psi(\sigma)}.$$

In particular, if χ and ψ are characters, then

$$\langle \chi, \psi \rangle = \frac{1}{n!} \sum_{\sigma \in \mathcal{S}_n} \chi(\sigma)\overline{\psi(\sigma)}. \tag{2.10}$$

Theorem 2.3. *If χ and ψ are irreducible characters, then we have character relation of the first kind*

$$\langle \chi, \psi \rangle = \delta_{\chi, \psi}, \tag{2.11}$$

where $\delta_{\chi, \psi}$ stands for Kronecker delta.

Since character is a class function, then we can rewrite (2.10) and (2.11) as

$$\langle \chi, \psi \rangle = \frac{1}{n!} \sum_K |K| \chi(K) \overline{\psi(K)} = \delta_{\chi, \psi},$$

where we mean by $\chi(K)$ and $\psi(K)$ that χ and ψ act in K respectively, $|K|$ denotes the number of K and the sum is over all conjugacy classes of $\mathcal{S}_n$. This implies that the modified character table

$$U = \left(\sqrt{\frac{|K|}{n!}} \chi(K) \right)_{\chi, K}$$

has orthonormal rows. Hence U, being a square, is a unitary matrix and has orthonormal columns. Thus we have proven the character relation of the second kind as follows.

Theorem 2.4.

$$\sum_\chi \chi(K) \chi(L) = \frac{n!}{|K|} \delta_{K, L}, \tag{2.12}$$

where the sum is take over all irreducible characters.

A partition of n is a sequence $(\lambda_1, \lambda_2, \cdots, \lambda_l)$ of non-increasing natural numbers such that

$$\sum_{i=1}^l \lambda_i = n,$$

where the λ_i's are called parts, l is called the length.

If $\lambda = (\lambda_1, \lambda_2, \cdots, \lambda_l)$ is a partition of n, then we write $\lambda \mapsto n$. We also use the notation $|\lambda| = \sum_{i=1}^l \lambda_i$. The cycle type of a permutation in $\mathcal{S}_n$ naturally gives a partition of n. Conversely, given a $\lambda \mapsto n$, let $r_k = \sharp\{i; \lambda_i = k\}$, then we have a cycle type $(1^{r_1}, 2^{r_2}, \cdots, n^{r_n})$. Thus there is a natural one-to-one correspondence between partitions of n and conjugacy classes of $\mathcal{S}_n$. As a consequence, the number of irreducible characters is equal to the number of partitions of n.

Let $\mathcal{P}_n$ be the set of all partitions of n. We need to find an ordering on $\mathcal{P}_n$. Since each partition is a sequence of integer numbers, then a natural ordering is the ordinary lexicographic order. Let $\lambda = (\lambda_1, \lambda_2, \cdots, \lambda_l)$ and $\mu = (\mu_1, \mu_2, \cdots, \mu_m)$ be partitions of n. Then $\lambda < \mu$ in lexicographic order if, for some index i,

$$\lambda_j = \mu_j \quad \text{for } j < i \quad \text{and} \quad \lambda_i < \mu_i.$$

This is a total ordering on $\mathcal{P}_n$. For instance, on $\mathcal{P}_6$ we have

$$\left(1^6\right) < \left(2,1^4\right) < \left(2^2,1^2\right) < \left(2^3\right) < \left(3,1^3\right) < \left(3,2,1\right)$$
$$< \left(3^2\right) < \left(4,1^2\right) < \left(4,2\right) < \left(5,1\right) < \left(6\right).$$

Another ordering is the following dominance order. If

$$\lambda_1 + \lambda_2 + \cdots + \lambda_i \geq \mu_1 + \mu_2 + \cdots + \mu_i$$

for all $i \geq 1$, then λ is said to dominate μ, written as $\lambda \trianglerighteq \mu$. The lexicographic order is a refinement of the dominance order in the sense that $\lambda \geq \mu$ if $\lambda, \mu \in \mathcal{P}_n$ satisfy $\lambda \trianglerighteq \mu$.

Next we shall describe a graphic representation of a partition. Suppose $\lambda = (\lambda_1, \lambda_2, \cdots, \lambda_l) \mapsto n$. The Young diagram (shape) of λ is an array of n boxes into l left-justified rows with row i containing λ_i boxes for $1 \leq i \leq l$. The box in row i and column j has coordinates (i, j), as in a matrix, see Figure 2.1.

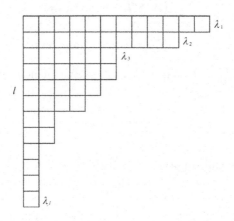

Fig. 2.1 Young diagram

A Young tableau of shape λ, t^λ, is an array obtained by putting the numbers $1, 2, \cdots, n$ into the boxes bijectively. A Young tableau t^λ is standard if the rows are increasing from left to right and the columns are increasing from top to bottom. Let $t_{i,j}$ stand for the entry of t in position (i, j). Clearly there are $n!$ Young tableau for any shape $\lambda \mapsto n$. Two tableaux t_1^λ and t_2^λ are row equivalent, $t_1^\lambda \sim t_2^\lambda$, if corresponding rows of the two tableaux contain the same elements. A tabloid of shape λ is

$$\{t^\lambda\} = \{t_1^\lambda, t_1^\lambda \sim t^\lambda\}.$$

The number of tableaux in any given equivalence class is $\lambda! := \lambda_1! \lambda_2! \cdots \lambda_l!$. Thus the number of tabloids of shape λ is just $n!/\lambda!$.

Given a $\sigma \in \mathcal{S}_n$, define

$$\sigma t^\lambda = (\sigma t_{i,j}).$$

To illustrate, if $\sigma = (1,2,3) \in \mathcal{S}_3$, $\lambda = (2,1) \mapsto 3$, $t^\lambda = \begin{pmatrix} 1 & 2 \\ 3 \end{pmatrix}$, then

$$\sigma t^\lambda = (\sigma t_{i,j}) = \begin{pmatrix} 2 & 3 \\ 1 \end{pmatrix}.$$

This induces an action on tabloids by letting

$$\sigma\{t^\lambda\} = \{\sigma t^\lambda\}.$$

Suppose that the tableau t^λ has columns $C_1, C_2, \cdots, C_{\lambda_1}$. Let

$$\kappa_{C_j} = \sum_{\sigma_j \in \mathcal{S}_{C_j}} sgn(\sigma_j)\sigma_j,$$

where $\mathcal{S}_{C_j}$ is a symmetric group of permutations of numbers from C_j. Let

$$\kappa_{t^\lambda} = \kappa_{C_1} \kappa_{C_2} \cdots \kappa_{C_{\lambda_1}}.$$

This is a linear combinations of elements of $\mathcal{S}_n$, so $\kappa_{t^\lambda} \in \mathbb{C}[\mathcal{S}_n]$. Now we can pass from tabloid t^λ to polytabloid

$$\mathbf{e}_{t^\lambda} = \kappa_{t^\lambda}\{t^\lambda\}.$$

Some basic properties are summarized in the following lemma.

Lemma 2.3. *(i) For any $\lambda \mapsto n$, $\{\mathbf{e}_{t^\lambda}, t^\lambda$ is a standard Young tableau$\}$ is independent;*
(ii) For any $\lambda \mapsto n$,

$$S^\lambda =: span\{\mathbf{e}_{t^\lambda}, t^\lambda \text{ is a standard Young tableau}\}$$
$$= span\{\mathbf{e}_{t^\lambda}, t^\lambda \text{ is a Young tableau}\};$$

(iii) $\{S^\lambda, \lambda \mapsto n\}$ form a complete list of irreducible $\mathcal{S}_n$-modules.

Let χ_λ be the character of matrix representation associated with S^λ, and d_λ the number of standard Young tableaux of shape λ, then we have

Theorem 2.5.

$$\chi_\lambda(1^n) = dim S^\lambda = d_\lambda$$

and

$$\sum_{\lambda \mapsto n} \chi_\lambda^2(1^n) = \sum_{\lambda \mapsto n} d_\lambda^2 = n!. \tag{2.13}$$

Formula (2.13) is often referred to as the Burnside identity. Some more information about partitions will be found in Chapters 4 and 5.

Consider the ring $\mathbb{Z}[x_1, \cdots, x_n]$ of polynomials in n independent variables $x_1, \cdots, x_n$ with rational integer coefficients. The symmetric group $\mathcal{S}_n$ acts on this ring by permuting the variables, and a polynomial is symmetric if it is invariant under this action. Let $\mathbf{\Lambda}_n := \mathbf{\Lambda}_n[x_1, \cdots, x_n]$ be the subring formed by the symmetric polynomials. We will list four classes of widely used symmetric polynomials, all indexed by partitions.

• Elementary symmetric polynomials. For each integer $r \geq 0$ the rth elementary symmetric polynomial $\mathbf{e}_r$ is the sum of all products of r distinct variables x_i, so that $\mathbf{e}_0 = 0$ and for $r \geq 1$

$$\mathbf{e}_r = \sum_{1 \leq i_1 < i_2 < \cdots < i_r \leq n} x_{i_1} x_{i_2} \cdots x_{i_r}.$$

For each partition $\lambda = (\lambda_1, \lambda_2, \cdots, \lambda_l)$ define

$$\mathbf{e}_\lambda = \mathbf{e}_{\lambda_1} \mathbf{e}_{\lambda_2} \cdots \mathbf{e}_{\lambda_l}.$$

$\mathbf{e}_r, r \geq 0$ are algebraically independent over $\mathbb{Z}$, and every element of $\mathbf{\Lambda}_n$ is uniquely expressible as a polynomial in the $\mathbf{e}_r$.

• Complete symmetric polynomials. For each integer $r \geq 0$ the rth complete symmetric polynomial h_r is the sum of all monomials of total degree r in the variables $x_1, x_2, \cdots, x_n$. In particular, $h_0 = 1$ and $h_1 = \mathbf{e}_1$. By convention, h_r and $\mathbf{e}_r$ are defined to be zero for $r < 0$. Define

$$h_\lambda = h_{\lambda_1} h_{\lambda_2} \cdots h_{\lambda_l}$$

for any partition $\lambda = (\lambda_1, \lambda_2, \cdots, \lambda_l)$. The $h_r, r \geq 0$ are algebraically independent over $\mathbb{Z}$, and

$$\mathbf{\Lambda}_n = \mathbb{Z}[h_1, h_2, \cdots, h_n]$$

• Schur symmetric polynomials. For each partition $\lambda = (\lambda_1, \lambda_2, \cdots, \lambda_l)$ with length $l \leq n$, consider the determinant

$$\det \left(x_i^{\lambda_j + n - j} \right)_{1 \leq i, j \leq n}.$$

This is divisible in $\mathbb{Z}[x_1, x_2, \cdots, x_n]$ by each of the differences $x_j - x_i$ ($1 \leq i < j \leq n$), and hence by their product $\prod_{1 \leq i < j \leq n}(x_j - x_i)$, which is the Vandermonde determinant $\det \left(x_i^{n-j} \right)_{1 \leq i, j \leq n}$. Define

$$s_\lambda := s_\lambda(x_1, \cdots, x_n) = \frac{\det(x_i^{\lambda_j + n - j})_{1 \leq i, j \leq n}}{\det(x_i^{n-j})_{1 \leq i, j \leq n}}, \tag{2.14}$$

where $s_\lambda = 0$ if the numbers $\lambda_j + n - j$ ($1 \leq j \leq n$) are not all distinct.

The quotient (2.14) is a symmetric and homogeneous polynomial of degree $|\lambda|$, that is, in Λ_n. It is called the Schur polynomial in the variable $x_1, x_2, \cdots, x_n$, corresponding to the partition λ. The Schur polynomial $s_\lambda(x_1, \cdots, x_n)$ with $l \leq n$ form a $\mathbb{Z}$-basis of Λ_n.

Each Schur polynomial can be expressed as a polynomial in the elementary symmetric polynomials e_r, and as a polynomial in the complete symmetric polynomial h_r. The formulas are:

$$s_\lambda = \det(h_{\lambda_i - i + j})_{1 \leq i, j \leq n}$$

where $l(\lambda) \leq n$, and

$$s_\lambda = \det(e_{\lambda'_i - i + j})_{1 \leq i, j \leq n}$$

where λ' is a conjugate partition with $l(\lambda') \leq n$.

In particular, we have

$$s_{(n)} = h_n, \quad s_{(1^n)} = e_n.$$

• Power sum polynomials. For each $r \geq 1$ the rth is

$$p_r := p_r(x_1, \cdots, x_n) = \sum_{i=1}^{n} x_i^r.$$

We define

$$p_\lambda = p_{\lambda_1} p_{\lambda_2} \cdots p_{\lambda_l}$$

for each partition $\lambda = (\lambda_1, \lambda_2, \cdots, \lambda_l)$. Note that p_λ, $\lambda \mapsto n$ do not form a $\mathbb{Z}$-basis of Λ_n. For instance,

$$h_2 = \frac{1}{2} p_2 + \frac{1}{2} p_1^2$$

does not have integral coefficients when expressed in terms of the p_λ. In general, for any partition $\lambda = (1^{r_1}, 2^{r_2}, \cdots)$ of n, define

$$z_\lambda = \prod_{i=1}^{\infty} i^{r_i} r_i!. \tag{2.15}$$

Then we can express h_n, e_n as linear combinations of the p_λ as follows

$$h_n = \sum_{\lambda \mapsto n} z_\lambda^{-1} p_\lambda$$

and

$$e_n = \sum_{\lambda \mapsto n} (-1)^{n - l(\lambda)} z_\lambda^{-1} p_\lambda.$$

The Schur polynomial s_λ can also be expressed as linear combinations of the p_λ

$$s_\lambda = \sum_{\rho \mapsto |\lambda|} z_\rho^{-1} \chi_\lambda(\rho) p_\rho,$$

where $\chi_\lambda(\rho)$ is the value of irreducible character χ_λ at permutations of cycle-type ρ. Conversely, we have the following inverse formula, which is called Schur-Weyl duality.

Theorem 2.6.

$$p_\lambda = \sum_{\rho \mapsto |\lambda|} \chi_\rho(\lambda) s_\rho. \tag{2.16}$$

We now define an inner product on Λ_n. Suppose $f \in \Lambda_n$, let

$$f(U_n) = f(e^{i\theta_1}, \cdots, e^{i\theta_n}),$$

where U_n is an $n \times n$ unitary matrix with eigenvalues $e^{i\theta_1}, \cdots, e^{i\theta_n}$. Thus $f : \mathcal{U}_n \mapsto \mathbb{C}$ is invariant under unitary transforms.

Suppose we are given two symmetric polynomials $f, g \in \Lambda_n$, their inner product is defined by

$$\langle f, g \rangle = \int_{\mathcal{U}_n} f(U_n) \overline{g(U_n)} d\mu_n.$$

It turns out that Schur polynomials are orthonormal with respect to this inner product, which is referred to as Schur orthonormality. In particular, we have

Theorem 2.7.

$$\langle s_\lambda, s_\tau \rangle = \delta_{\lambda,\tau} \mathbf{1}_{(l(\lambda) \leq n)}. \tag{2.17}$$

Proof. According to (2.1), we have

$$\langle f, g \rangle = \int_{\mathcal{U}_n} f(U_n) \overline{g(U_n)} d\mu_n$$

$$= \frac{1}{(2\pi)^n n!} \int_{[0, 2\pi]^n} f(e^{i\theta_1}, \cdots, e^{i\theta_n}) g(e^{-i\theta_1}, \cdots, e^{-i\theta_n})$$

$$\cdot \prod_{1 \leq j < k \leq n} |e^{i\theta_j} - e^{i\theta_k}|^2 d\theta_1 \cdots d\theta_n.$$

If λ and τ are two partitions of lengths $\leq n$, then by (2.14)

$$\langle s_\lambda, s_\tau \rangle = \frac{1}{(2\pi)^n n!} \int_{[0,2\pi]^n} s_\lambda \left(e^{i\theta_1}, \cdots, e^{i\theta_n} \right) s_\tau \left(e^{-i\theta_1}, \cdots, e^{-i\theta_n} \right)$$

$$\cdot \prod_{1 \leq j < k \leq n} \left| e^{i\theta_j} - e^{i\theta_k} \right|^2 d\theta_1 \cdots d\theta_n$$

$$= \frac{1}{(2\pi)^n n!} \int_{[0,2\pi]^n} \det \left(e^{i(\lambda_k + n - k)\theta_j} \right) \det \left(e^{-i(\tau_k + n - k)\theta_j} \right) d\theta_1 \cdots d\theta_n$$

$$= \frac{1}{n!} \left[\det(e^{i(\lambda_k + n - k)\theta_j}) \det(e^{-i(\tau_k + n - k)\theta_j}) \right]_1, \tag{2.18}$$

where $[f]_1$ denotes the constant term of f.

A simple algebra shows

$$\left[\det(e^{i(\lambda_k + n - k)\theta_j}) \det(e^{-i(\tau_k + n - k)\theta_j}) \right]_1 = n! \delta_{\lambda, \tau},$$

which together with (2.18) implies

$$\langle s_\lambda, s_\tau \rangle = \delta_{\lambda, \tau}.$$

We conclude the proof. $\qquad\square$

Having Schur orthonormality (2.17), we can further compute the inner product of power sum polynomials. For any partitions μ and ν, applying the Schur-Weyl duality (2.16) immediately yields

$$\langle p_\mu, p_\nu \rangle = \sum_{\rho \mapsto |\mu|} \sum_{\sigma \mapsto |\nu|} \chi_\rho(\mu) \chi_\sigma(\nu) \langle s_\rho, s_\sigma \rangle$$

$$= \sum_{\rho \mapsto |\mu|} \sum_{\sigma \mapsto |\nu|} \chi_\rho(\mu) \chi_\sigma(\nu) \delta_{\rho, \sigma} \mathbf{1}_{(l(\rho) \leq n)}$$

$$= \delta_{|\mu|, |\nu|} \sum_{\rho \mapsto |\mu|} \chi_\rho(\mu) \chi_\rho(\nu) \mathbf{1}_{(l(\rho) \leq n)}.$$

When $|\mu| = |\nu| \leq n$, the sum is taken over all partitions of $|\mu|$, and so the character relation (2.12) of the second kind shows

$$\langle p_\mu, p_\nu \rangle = z_\mu \delta_{\mu, \nu}, \tag{2.19}$$

where z_μ is defined as in (2.15).

2.3 Linear functionals of eigenvalues

Let $f : \mathbb{T} \mapsto \mathbb{R}$ be a square integrable function, that is, $f \in L^2(\mathbb{T}, d\mu)$. Define the Fourier coefficients by

$$\hat{f}_l = \frac{1}{2\pi} \int_0^{2\pi} f(e^{i\theta}) e^{-il\theta} d\theta, \quad -\infty < l < \infty,$$

so that $\hat{f}_0$ is the average of f over $\mathbb{T}$. Since f is real, then $\hat{f}_{-l} = \hat{f}_l$.

In this section we shall focus on the fluctuation of linear eigenvalue statistic $\sum_{k=1}^{n} f(e^{i\theta_k})$ around the average $n\hat{f}_0$.

Theorem 2.8. *If $f \in L^2(\mathbb{T}, d\mu)$ is such that $\sum_{l=1}^{\infty} l|\hat{f}_l|^2 < \infty$, then*

$$\sum_{k=1}^{n} f(e^{i\theta_k}) - n\hat{f}_0 \xrightarrow{d} N(0, \sigma_f^2), \quad n \to \infty \qquad (2.20)$$

where $\sigma_f^2 = 2\sum_{l=1}^{\infty} l|\hat{f}_l|^2$.

This theorem goes back to Szegö as early as in 1950s, and is now known as Szegö's strong limit theorem. There exist at least six different proofs with slight different assumptions on f in literature, and the most classical one uses the orthogonal polynomials on the unit circle $\mathbb{T}$. Here we prefer to prove the theorem using the moment method of Diaconis and Evans (2001), Diaconis (2003). The interested reader is referred to Simon (2004) for other five proofs. See also a recent survey of Deift, Its and Krasovsky (2012) for extensions and applications.

Lemma 2.4. *Suppose that $Z = X + iY$ is a complex standard normal random variable, then for any non-negative integers a and b*

$$EZ^a \bar{Z}^b = a!\delta_{a,b}.$$

Proof. Z can clearly be written in polar coordinates as follows:

$$Z = \gamma e^{i\theta},$$

where γ and θ are independent, θ is uniform over $[0, 2\pi]$, and γ has density function $2re^{-r^2}$, $r \geq 0$. It easily follows

$$\begin{aligned}
EZ^a \bar{Z}^b &= E\gamma^{a+b} e^{i\theta(a-b)} \\
&= E\gamma^{a+b} E e^{i\theta(a-b)} \\
&= E\gamma^{2a} \delta_{a,b} = a!\delta_{a,b},
\end{aligned}$$

as desired. $\qquad\qquad\qquad\qquad\qquad\qquad\qquad\qquad\qquad\qquad\qquad\qquad\Box$

Lemma 2.5. *(i) Suppose that $Z_l, l \geq 1$ is a sequence of i.i.d. complex standard normal random variables. Then for any $m \geq 1$ and any non-negative integers $a_1, a_2, \cdots, a_m$ and $b_1, b_2, \cdots, b_m$*

$$\begin{aligned}
&E\prod_{l=1}^{m}\left(\sum_{k=1}^{n} e^{il\theta_k}\right)^{a_l} \prod_{l=1}^{m}\left(\sum_{k=1}^{n} e^{-il\theta_k}\right)^{b_l} \\
&= E\prod_{l=1}^{m}\left(\sqrt{l}Z_l\right)^{a_l} \prod_{l=1}^{m}\left(\sqrt{l}\bar{Z}_l\right)^{b_l}
\end{aligned} \qquad (2.21)$$

whenever $n \geq \max\left\{\sum_{l=1}^{m} la_l, \sum_{l=1}^{m} lb_l\right\}$.

(ii) For any integer $j, l \geq 1$,

$$E \sum_{k=1}^{n} e^{il\theta_k} \sum_{k=1}^{n} e^{-ij\theta_k} = \delta_{j,l} \min(l, n). \tag{2.22}$$

Proof. Recall the lth power sum polynomial

$$p_l\left(e^{i\theta_1}, \cdots, e^{i\theta_n}\right) = \sum_{k=1}^{n} e^{il\theta_k}.$$

Then it follows

$$\prod_{l=1}^{m}\left(\sum_{k=1}^{n} e^{il\theta_k}\right)^{a_l} = p_\lambda\left(e^{i\theta_1}, \cdots, e^{i\theta_n}\right)$$

and

$$\prod_{l=1}^{m}\left(\sum_{k=1}^{n} e^{-il\theta_k}\right)^{b_l} = \overline{p_\mu\left(e^{i\theta_1}, \cdots, e^{i\theta_n}\right)}$$

where $\lambda = \left(1^{a_1}, 2^{a_2}, \cdots, m^{a_m}\right)$ and $\mu = \left(1^{b_1}, 2^{b_2}, \cdots, m^{b_m}\right)$.

According to the orthogonality relation in (2.19), we have

$$E \prod_{l=1}^{m}\left(\sum_{k=1}^{n} e^{il\theta_k}\right)^{a_l} \prod_{l=1}^{m}\left(\sum_{k=1}^{n} e^{-il\theta_k}\right)^{b_l}$$

$$= \langle p_\lambda, p_\mu \rangle$$

$$= \delta_{\lambda,\mu} \prod_{l=1}^{m} l^{a_l} a_l! \tag{2.23}$$

whenever $|\lambda|, |\mu| \leq n$. Now we can get the identity (2.21) using Lemma 2.4.

Turn to (2.22). We immediately know from (2.23) that the expectation is zero if $j \neq l$, while the case $j = l$ has been proven in (2.6). $\square$

As an immediate consequence, we have the following

Theorem 2.9. *For each integer* $m \geq 1$

$$\left(\sum_{k=1}^{n} e^{il\theta_k}, \quad 1 \leq l \leq m\right) \xrightarrow{d} \left(\sqrt{l}\, Z_l, \quad 1 \leq l \leq m\right), \quad n \to \infty.$$

In particular, it holds

$$\sum_{l=-m}^{m} \hat{f}_l \sum_{k=1}^{n} e^{-il\theta_k} - n\hat{f}_0 \xrightarrow{d} N\left(0, \sigma_{m,f}^2\right), \quad n \to \infty$$

where $\sigma_{m,f}^2 = 2\sum_{l=1}^{m} l|\hat{f}_l|^2$.

Proof of Theorem 2.8. Since $\sum_{l=-\infty}^{\infty} |\hat{f}_l|^2 < \infty$, then we can express f in terms of Fourier series

$$f(e^{i\theta}) = \sum_{l=-\infty}^{\infty} \hat{f}_l e^{-il\theta},$$

from which it follows

$$\sum_{k=1}^{n} f(e^{i\theta_k}) = \sum_{l=-\infty}^{\infty} \hat{f}_l \sum_{k=1}^{n} e^{-il\theta_k}$$

$$= n\hat{f}_0 + \sum_{l=1}^{\infty} \hat{f}_l \sum_{k=1}^{n} e^{-il\theta_k} + \overline{\sum_{l=1}^{\infty} \hat{f}_l \sum_{k=1}^{n} e^{-il\theta_k}}.$$

It is sufficient for us to establish the following statement: there exists a sequence of numbers m_n with $m_n \to \infty$ and $m_n/n \to 0$ such that
(i)

$$\sum_{l=m_n+1}^{\infty} \hat{f}_l \sum_{k=1}^{n} e^{-il\theta_k} \xrightarrow{P} 0; \qquad (2.24)$$

(ii)

$$\sum_{l=1}^{m_n} \hat{f}_l \sum_{k=1}^{n} e^{-il\theta_k} \xrightarrow{d} \left(\sum_{l=1}^{\infty} l|\hat{f}_l|^2 \right)^{1/2} N_{\mathbb{C}}(0,1), \qquad (2.25)$$

where $N_{\mathbb{C}}(0,1)$ denotes a complex standard normal random variable.

Indeed, for any sequence of numbers m_n with $m_n \to \infty$, we have by (2.22)

$$E \left| \sum_{l=m_n+1}^{\infty} \hat{f}_l \sum_{k=1}^{n} e^{-il\theta_k} \right|^2 = \sum_{l=m_n+1}^{\infty} \min(l,n)|\hat{f}_l|^2$$

$$\leq \sum_{l=m_n+1}^{\infty} l|\hat{f}_l|^2$$

$$\to 0,$$

which directly implies (2.24) using the Markov inequality.

For (2.25), note the moment identity (2.21) is applicable to yield

$$E \left(\sum_{l=1}^{m_n} \hat{f}_l \sum_{k=1}^{n} e^{-il\theta_k} \right)^a \left(\sum_{l=1}^{m_n} \bar{\hat{f}}_l \sum_{k=1}^{n} e^{il\theta_k} \right)^b$$

$$= E \left(\sum_{l=1}^{m_n} \hat{f}_l \sqrt{l}\, Z_l \right)^a \left(\sum_{l=1}^{m_n} \bar{\hat{f}}_l \sqrt{l}\, \bar{Z}_l \right)^b$$

$$= \left(\sum_{l=1}^{m_n} l|\hat{f}_l|^2 \right)^a a! \delta_{a,b} \qquad (2.26)$$

whenever $n \geq m_n(a+b)$.

If $m_n/n \to 0$, then for any non-negative integer numbers a and b, the assumption $n \geq m_n(a+b)$ can be guaranteed for sufficiently large n. Now we can conclude the claim (2.25) by letting $n \to \infty$ in (2.26). The proof is now complete. $\qquad\qquad\qquad\qquad\qquad\qquad\qquad\qquad\qquad\qquad\qquad\qquad\qquad$ $\square$

Remark 2.1. A remarkable feature in Theorem 2.8 is that there is no normalizing constant in left hand side of (2.20). Recall that there is a normalizing constant $1/\sqrt{n}$ in the central limit theorem for sums of i.i.d. random variables with finite variance. This further manifests that the eigenvalues of the CUE spread out more regularly on the unit circle $\mathbb{T}$ than independent uniform points. This phenomena also appears in the central limit theorem for linear functional of eigenvalues of the Gaussian Unitary Ensemble (GUE), see Chapter 3 below.

The following result shows that even when

$$\sum_{l=1}^{\infty} l|\hat{f}_l|^2 = \infty,$$

the central limit theorem for $\sum_{k=1}^{n} f(e^{i\theta_k})$ after properly scaled still holds under a weak additional assumption. Recall that a positive sequence $\{c_k\}$ is said to be slowly varying if for any $\alpha > 0$

$$\lim_{k \to \infty} \frac{c_{\lfloor \alpha k \rfloor}}{c_k} = 1.$$

Theorem 2.10. *Suppose that $f \in L^2(\mathbb{T}, d\mu)$ is such that*

$$B_n := \sum_{l=1}^{n} l|\hat{f}_l|^2, \quad n \geq 1$$

is slowly varying. Then

$$\frac{1}{\sqrt{2B_n}} \Big(\sum_{k=1}^{n} f(e^{i\theta_k}) - n\hat{f}_0 \Big) \xrightarrow{d} N(0,1).$$

Proof. As in the proof of Theorem 2.8, it follows

$$\sum_{k=1}^{n} f(e^{i\theta_k}) - n\hat{f}_0 = \sum_{l=1}^{\infty} \hat{f}_l \sum_{k=1}^{n} e^{-il\theta_k} + \sum_{l=1}^{\infty} \bar{\hat{f}}_l \sum_{k=1}^{n} e^{il\theta_k}.$$

It is enough to prove

$$\frac{1}{\sqrt{B_n}} \sum_{l=1}^{\infty} \hat{f}_l \sum_{k=1}^{n} e^{-il\theta_k} \xrightarrow{d} N_{\mathbb{C}}(0,1).$$

Because B_n, $n \geq 1$ is slowly varying, there must be a sequence of integers m_n such that as $n \to \infty$

$$m_n \to \infty, \quad \frac{m_n}{n} \to 0$$

and

$$\frac{B_{m_n}}{B_n} \to 1. \tag{2.27}$$

We shall establish the following statements:

(i)

$$\frac{1}{\sqrt{B_n}} \sum_{l=m_n+1}^{\infty} \hat{f}_l \sum_{k=1}^{n} e^{-il\theta_k} \xrightarrow{P} 0; \tag{2.28}$$

(ii)

$$\frac{1}{\sqrt{B_n}} \sum_{l=1}^{m_n} \hat{f}_l \sum_{k=1}^{n} e^{-il\theta_k} \xrightarrow{d} N_{\mathbb{C}}(0,1). \tag{2.29}$$

According to (2.22),

$$E\left| \sum_{l=m_n+1}^{\infty} \hat{f}_l \sum_{k=1}^{n} e^{-il\theta_k} \right|^2 = \sum_{l=m_n+1}^{\infty} |\hat{f}_l|^2 \min(l,n)$$

$$= \sum_{l=m_n+1}^{n} l|\hat{f}_l|^2 + n \sum_{l=n+1}^{\infty} |\hat{f}_l|^2.$$

Summing by parts,

$$2 \sum_{l=n+1}^{\infty} |\hat{f}_l|^2 = \sum_{l=n}^{\infty} (B_{l+1} - B_l) \frac{1}{l+1}$$

$$= \sum_{l=n}^{\infty} \frac{B_l}{l(l+1)} - \frac{B_n}{n+1}.$$

Since B_n, $n \geq 1$ is slowly varying,

$$\frac{n}{B_n} \sum_{l=n}^{\infty} \frac{B_l}{l(l+1)} \to 1.$$

Putting these together implies

$$\frac{1}{B_n} E\left| \sum_{l=m_n+1}^{\infty} \hat{f}_l \sum_{k=1}^{n} e^{-il\theta_k} \right|^2 \to 0,$$

which in turn implies (2.28).

Turn to (2.29). Fix non-negative integers a and b. Since $m_n/n \to 0$, then $n \geq m_n(a+b)$ for sufficiently large n. So (2.21) is applicable to yield

$$E\left(\sum_{l=1}^{m_n} \hat{f}_l \sum_{k=1}^{n} e^{-il\theta_k}\right)^a \left(\sum_{l=1}^{m_n} \bar{\hat{f}}_l \sum_{k=1}^{n} e^{il\theta_k}\right)^b$$

$$= \left(\sum_{l=1}^{m_n} l|\hat{f}_l|^2\right)^a a! \delta_{a,b}. \tag{2.30}$$

Thus (2.29) is valid from (2.27) and (2.30). The proof is now complete. □

To conclude this section, we shall look at two interesting examples. The first one is the distribution of values taken by the logarithm of characteristic polynomial of a random unitary matrix. Recall the characteristic polynomial of a matrix U_n is defined by the determinant

$$\det(zI_n - U_n).$$

Fix $z = e^{i\theta_0}$ and assume U_n is from the CUE. Since $e^{i\theta_0}$ is almost surely not an eigenvalue of U_n, then

$$\det\left(e^{i\theta_0}I_n - U_n\right) = \prod_{k=1}^{n}\left(e^{i\theta_0} - e^{i\theta_k}\right) \neq 0.$$

It is fascinating that the logarithm of $\det\left(e^{i\theta_0}I_n - U_n\right)$ after properly scaled weakly converges to a normal distribution, analogous to Selberg's result on the normal distribution of values of the logarithm of the Riemann zeta function. This was first observed by Keating and Snaith (2000), which argued that the Riemann zeta function on the critical line could be modelled by the characteristic polynomial of a random unitary matrix.

Theorem 2.11. *As $n \to \infty$,*

$$\frac{1}{\sqrt{\log n}}\left(\log \det(e^{i\theta_0}I_n - U_n) - in\theta_0\right) \xrightarrow{d} N_{\mathbb{C}}(0,1),$$

where $\log$ *denotes the usual branch of the logarithm defined on* $\mathbb{C} \setminus \{z : Re(z) \leq 0\}$.

Proof. First observe

$$\log \det\left(e^{i\theta_0}I_n - U_n\right) - in\theta_0 = \sum_{k=1}^{n} \log\left(1 - e^{i(\theta_k - \theta_0)}\right).$$

According to Weyl's formula,

$$\left(e^{i(\theta_k - \theta_0)}, \quad 1 \leq k \leq n\right) \stackrel{d}{=} \left(e^{i\theta_k}, \quad 1 \leq k \leq n\right).$$

Hence it suffices to prove

$$\frac{1}{\sqrt{\log n}} \sum_{k=1}^{n} \log\left(1 - e^{i\theta_k}\right) \xrightarrow{d} N_{\mathbb{C}}(0,1).$$

Note that it follows for any $n \geq 1$

$$\sum_{k=1}^{n} \log\left(1 - e^{i\theta_k}\right) = -\sum_{l=1}^{\infty} \frac{1}{l} \sum_{k=1}^{n} e^{il\theta_k}, \quad \text{a.s.} \tag{2.31}$$

Indeed, for any real $r > 1$

$$\log\left(1 - \frac{e^{i\theta_k}}{r}\right) = -\sum_{l=1}^{\infty} \frac{1}{lr^l} e^{il\theta_k},$$

and so

$$\sum_{k=1}^{n} \log\left(1 - \frac{e^{i\theta_k}}{r}\right) = -\sum_{l=1}^{\infty} \frac{1}{lr^l} \sum_{k=1}^{n} e^{il\theta_k}.$$

Thus we have by virtue of (2.22)

$$E\left|\sum_{l=1}^{\infty} \frac{1}{l}\left(\frac{1}{r^l} - 1\right) \sum_{k=1}^{n} e^{il\theta_k}\right|^2 = \sum_{l=1}^{\infty} \frac{1}{l^2}\left(\frac{1}{r^l} - 1\right)^2 \min(l,n)$$

$$= \sum_{l=1}^{n} \frac{1}{l}\left(\frac{1}{r^l} - 1\right)^2 + \sum_{l=n+1}^{\infty} \frac{1}{l^2}\left(\frac{1}{r^l} - 1\right)^2.$$

Letting $r \to 1+$ easily yields

$$E\left|\sum_{l=1}^{\infty} \frac{1}{l}\left(\frac{1}{r^l} - 1\right) \sum_{k=1}^{n} e^{il\theta_k}\right|^2 \to 0,$$

which in turn implies (2.31).

Now we need only prove

$$\frac{1}{\sqrt{\log n}} \sum_{l=1}^{\infty} \frac{1}{l} \sum_{k=1}^{n} e^{il\theta_k} \xrightarrow{d} N_{\mathbb{C}}(0,1).$$

The proof is very similar to that of Theorem 2.8. Let $m_n = n/\log n$ so that $m_n \to \infty$ and $m_n/n \to 0$. We shall establish the following statements:
(i)

$$\frac{1}{\sqrt{\log n}} \sum_{l=m_n+1}^{\infty} \frac{1}{l} \sum_{k=1}^{n} e^{il\theta_k} \xrightarrow{P} 0; \tag{2.32}$$

(ii)

$$\frac{1}{\sqrt{\log n}} \sum_{l=1}^{m_n} \frac{1}{l} \sum_{k=1}^{n} e^{il\theta_k} \xrightarrow{d} N_{\mathbb{C}}(0,1). \tag{2.33}$$

According to (2.22), it holds

$$E\left| \sum_{l=m_n+1}^{\infty} \frac{1}{l} \sum_{k=1}^{n} e^{il\theta_k} \right|^2 = \sum_{l=m_n+1}^{n} \frac{1}{l} + n \sum_{l=n+1}^{\infty} \frac{1}{l^2}$$
$$= O(\log\log n),$$

which together with the Markov inequality directly implies (2.32).

To prove (2.33), note for any non-negative integers a and b,

$$E\left(\sum_{l=1}^{m_n} \frac{1}{l} \sum_{k=1}^{n} e^{il\theta_k} \right)^a \left(\sum_{l=1}^{m_n} \frac{1}{l} \sum_{k=1}^{n} e^{-il\theta_k} \right)^b$$
$$= \left(\sum_{l=1}^{m_n} \frac{1}{l} \right)^a a! \delta_{a,b}$$
$$= (1 + o(1))(\log n)^a a! \delta_{a,b},$$

as desired. □

The second example of interest is the numbers of eigenvalues lying in an arc. For $0 \le a < b < 2\pi$, write $N_n(a,b)$ for the number of eigenvalues $e^{i\theta_k}$ with $\theta_k \in [a,b]$. Particularly speaking,

$$N_n(a,b) = \sum_{k=1}^{n} \mathbf{1}_{(a,b)}(\theta_k).$$

Since each eigenvalue $e^{i\theta_k}$ is uniform over $\mathbb{T}$, then

$$EN_n(a,b) = \frac{n(b-a)}{2\pi}.$$

The following theorem, due to Wieand (1998) (see also Diaconis and Evans (2001)), shows that the fluctuation of $N_n(a,b)$ around the mean is asymptotically normal. It is worth mentioning that the asymptotic variance $\log n$ (up to a constant) is very typical in the study of numbers of points like eigenvalues in an interval. The reader will again see it in the study of GUE and random Plancherel partitions.

Theorem 2.12. *For $0 \le a < b < 2\pi$, as $n \to \infty$*

$$\frac{N_n(a,b) - \frac{n(b-a)}{2\pi}}{\frac{1}{\pi}\sqrt{\log n}} \xrightarrow{d} N(0,1). \tag{2.34}$$

Proof.   (2.34) is actually a direct consequence of Theorem 2.10. Indeed, set

$$f\left(e^{i\theta}\right) = 1_{(a,\,b)}(\theta).$$

Then a simple calculation shows

$$\hat{f}_0 = \frac{b-a}{2\pi}, \quad \hat{f}_l = \frac{1}{l2\pi \mathbf{i}}\left(e^{-ila} - e^{-ilb}\right), \quad l \neq 0$$

and so

$$B_n := \sum_{l=1}^{n} l|\hat{f}_l|^2$$

$$= \frac{1}{4\pi^2}\sum_{l=1}^{n}\frac{1}{l}\left|e^{-ila} - e^{-ilb}\right|^2$$

$$= \frac{1}{4\pi^2}\sum_{l=1}^{n}\frac{1}{l}\left(2 + 2\cos l(b-a)\right).$$

On the other hand, an elementary calculus shows

$$\frac{1}{\log n}\sum_{l=1}^{n}\frac{\cos l(b-a)}{l} \to 0, \quad n \to \infty.$$

Hence B_n is a slowly varying and

$$\frac{B_n}{\log n} \to \frac{1}{2\pi^2}, \quad n \to \infty.$$

The proof is complete.   □

The above theorem deals only with the number of eigenvalues in a single arc. In a very similar way, employing the Cramér-Wald device, one may prove a finite dimensional normal convergence for multiple arcs.

Theorem 2.13. *As $n \to \infty$, the finite dimensional distribution of the processes*

$$\frac{N_n(a,b) - \frac{n(b-a)}{2\pi}}{\frac{1}{\pi}\sqrt{\log n}}, \quad 0 \leq a < b < 2\pi$$

converges to those of a centered Gaussian process $\{Z(a,b) : 0 \leq a < b < 2\pi\}$ with the covariance structure

$$EZ(a,b)Z(a',b') = \begin{cases} 1, & \textit{if } a = a' \textit{ and } b = b', \\ \frac{1}{2}, & \textit{if } a = a' \textit{ and } b \neq b', \\ \frac{1}{2}, & \textit{if } a \neq a' \textit{ and } b = b', \\ -\frac{1}{2}, & \textit{if } b = a', \\ 0, & \textit{otherwise.} \end{cases}$$

Proof.   See Theorem 6.1 of Diaconis (2001).   □

2.4   Five diagonal matrix models

This section is aimed to establish a five diagonal sparse matrix model for the CUE and to provide an alternate approach to asymptotic normality of the characteristic polynomials and the number of eigenvalues lying in an arc. We first introduce basic notions of orthogonal polynomials and Verblunsky coefficients associated to a finitely supported measure on the unit circle, and quickly review some well-known facts, including the Szegö recurrence equations and Verblunsky's theorem. The measure we will be concerned with is the spectral measure induced by a unitary matrix and a cyclic vector. Two matrices of interest to us are upper triangular Hessenberg matrix and CMV five diagonal matrix, whose Verblunsky coefficients can be expressed in a simple way. Then we turn to a random unitary matrix distributed with Haar measure. Particularly interesting, the associated Verblunsky coefficients are independent Θ_v-distributed complex random variables. Thus as a consequence of Verblunsky's theorem, we naturally get a five diagonal matrix model for the CUE. Lastly, we rederive Theorems 2.11 and 2.12 via a purely probabilistic approach: use only the classical central limit theorems for sums of independent random variables and martingale difference sequences.

Assume we are given a finitely supported probability measure $d\nu$ on exactly n points $e^{i\theta_1}, e^{i\theta_2}, \cdots, e^{i\theta_n}$ with masses $\nu_1, \nu_2, \cdots, \nu_n$, where $\nu_i > 0$ and $\sum_{i=1}^{n} \nu_i = 1$. Let $L^2(\mathbb{T}, d\nu)$ be the of square integrable functions on $\mathbb{T}$ with respective to $d\nu$ with the inner product given by

$$\langle f, g \rangle = \int_{\mathbb{T}} f(e^{i\theta}) \overline{g(e^{i\theta})} d\nu.$$

Applying the Gram-Schmidt algorithm to the ordered set $\{1, z, \cdots, z^{n-1}\}$, we can get a sequence of orthogonal polynomials $\Phi_0, \Phi_1, \cdots, \Phi_{n-1}$, where

$$\Phi_0(z) = 1, \quad \Phi_k(z) = z^k + \text{lower order}.$$

Define the Szegö dual by

$$\Phi_k^*(z) = z^k \overline{\Phi_k(\bar{z}^{-1})}.$$

Namely,

$$\Phi_k(z) = \sum_{l=0}^{k} c_l z^l \quad \Rightarrow \quad \Phi_k^*(z) = \sum_{l=0}^{k} \bar{c}_{k-l} z^l.$$

As Szegö discovered, there exist complex constants $\alpha_0, \alpha_1, \cdots, \alpha_{n-2} \in \mathbb{D}$, where $\mathbb{D} := \{z \in \mathbb{C}, |z| < 1\}$, such that for $0 \leq k \leq n-2$

$$\Phi_{k+1}(z) = z\Phi_k(z) - \bar{\alpha}_k \Phi_k^*(z) \tag{2.35}$$

and

$$\Phi_{k+1}^*(z) = \Phi_k^*(z) - \alpha_k z \Phi_k(z). \tag{2.36}$$

Expanding z^n in this basis shows that there exists an α_{n-1}, say $\alpha_{n-1} = e^{i\eta} \in \mathbb{T}$, $0 \le \eta < 2\pi$, such that if letting

$$\Phi_n(z) = z\Phi_{n-1}(z) - \bar{\alpha}_{n-1}\Phi_{n-1}^*(z), \tag{2.37}$$

then

$$\Phi_n(z) = 0 \quad \text{in } L^2(\mathbb{T}, d\nu).$$

Define

$$\rho_k = \sqrt{1 - |\alpha_k|^2}, \quad 0 \le k \le n - 1,$$

then it follows from recurrence relations (2.35) and (2.36)

$$\|\Phi_0\| = 1, \quad \|\Phi_k\| = \prod_{l=0}^{k-1} \rho_l, \quad k \ge 1. \tag{2.38}$$

The orthonormal polynomial ϕ_k is defined by

$$\phi_k(z) = \frac{\Phi_k(z)}{\|\Phi_k\|}.$$

We call α_k, $0 \le k \le n - 1$ the Verblunsky coefficients associated to the measure $d\nu$, which play an important role in the study of unitary matrices. We sometimes write $\alpha_k(d\nu)$ for α_k to emphasize the dependence on the underlying measure $d\nu$. A basic fact we need below is

Theorem 2.14. *There is a one-to-one correspondence between the finitely supported probability measure $d\nu$ on $\mathbb{T}$ and complex numbers $\alpha_0, \alpha_1, \cdots,$ α_{n-1} with $\alpha_0, \alpha_1, \cdots, \alpha_{n-2} \in \mathbb{D}$ and $\alpha_{n-1} \in \mathbb{T}$.*

This theorem is now called Verblunsky's theorem (also called Favard's theorem for the circle). The reader is referred to Simon (2004) for the proof (at least four proofs are presented).

It is very expedient to encode the Szegö recurrence relation (2.35) and (2.36). Let

$$B_k(z) = z\frac{\Phi_k(z)}{\Phi_k^*(z)}.$$

It easily follows

$$B_0(z) = z, \quad B_{k+1}(z) = zB_k(z)\frac{1 - \bar{\alpha}_k\bar{B}_k(z)}{1 - \alpha_k B_k(z)}, \quad z \in \mathbb{T}, \tag{2.39}$$

which shows that the $B_k(z)$ can be completely expressed in terms of the Verblunsky coefficients α_k.

Note in view of (2.39), B_k is a finite Blaschke product of degree $k+1$. Define a continuous function $\psi_k(\theta) : [0, 2\pi] \mapsto \mathbb{R}$ via

$$B_k\left(e^{i\theta}\right) = e^{i\psi_k(\theta)}. \tag{2.40}$$

$\psi_k(\theta)$ is the absolute Präfer phase of B_k, so the set of points

$$\left\{e^{i\theta} : B_{n-1}(e^{i\theta}) = \bar{\alpha}_{n-1}\right\} = \left\{e^{i\theta} : \psi_{n-1}(\theta) \in 2\pi\mathbb{Z} + \eta\right\}$$

is the support of $d\nu$.

Also, $\psi_k(\theta)$ is a strictly increasing function of θ. To avoid ambiguity, we may choose a branch of the logarithm in (2.39) so that

$$\psi_0(\theta) = \theta, \quad \psi_{k+1}(\theta) = \psi_k(\theta) + \theta + \Upsilon(\psi_k, \alpha_k), \tag{2.41}$$

where $\Upsilon(\psi, \alpha) = -2Im \log(1 - \alpha e^{i\psi})$.

Let $\mathbb{C}[z]$ be the vector space of complex polynomials in the variable z. Consider the multiplication operator $\Pi : f(z) = zf(z)$ in $\mathbb{C}(z)$. We easily obtain an explicit expression of Π in the basis of orthonormal polynomials ϕ_k, $0 \le k \le n-1$. In particular,

$$\Pi \begin{pmatrix} \phi_0 \\ \phi_1 \\ \vdots \\ \phi_{n-2} \\ \phi_{n-1} \end{pmatrix} = H_n^L \begin{pmatrix} \phi_0 \\ \phi_1 \\ \vdots \\ \phi_{n-2} \\ \phi_{n-1} \end{pmatrix} + \begin{pmatrix} 0 \\ 0 \\ \vdots \\ 0 \\ \frac{\Phi_n}{\|\Phi_{n-1}\|} \end{pmatrix},$$

where $H_n^L = \left(H_{ij}^L\right)_{0 \le i,j \le n-1}$ is an lower triangular Hessenberg matrix given by

$$H_{ij}^L = \begin{cases} -\alpha_{j-1}\bar{\alpha}_i \prod_{l=j+1}^i \rho_l, & j \le i-1, \\ -\alpha_{i-1}\bar{\alpha}_i, & j = i, \\ \rho_i, & j = i+1, \\ 0, & j > i+1. \end{cases}$$

A simple algebra further shows that the characteristic polynomial of H_n^L is equal to the nth polynomial $\Phi_n(z)$ defined in (2.37). Namely,

$$\det(zI_n - H_n^L) = \Phi_n(z), \tag{2.42}$$

which implies $e^{i\theta_1}, \cdots, e^{i\theta_n}$ are the spectrum of H_n^L. So, the spectral analysis of H_n^L can give relations between the zeros of orthogonal polynomials and the Verblunsky coefficients. However, H_n^L is a far from sparse matrix and the entries H_{ij}^L depend on the Verblunsky coefficients α_k and ρ_k in a complicated way. This makes difficult this task. Moreover, the numerical computations of zeros of high degree orthogonal polynomials becomes a nontrivial problem due to the Hessenberg structure of H_n^L.

To overcome this difficulty, Cantero, Moral, and Velázquez (2003) used a simple and ingenious idea. Applying the Gram-Schmidt procedure to the first n of the ordered set $\{1, z, z^{-1}, z^2, z^{-2}, \cdots\}$ rather than $\{1, z, \cdots, z^{n-1}\}$, we can get a sequence of orthogonal Laurent polynomials, denoted by $\chi_k(z)$, $0 \leq k \leq n-1$. We will refer to the χ_k as the standard right orthonormal L-polynomial with respect to the measure $d\nu$. Interestingly, the χ_k can be expressed in terms of the orthonormal polynomial ϕ_k and its Szegö dual ϕ_k^* as follows:

$$\chi_{2k+1}(z) = z^{-k}\phi_{2k+1}(z),$$
$$\chi_{2k}(z) = z^{-k}\phi_{2k}^*(z).$$

Similarly, applying the Gram-Schmidt procedure to the first n of the ordered set of $\{1, z^{-1}, z, z^{-2}, z^2, \cdots\}$, we can get another sequence of orthogonal Laurent polynomial, denoted by χ_{k*}. We call the χ_{k*} the standard left orthogonal L-polynomial. It turns out that the χ_k and χ_{k*} are closely related to each other through the equation:

$$\chi_{k*}(z) = \bar{\chi}_k(z^{-1}).$$

Define

$$\Xi_k = \begin{pmatrix} \bar{\alpha}_k & \rho_k \\ \rho_k & -\alpha_k \end{pmatrix}$$

for $0 \leq k \leq n-2$, while $\Xi_{-1} = (1)$ and $\Xi_{n-1} = (\bar{\alpha}_{n-1})$ are 1×1 matrices. Then it readily follows from the Szegö recurrence relation that

$$\begin{pmatrix} \chi_{2k}(z) \\ \chi_{2k*}(z) \end{pmatrix} = \frac{1}{\rho_{2k-1}} \begin{pmatrix} -\alpha_{2k-1} & 1 \\ 1 & -\bar{\alpha}_{2k-1} \end{pmatrix} \begin{pmatrix} \chi_{2k-1}(z) \\ \chi_{2k-1*}(z) \end{pmatrix},$$
$$\begin{pmatrix} \chi_{2k-1}(z) \\ \chi_{2k}(z) \end{pmatrix} = \Xi_{2k-1} \begin{pmatrix} \chi_{2k-1*}(z) \\ \chi_{2k*}(z) \end{pmatrix},$$
$$z \begin{pmatrix} \chi_{2k*}(z) \\ \chi_{2k+1*}(z) \end{pmatrix} = \Xi_{2k} \begin{pmatrix} \chi_{2k}(z) \\ \chi_{2k+1}(z) \end{pmatrix}.$$

It can be further written as a five term recurrence equation:

$$z\chi_0(z) = \bar{\alpha}_0\chi_0(z) + \rho_0\chi_1(z),$$

$$z \begin{pmatrix} \chi_{2k-1}(z) \\ \chi_{2k}(z) \end{pmatrix} = \begin{pmatrix} \rho_{2k-2}\bar{\alpha}_{2k-1} & -\alpha_{2k-2}\bar{\alpha}_{2k-1} \\ \rho_{2k-2}\rho_{2k-1} & -\alpha_{2k-2}\rho_{2k-1} \end{pmatrix} \begin{pmatrix} \chi_{2k-2}(z) \\ \chi_{2k-1}(z) \end{pmatrix}$$

$$+ \begin{pmatrix} \rho_{2k-1}\bar{\alpha}_{2k} & \rho_{2k-1}\rho_{2k} \\ -\alpha_{2k-1}\bar{\alpha}_{2k} & -\alpha_{2k-1}\rho_{2k} \end{pmatrix} \begin{pmatrix} \chi_{2k}(z) \\ \chi_{2k+1}(z) \end{pmatrix}.$$

Construct now the $n \times n$ block diagonal matrices

$$\mathcal{L} = diag(\Xi_0, \Xi_2, \Xi_4, \cdots), \quad \mathcal{M} = diag(\Xi_{-1}, \Xi_1, \Xi_3, \cdots)$$

and define

$$\mathcal{C}_n = \mathcal{ML}, \quad \mathcal{C}_n^\tau = \mathcal{LM}. \tag{2.43}$$

It is easy to check that $\mathcal{L}$ and $\mathcal{M}$ are symmetric unitary matrices, and so both $\mathcal{C}_n$ and $\mathcal{C}_n^\tau$ are unitary.

A direct manipulation of matrix product shows that $\mathcal{C}_n$ is a five diagonal sparse matrix. Specifically speaking, if $n = 2k$, then $\mathcal{C}_n$ is equal to

$$\begin{pmatrix} \bar{\alpha}_0 & \rho_0 & 0 & 0 & 0 & \cdots & 0 & 0 \\ \rho_0\bar{\alpha}_1 & -\alpha_0\bar{\alpha}_1 & \rho_1\bar{\alpha}_2 & \rho_1\rho_2 & 0 & \cdots & 0 & 0 \\ \rho_0\rho_1 & -\alpha_0\rho_1 & -\alpha_1\bar{\alpha}_2 & -\alpha_1\rho_2 & 0 & \cdots & 0 & 0 \\ 0 & 0 & \rho_2\bar{\alpha}_3 & -\alpha_2\bar{\alpha}_3 & \rho_3\bar{\alpha}_4 & \cdots & 0 & 0 \\ 0 & 0 & \rho_2\rho_3 & -\alpha_2\rho_3 & -\alpha_3\alpha_4 & \cdots & 0 & 0 \\ \vdots & \vdots & \vdots & \vdots & \vdots & \ddots & \vdots & \vdots \\ 0 & 0 & 0 & 0 & 0 & \cdots & -\alpha_{n-3}\bar{\alpha}_{n-2} & -\alpha_{n-3}\rho_{n-2} \\ 0 & 0 & 0 & 0 & 0 & \cdots & \rho_{n-2}\bar{\alpha}_{n-1} & -\alpha_{n-2}\bar{\alpha}_{n-1} \end{pmatrix},$$

while if $n = 2k + 1$, then $\mathcal{C}_n$ is equal to

$$\begin{pmatrix} \bar{\alpha}_0 & \rho_0 & 0 & 0 & \cdots & 0 & 0 & 0 \\ \rho_0\bar{\alpha}_1 & -\alpha_0\bar{\alpha}_1 & \rho_1\bar{\alpha}_2 & \rho_1\rho_2 & \cdots & 0 & 0 & 0 \\ \rho_0\rho_1 & -\alpha_0\rho_1 & -\alpha_1\bar{\alpha}_2 & -\alpha_1\rho_2 & \cdots & 0 & 0 & 0 \\ 0 & 0 & \rho_2\bar{\alpha}_3 & -\alpha_2\bar{\alpha}_3 & \cdots & 0 & 0 & 0 \\ 0 & 0 & \rho_2\rho_3 & -\alpha_2\rho_3 & \cdots & 0 & 0 & 0 \\ \vdots & \vdots & \vdots & \vdots & \ddots & \vdots & \vdots & \vdots \\ 0 & 0 & 0 & 0 & \cdots & \rho_{n-3}\bar{\alpha}_{n-2} & -\alpha_{n-3}\bar{\alpha}_{n-2} & \rho_{n-2}\bar{\alpha}_{n-1} \\ 0 & 0 & 0 & 0 & \cdots & \rho_{n-3}\rho_{n-2} & -\alpha_{n-3}\rho_{n-2} & -\alpha_{n-2}\bar{\alpha}_{n-1} \end{pmatrix}.$$

The multiplication operator $\Pi : f(z) \mapsto zf(z)$ can be explicitly expressed in the basis of χ_k, $0 \le k \le n - 1$ as follows. If $n = 2k$, then

$$\Pi \begin{pmatrix} \chi_0 \\ \chi_1 \\ \vdots \\ \chi_{n-2} \\ \chi_{n-1} \end{pmatrix} = \mathcal{C}_n \begin{pmatrix} \chi_0 \\ \chi_1 \\ \vdots \\ \chi_{n-2} \\ \chi_{n-1} \end{pmatrix} + \begin{pmatrix} 0 \\ 0 \\ \vdots \\ 0 \\ \frac{\Phi_n}{z^{k-1}\|\Phi_{n-1}\|} \end{pmatrix}, \tag{2.44}$$

while if $n = 2k + 1$, then

$$
\Pi
\begin{pmatrix}
\chi_0 \\
\chi_1 \\
\vdots \\
\chi_{n-3} \\
\chi_{n-2} \\
\chi_{n-1}
\end{pmatrix}
= \mathcal{C}_n
\begin{pmatrix}
\chi_0 \\
\chi_1 \\
\vdots \\
\chi_{n-3} \\
\chi_{n-2} \\
\chi_{n-1}
\end{pmatrix}
+
\begin{pmatrix}
0 \\
0 \\
\vdots \\
0 \\
\rho_{n-2}\frac{\Phi_n}{z^{k-1}\|\Phi_{n-1}\|} \\
-\alpha_{n-2}\frac{\Phi_n}{z^{k-1}\|\Phi_{n-1}\|}
\end{pmatrix}.
\tag{2.45}
$$

The analog in the basis of $\chi_{k*}, 0 \le k \le n-1$ holds with $\mathcal{C}_n$ replaced by $\mathcal{C}_n^\tau$. Call $\mathcal{C}_n$ and $\mathcal{C}_n^\tau$ the CMV matrices associated to $\alpha_0, \alpha_1, \cdots, \alpha_{n-1}$.

Similarly to the equation (2.42), we have

Lemma 2.6. *In the above notations,*

$$
\det(zI_n - \mathcal{C}_n) = \Phi_n(z).
$$

Proof. If $n = 2k$, then by (2.44)

$$
(zI_n - \mathcal{C}_n)
\begin{pmatrix}
\chi_0 \\
\chi_1 \\
\vdots \\
\chi_{n-2} \\
\chi_{n-1}
\end{pmatrix}
=
\begin{pmatrix}
0 \\
0 \\
\vdots \\
0 \\
\frac{\Phi_n(z)}{z^{k-1}\|\Phi_{n-1}\|}
\end{pmatrix}.
$$

Denote by $\mathcal{C}_{n,k}$ the $k \times k$ subminor matrix of $\mathcal{C}_n$. Applying to solve the above system with respect to $\chi_{n-1}(z)$, we get

$$
\chi_{n-1}(z) = \frac{1}{\det(zI_n - \mathcal{C}_n)} \det
\begin{pmatrix}
& & & 0 \\
zI_{n-1} - \mathcal{C}_{n,n-1} & & & \vdots \\
& & & 0 \\
& & & \frac{\Phi_n(z)}{z^k\|\Phi_{n-1}\|} \\
& \cdots & &
\end{pmatrix}
$$

$$
= \frac{\Phi_n(z)}{z^{k-1}\|\Phi_{n-1}\|} \frac{\det(zI_{n-1} - \mathcal{C}_{n,n-1})}{\det(zI_n - \mathcal{C}_n)},
$$

which implies

$$
\frac{\det(zI_n - \mathcal{C}_n)}{\det(zI_{n-1} - \mathcal{C}_{n,n-1})} = \frac{\Phi_n(z)}{\Phi_{n-1}(z)}.
$$

Similarly, if $n = 2k + 1$, applying Cramér's rule to solve the initial system (2.45) with respect to $\chi_{n-2}(z)$ gives

$$\chi_{n-2}(z) = \frac{1}{\det(zI_n - \mathcal{C}_n)}$$

$$\cdot \det \begin{pmatrix} & & 0 & 0 \\ zI_{n-2} - \mathcal{C}_{n,n-2} & \vdots & \vdots \\ & 0 & 0 \\ \cdots & \rho_{n-2}\frac{\Phi_n(z)}{z^{k-1}\|\Phi_{n-1}\|} & -\rho_{n-2}\bar{\alpha}_{n-1} \\ \cdots & -\alpha_{n-2}\frac{\Phi_n(z)}{z^{k-1}\|\Phi_{n-1}\|} & z + \alpha_{n-2}\bar{\alpha}_{n-1} \end{pmatrix}$$

$$= z\rho_{n-2}\frac{\Phi_n(z)}{z^{k-1}\|\Phi_{n-1}\|}\frac{\det(zI_{n-2} - \mathcal{C}_{n,n-2})}{\det(zI_n - \mathcal{C}_n)}$$

and so

$$\frac{\det(zI_n - \mathcal{C}_n)}{\det(zI_{n-2} - \mathcal{C}_{n,n-2})} = \frac{\Phi_n(z)}{\Phi_{n-2}(z)}.$$

Thus we find by induction that for $n \geq 1$

$$\frac{\det(zI_n - \mathcal{C}_n)}{\det(z - \mathcal{C}_{n,1})} = \frac{\Phi_n(z)}{\Phi_1(z)}.$$

Since

$$\Phi_1(z) = z - \bar{\alpha}_0 = \det(z - \mathcal{C}_{n,1}),$$

then the claim follows. $\qquad\qquad\square$

In what follows we will be concerned with the spectral measure of a unitary matrix. Let U_n be a unitary matrix from $\mathcal{U}_n$, $\mathbf{e}_1 = (1, 0, \cdots, 0)^\tau$ a cyclic vector. Construct a probability measure $d\nu$ on $\mathbb{T}$ such that

$$\int_{\mathbb{T}} z^m d\nu = \langle U_n^m \mathbf{e}_1, \mathbf{e}_1 \rangle, \quad m \geq 0.$$

Note that $d\nu$ is of finite support. Indeed, let $e^{i\theta_1}, e^{i\theta_2}, \cdots, e^{i\theta_n}$ be the eigenvalues of U_n, then there must exist a unitary matrix V_n such that

$$U_n = V_n \begin{pmatrix} e^{i\theta_1} & 0 & \cdots & 0 \\ 0 & e^{i\theta_2} & \cdots & 0 \\ \vdots & \vdots & \ddots & \vdots \\ 0 & 0 & \cdots & e^{i\theta_n} \end{pmatrix} V_n^*.$$

Furthermore, V_n may be chosen to consist of eigenvectors $\mathbf{v}_1, \mathbf{v}_2, \cdots, \mathbf{v}_n$. If we further require that $v_{11} := q_1 > 0$, $v_{12} := q_2 > 0$, $\cdots$, $v_{1n} := q_n > 0$, then V_n is uniquely determined. In addition, it easily follows

$$q_1^2 + q_2^2 + \cdots + q_n^2 = 1$$

because of orthonormality of V_n. Thus it follows

$$\langle U_n^m \mathbf{e}_1, \mathbf{e}_1 \rangle = \sum_{j=1}^{n} q_j^2 e^{im\theta_j}, \quad m \geq 0.$$

So $d\nu$ is supported by $\{e^{i\theta_1}, e^{i\theta_2}, \cdots, e^{i\theta_n}\}$ and

$$\nu(e^{i\theta_j}) = q_j^2, \quad j = 1, 2, \cdots, n.$$

Having the measure $d\nu$ on $\mathbb{T}$, we can produce the Verblunsky coefficient. s $\alpha_k(d\nu)$ We shall below write $\alpha_k(U_n, \mathbf{e}_1)$ for $\alpha_k(d\nu)$ to indicate the underlying matrix and cyclic vector. The following lemmas provide us with two nice examples of unitary matrices, whose proofs can be found in Simon (2004).

Lemma 2.7. *Given a sequence of complex numbers $\alpha_0, \alpha_1, \cdots, \alpha_{n-2} \in \mathbb{D}$ and $\alpha_{n-1} \in \mathbb{T}$, construct an upper triangular Hessenberg matrix $H_n^U = \left(H_{ij}^U\right)_{0 \leq i,j \leq n-1}$ by letting*

$$H_{ij}^U = \begin{cases} -\alpha_{i-1}\bar{\alpha}_j \prod_{l=i}^{j-1} \rho_l, & i < j, \\ -\alpha_{i-1}\bar{\alpha}_i, & i = j, \\ \rho_j, & i = j+1, \\ 0, & i > j+1. \end{cases} \tag{2.46}$$

Then $\alpha_k\left(H_n^U, \mathbf{e}_1\right) = \alpha_k$, $0 \leq k \leq n-1$.

Lemma 2.8. *Given a sequence of complex numbers $\alpha_0, \alpha_1, \cdots, \alpha_{n-2} \in \mathbb{D}$ and $\alpha_{n-1} \in \mathbb{T}$, construct a CMV matrix $\mathcal{C}_n$ as in (2.43). Then $\alpha_k(\mathcal{C}_n, \mathbf{e}_1) = \alpha_k$, $0 \leq k \leq n-1$.*

What is the distribution of the $\alpha_k(\mathcal{C}_n, \mathbf{e}_1)$ if U_n is chosen at random from the CUE? To answer this question, we need to introduce a notion of Θ_v-distributed random variable. A complex random variable Z is said to be Θ_v-distributed ($v > 1$) if for any f

$$Ef(Z) = \frac{v-1}{2\pi} \int \int_{\mathbb{D}} f(z)\left(1 - |z|^2\right)^{(v-3)/2} dz.$$

For $v \geq 2$ an integer, there is an intuitive geometric interpretation for Z.

Lemma 2.9. *If $\mathbf{X} = (X_1, \cdots, X_n, X_{n+1}) \in \mathbb{R}^{n+1}$ is uniform over the n-dimensional unit sphere $\mathbb{S}^n$, then for any $1 \leq k \leq n$,*

$$Ef(X_1, \cdots, X_k) = \frac{\Gamma(\frac{n+1}{2})}{2\pi^{k/2}\Gamma(\frac{n-k+1}{2})} \int_{\mathbb{B}^k} f(x_1, \cdots, x_k)$$

$$\cdot \left(1 - x_1^2 - \cdots - x_k^2\right)^{(n-k-1)/2} dx_1 \cdots dx_k, \tag{2.47}$$

where $\mathbb{B}^k = \{(x_1, \cdots, x_k) : x_1^2 + \cdots + x_k^2 < 1\}$.

In particular, X_1 *is Beta(1/2, n/2)-distributed and* $X_1 + iX_2$ *is* Θ_n-*distributed.*

Proof. (2.47) is actually a direct consequence of the following change-of-variables formula using matrix volume:

$$\int_V f(\mathbf{v})d\mathbf{v} = \int_U (f \circ \phi)(\mathbf{u})|J_\phi(\mathbf{u})|d\mathbf{u} \tag{2.48}$$

where $V \subseteq \mathbb{R}^n$ and $U \subseteq \mathbb{R}^m$ with $n \geq m$, f is integrable on V, $\phi : U \mapsto V$ is sufficiently well-behaved function, $d\mathbf{v}$ and $d\mathbf{u}$ denote respectively the volume element in $V \subseteq \mathbb{R}^n$, and $|J_\phi(\mathbf{u})|$ is the volume of Jacobian matrix $J_\phi(\mathbf{u})$.

To apply (2.48) in our setting, let

$$\phi_k(x_1, \cdots, x_n) = x_k, \quad 1 \leq k \leq n$$

and

$$\phi_{n+1}(x_1, \cdots, x_n) = (1 - x_1^2 - \cdots - x_n^2)^{1/2}.$$

So the $\mathbb{S}^n$ is the graph of $\mathbb{B}^n$ under the mapping $\phi = (\phi_1, \cdots, \phi_{n+1})$. The Jacobian matrix of ϕ is

$$J_\phi = \begin{pmatrix} 1 & 0 & 0 & \cdots & 0 & 0 \\ 0 & 1 & 0 & \cdots & 0 & 0 \\ 0 & 0 & 0 & \ddots & 1 & 0 \\ 0 & 0 & 0 & \cdots & 0 & 1 \\ \frac{\partial \phi_{n+1}}{\partial x_1} & \frac{\partial \phi_{n+1}}{\partial x_2} & \frac{\partial \phi_{n+1}}{\partial x_3} & \cdots & \frac{\partial \phi_{n+1}}{\partial x_{n-1}} & \frac{\partial \phi_{n+1}}{\partial x_n} \end{pmatrix}.$$

This is an $n + 1 \times n$ rectangular matrix, whose volume is computed by

$$|J_\phi| = \sqrt{\det(J_\phi^\tau J_\phi)}$$

$$= (1 - x_1^2 - \cdots - x_n^2)^{-1/2}.$$

Hence according to (2.48), we have

$$Ef(X_1, \cdots, X_k)$$

$$= \int_{\mathbb{S}^n} f(x_1, \cdots, x_k)d\mathbf{s}$$

$$= \frac{\Gamma(\frac{n+1}{2})}{2\pi^{(n+1)/2}} \int_{\mathbb{B}^n} f(x_1, \cdots, x_k)(1 - x_1^2 - \cdots - x_n^2)^{-1/2}dx_1 \cdots dx_n$$

$$= \frac{\Gamma(\frac{n+1}{2})}{2\pi^{(n+1)/2}} \int_{\mathbb{B}^k} f(x_1, \cdots, x_k)dx_1 \cdots dx_k$$

$$\cdot \int_{x_{k+1}^2 + \cdots + x_n^2 < 1 - x_1^2 - \cdots - x_k^2} (1 - x_1^2 - \cdots - x_n^2)^{-1/2}dx_{k+1} \cdots dx_n \tag{2.49}$$

where ds denotes the uniform measure on $\mathbb{S}^n$.

On the other hand, it is easy to compute

$$\int_{x_{k+1}^2+\cdots+x_n^2<1-x_1^2-\cdots-x_k^2} \left(1-x_1^2-\cdots-x_n^2\right)^{-1/2} dx_{k+1}\cdots dx_n$$

$$= \left(1-x_1^2-\cdots-x_k^2\right)^{(n-k-1)/2}$$

$$\int_{\mathbb{B}^{n-k}} \left(1-x_1^2-\cdots-x_{n-k}^2\right)^{-1/2} dx_1\cdots dx_{n-k}$$

$$= \frac{\pi^{(n-k+1)/2}}{\Gamma\left(\frac{n-k+1}{2}\right)}\left(1-x_1^2-\cdots-x_k^2\right)^{(n-k-1)/2}. \tag{2.50}$$

Substituting (2.50) into (2.49) immediately get (2.47).   □

Remark 2.2. Lemma 2.9 can also be proved by using the following well-known fact: let $g_1,\cdots,g_{n+1}$ be a sequence of i.i.d. standard normal random variables, then

$$(X_1,\cdots,X_{n+1}) \stackrel{d}{=} \frac{1}{(g_1^2+\cdots+g_{n+1}^2)^{1/2}}(g_1,\cdots,g_{n+1}).$$

To keep notation consistent, Z is said to be Θ_1-distributed if Z is uniform on the unit circle.

Theorem 2.15. *Assume that U_n is a unitary matrix chosen from $\mathcal{U}_n$ at random according to the Haar measure $d\mu_n$. Then the Verblunsky coefficients $\alpha_k(U_n,\mathbf{e}_1)$ are independent $\Theta_{2(n-k-1)+1}$-distributed complex random variables.*

The key to proving Theorem 2.15 is the Householder transform, which will transfer unitary matrix into an upper triangular Hessenberg form. Write $U_n = (u_{ij})_{n\times n}$. Let $\mathbf{w} = (w_1, w_2,\cdots,w_n)^{\tau}$ where

$$w_1 = 0, \quad w_2 = -\frac{u_{21}}{|u_{21}|}\left(\frac{1}{2}-\frac{|u_{21}|}{2\alpha}\right)^{1/2},$$

$$w_l = -\frac{u_{l1}}{(2\alpha^2-2\alpha|u_{21}|)^{1/2}}, \quad l\geq 3$$

where $\alpha > 0$ and

$$\alpha^2 := |u_{21}|^2 + |u_{31}|^2 + \cdots + |u_{n1}|^2$$
$$= 1 - |u_{11}|^2.$$

Trivially, it follows

$$\mathbf{w}^*\mathbf{w} = 1.$$

Define

$$R_n = I_n - 2\mathbf{w}\mathbf{w}^* = \begin{pmatrix} 1 & 0 & \cdots & 0 \\ 0 & & & \\ \vdots & & V_{n-1} & \\ 0 & & & \end{pmatrix}.$$

This is a reflection through the plane perpendicular to $\mathbf{w}$. It is easy to check

$$R_n^{-1} = R_n^* = R_n$$

and

$$R_n U_n R_n = \begin{pmatrix} u_{11} & (u_{12}, u_{13}, \cdots, u_{1n})V_{n-1} \\ & \\ V_{n-1}\begin{pmatrix} u_{21} \\ u_{31} \\ \vdots \\ u_{n1} \end{pmatrix} & V_{n-1}U_{n,n-1}V_{n-1} \end{pmatrix},$$

where $U_{n,n-1}$ is the $n-1 \times n-1$ submatrix of U_n by deleting the first row and the first column.

Take a closer look at the first column. The first element of $R_n U_n R_n$ is unchanged, u_{11}; while the second is

$$\begin{pmatrix} 1 - 2w_2 w_2^*, -2w_2 w_3^*, \cdots, -2w_2 w_n^* \end{pmatrix} \begin{pmatrix} u_{21} \\ u_{32} \\ \vdots \\ u_{n1} \end{pmatrix} = \alpha \frac{u_{21}}{|u_{21}|},$$

and the third and below are zeros. So far we have described one step of the usual Householder algorithm. To make the second entry nonnegative, we need to add one further conjugation. Let D_n differ from the identity matrix by having $(2,2)$-entry $e^{-i\phi}$ with ϕ chosen appropriately and form $D_n R_n U_n R_n D_n^*$. Then we get the desired matrix

$$\begin{pmatrix} u_{11} & (u_{12}, u_{13}, \cdots, u_{1n})V_{n-1} \\ \sqrt{1 - |u_{11}|^2} & \\ 0 & \\ \vdots & V_{n-1}U_{n,n-1}V_{n-1} \\ 0 & \end{pmatrix}.$$

Proof of Theorem 2.15. We shall apply the above refined Householder algorithm to a random unitary matrix U_n. To do this, we need the following realization of Haar measure: choose the first column at random from the unit sphere $\mathbb{S}^{n-1}$; then choose the second column from the unit sphere of vectors orthogonal to the first; then the third column and so forth. In this way we get a Haar matrix because it is invariant under left multiplication by any unitary matrix.

Now the first column of U_n is a random vector from the unit sphere $\mathbb{S}^{n-1}$. After applying the above refined Householder algorithm, the new first column take the form $(\bar{\alpha}_0, \rho_0, 0, \cdots, 0)^\tau$ where $\bar{\alpha}_0 = u_{11}$ the original $(1,1)$ entry of U_n and so is by Lemma 2.9 Θ_{2n-1}-distributed, while $\rho_0 = \sqrt{1 - |\alpha_0|^2}$ as desired. The other columns are still orthogonal to the first column and form a random orthogonal basis for the orthogonal complement of the first column. Remember Haar measure is invariant under both right and left multiplication by a unitary.

For the subsequent columns the procedure is similar. Assume the $(k - 1)$th column is

$$
\begin{pmatrix}
\bar{\alpha}_{k-2}\rho_0\rho_1 \cdots \rho_{k-3} \\
-\bar{\alpha}_{k-2}\alpha_0\rho_1 \cdots \rho_{k-3} \\
\vdots \\
-\bar{\alpha}_{k-2}\alpha_{k-3} \\
\rho_{k-2} \\
0 \\
\vdots \\
0
\end{pmatrix}.
$$

Let

$$
X =
\begin{pmatrix}
\rho_0\rho_1 \cdots \rho_{k-2} \\
-\alpha_0\rho_1 \cdots \rho_{k-2} \\
\vdots \\
-\alpha_{k-3}\rho_{k-2} \\
-\alpha_{k-2} \\
0 \\
\vdots \\
0
\end{pmatrix},
$$

then X is a unit vector orthogonal to the first $k - 1$ columns. Namely X is an element of the linear vector space spanned by the last $n-k+1$ columns in U_n. Its inner product with the kth column, denoted by $\bar{\alpha}_{k-1}$, is distributed as the entry of a random vector from $2(n-k+1)$-sphere and is independent of $\alpha_0, \alpha_1, \cdots, \alpha_{k-2}$. This implies that α_{k-1} is $\Theta_{2(n-k)+1}$-distributed.

We now multiply the matrix at hand from left by the appropriate reflection and rotation to bring the kth columns into the desired form. Note neither these operations alter the top k rows and so the inner product of the kth column with X is unchanged. But now the kth column is uniquely determined; it must be $\bar{\alpha}_{k-1}X + \rho_{k-1}\mathbf{e}_{k+1}$, where $\mathbf{e}_{k+1} = (0, \cdots, 0, 1, 0 \cdots 0)^\tau$.

We then multiply on the right by RD^*, but this leaves the first k column unchanged, while orthogonally intermixing the other columns. In this way, we obtain a matrix whose first k column confirm to the structure H_n^U. While the remaining columns form a random basis for the orthogonal complements of the span of those k columns.

In this way, we can proceed inductively until we reach the last column. It is obliged to be a random orthonormal basis for the one-dimensional space orthogonal to the preceding $n-1$ columns and hence a random unimodular multiple, say $\bar{\alpha}_{n-1}$, of X. This is why the last Verblunsky coefficient is Θ_1-distributed.

We have now conjugated U_n to a matrix in the form of Hessenberg as in Lemma 2.7. Note the vector $\mathbf{e}_1$ is unchanged under the action of each of the conjugating matrices, then

$$\alpha_k(U_n, \mathbf{e}_1) = \alpha_k(H_n^U, \mathbf{e}_1) = \alpha_k.$$

We conclude the proof. $\qquad\qquad\qquad\qquad\qquad\qquad\qquad\qquad\square$

Combining Lemma 2.8 and Theorem 2.15 together, we immediately have

Theorem 2.16. *Let $\alpha_0, \alpha_1, \cdots, \alpha_{n-1}$ be a sequence of independent complex random variables and α_k is $\Theta_{2(n-k-1)+1}$-distributed. Define the CMV matrix C_n as in (2.43), then its eigenvalues are distributed according to (2.1).*

C_n is called a five diagonal matrix model of the CUE. It first appeared in the work of Killip and Nenciu (2004).

The rest of this section will be used to rederive Theorems 2.11 and 2.12 with help of the Verblunsky coefficients α_k and the Präfer phase ψ_k introduced above.

Start by an identity in law due to Bourgade, Hughes, Nikeghbali and Yor (2008).

Lemma 2.10. *Let $V_n \in \mathcal{U}_n$ be a random matrix with the first column $\mathbf{v}_1$ uniformly distributed on the n-dimensional unit complex sphere. If $U_{n-1} \in \mathcal{U}_{n-1}$ is distributed with Haar measure $d\mu_{n-1}$ and is independent of V_n,* *then*

$$U_n := V_n \begin{pmatrix} 1 & 0 \\ 0 & U_{n-1} \end{pmatrix} \tag{2.51}$$

is distributed with Haar measure $d\mu_n$ on $\mathcal{U}_n$.

Proof.   We shall prove for a fixed matrix $M \in \mathcal{U}_n$

$$MU_n \overset{d}{=} U_n.$$

Namely,

$$MV_n \begin{pmatrix} 1 & 0 \\ 0 & U_{n-1} \end{pmatrix} \overset{d}{=} V_n \begin{pmatrix} 1 & 0 \\ 0 & U_{n-1} \end{pmatrix}.$$

Write $V_n = (\mathbf{v}_1, \mathbf{v}_2, \cdots, \mathbf{v}_n)$. Since $\mathbf{v}_1$ is uniform, then so is $M\mathbf{v}_1$. By conditioning on $\mathbf{v}_1 = v$ and $M\mathbf{v}_1 = v$, it suffices to show

$$(v, M\mathbf{v}_2, \cdots, M\mathbf{v}_n) \begin{pmatrix} 1 & 0 \\ 0 & U_{n-1} \end{pmatrix} \overset{d}{=} (v, \mathbf{v}_2, \cdots, \mathbf{v}_n) \begin{pmatrix} 1 & 0 \\ 0 & U_{n-1} \end{pmatrix}. \tag{2.52}$$

Choose a unitary matrix A such that $Av = \mathbf{e}_1$. Since $A(v, M\mathbf{v}_2, \cdots, M\mathbf{v}_n)$ is unitary, then it must be equal to

$$\begin{pmatrix} 1 & 0 \\ 0 & X_{n-1} \end{pmatrix}$$

for some $X_{n-1} \in \mathcal{U}_{n-1}$. Similarly,

$$A(v, \mathbf{v}_2, \cdots, \mathbf{v}_n) = \begin{pmatrix} 1 & 0 \\ 0 & Y_{n-1} \end{pmatrix}$$

for some $Y_{n-1} \in \mathcal{U}_{n-1}$.

It is now easy to see

$$\begin{pmatrix} 1 & 0 \\ 0 & X_{n-1} \end{pmatrix} \begin{pmatrix} 1 & 0 \\ 0 & U_{n-1} \end{pmatrix} \overset{d}{=} \begin{pmatrix} 1 & 0 \\ 0 & Y_{n-1} \end{pmatrix} \begin{pmatrix} 1 & 0 \\ 0 & U_{n-1} \end{pmatrix}$$

since $X_{n-1}U_{n-1} \overset{d}{=} U_{n-1}$ and $Y_{n-1}U_{n-1} \overset{d}{=} U_{n-1}$ by rotation invariance of Haar measure. Thus by virtue of invertibility of A, (2.52) immediately follows, which concludes the proof. □

Lemma 2.11. *Let U_n be a random matrix from the CUE with the Verblunsky coefficients $\alpha_0, \alpha_1, \cdots, \alpha_{n-1}$. Then*

$$\det(I_n - U_n) \overset{d}{=} \prod_{k=0}^{n-1} (1 - \alpha_k). \tag{2.53}$$

Proof. To apply Lemma 2.10, we particularly choose a V_n as follows. Let $\mathbf{v}_1$ be a random vector uniformly distributed on the n-dimensional unit complex sphere. Define

$$V_n = \left(\mathbf{v}_1, \mathbf{e}_2 + a_2(\mathbf{v}_1 - \mathbf{e}_1), \cdots, \mathbf{e}_n + a_n(\mathbf{v}_1 - \mathbf{e}_1)\right),$$

where $\mathbf{e}_1, \cdots, \mathbf{e}_n$ are classic base in $\mathbb{R}^n$, $a_2, \cdots, a_n$ are such that V_n is unitary, that is,

$$a_k = \frac{\langle \mathbf{v}_1, \mathbf{e}_k \rangle}{\langle \mathbf{v}_1 - \mathbf{e}_1, \mathbf{e}_1 \rangle}, \quad k = 2, 3, \cdots, n.$$

According to Lemma 2.10, it follows

$$\det(I_n - U_n) \overset{d}{=} \det\left(I_n - V_n \begin{pmatrix} 1 & \mathbf{0} \\ \mathbf{0} & U_{n-1} \end{pmatrix}\right), \tag{2.54}$$

where U_{n-1} is distributed with Haar measure $d\mu_{n-1}$ independently of $\mathbf{v}_1$.

It remains to computing the determinant on the right hand side of (2.54). Note

$$\det\left(I_n - V_n \begin{pmatrix} 1 & \mathbf{0} \\ \mathbf{0} & U_{n-1} \end{pmatrix}\right) = \det\left(\begin{pmatrix} 1 & \mathbf{0} \\ \mathbf{0} & U_{n-1}^* \end{pmatrix} - V_n\right) \det U_{n-1}. \tag{2.55}$$

Set $U_{n-1}^* = (\mathbf{u}_2, \mathbf{u}_3, \cdots, \mathbf{u}_n)$ and $\mathbf{w}_1 = \mathbf{v}_1 - \mathbf{e}_1$. Then

$$\begin{pmatrix} 1 & \mathbf{0} \\ \mathbf{0} & U_{n-1}^* \end{pmatrix} - V_n$$

$$= \left(-\mathbf{w}_1, \begin{pmatrix} 0 \\ \mathbf{u}_2 \end{pmatrix} - (\mathbf{e}_2 + a_2\mathbf{w}_1), \cdots, \begin{pmatrix} 0 \\ \mathbf{u}_n \end{pmatrix} - (\mathbf{e}_n + a_n\mathbf{w}_1)\right).$$

So by the multi-linearity property,

$$\det\left(\begin{pmatrix} 1 & \mathbf{0} \\ \mathbf{0} & U_{n-1}^* \end{pmatrix} - V_n\right) = \det \begin{pmatrix} -w_{11} & & \mathbf{0} \\ -w_{21} & & \\ \vdots & & U_{n-1}^* - I_{n-1} \\ -w_{n1} & & \end{pmatrix}$$

$$= -w_{11} \det(U_{n-1}^* - I_{n-1}). \tag{2.56}$$

Substituting (2.56) in (2.55), we get

$$\det\left(I_n - V_n \begin{pmatrix} 1 & \mathbf{0} \\ \mathbf{0} & U_{n-1} \end{pmatrix}\right) = -w_{11} \det(I_{n-1} - U_{n-1}).$$

Observe $w_{11} = v_{11} - 1$ and $v_{11} \sim \Theta_{2n-1}$-distributed. Thus it follows

$$\det((I_n - U_n) \overset{d}{=} (1 - \alpha_0) \det(I_{n-1} - U_{n-1}).$$

Proceeding in this manner, we have

$$\det(I_n - U_n) \overset{d}{=} \prod_{k=0}^{n-1} (1 - \alpha_k),$$

as required. $\square$

Remark 2.3. The identity (2.53) can also be deduced using the recurrence relation of orthogonal polynomials $\Phi_k(z)$. Indeed, according to Theorem 2.16,

$$\det(I_n - U_n) \stackrel{d}{=} \det(I_n - C_n). \tag{2.57}$$

On the other hand, by Lemma 2.6 and (2.37)

$$\det(I_n - C_n) = \Phi_n(1)$$
$$= \Phi_{n-1}(1) - \bar{\alpha}_{n-1}\Phi_{n-1}^*(1).$$

Note by (2.40)

$$\frac{\Phi_{n-1}^*(1)}{\Phi_{n-1}(1)} = e^{-i\psi_{n-1}(0)}.$$

Hence we have

$$\det(I_n - C_n) = \Phi_{n-1}(1)\Big(1 - \bar{\alpha}_{n-1}e^{-i\psi_{n-1}(0)}\Big).$$

Inductively, using (2.35) we get

$$\det(I_n - C_n) = \prod_{k=0}^{n-1}\Big(1 - \bar{\alpha}_k e^{-i\psi_k(0)}\Big).$$

Observe that $\psi_0(0) = 0$ and $\psi_k(0)$ depends only on $\alpha_0, \alpha_1, \cdots, \alpha_{k-1}$. Using the conditioning argument and the rotation invariance of α_k, we can get

$$\det(I_n - C_n) \stackrel{d}{=} \prod_{k=0}^{n-1}(1 - \alpha_k), \tag{2.58}$$

which together with (2.57) implies (2.53). It is worth mentioning that the identity (2.58) is still valid for the CβE discussed in next section.

We also need some basic estimates about the moments of Θ_v-distributed random variables.

Lemma 2.12. *Assume Z is Θ_v-distributed for some $v \geq 1$.*
(i) $|Z|$ and $\arg Z$ are independent real random variables. Moreover, $\arg Z$ is uniform over $(0, 2\pi)$ and $|Z|$ is distributed with density function

$$p_{|Z|}(r) = (v - 1)r(1 - r^2)^{(v-3)/2}, \quad 0 < r < 1.$$

(ii)

$$EZ = EZ^2 = 0, \quad E|Z|^2 = \frac{2}{v+1}, \quad E|Z|^4 = \frac{8}{(v+1)(v+3)}.$$

Proof. (i) directly follows from (2.47), while a simple computation easily yields (ii). □

Proof of Theorem 2.11. Without loss of generality, we may and do assume $\theta_0 = 0$. According to Lemma 2.11, it suffices to prove the following asymptotic normality

$$\frac{1}{\sqrt{\log n}} \sum_{k=0}^{n-1} \log(1 - \alpha_k) \xrightarrow{d} N_{\mathbb{C}}(0, 1). \qquad (2.59)$$

Since $|\alpha_k| < 1$ almost surely for $k = 0, 1, \cdots, n - 2$, then

$$\log(1 - \alpha_k) = -\sum_{l=1}^{\infty} \frac{1}{l} \alpha_k^l.$$

Taking summation over k, we have the following

$$\sum_{k=0}^{n-1} \log(1 - \alpha_k) = -\sum_{l=1}^{\infty} \sum_{k=0}^{n-2} \frac{1}{l} \alpha_k^l + \log(1 - \alpha_{n-1})$$

$$= -\sum_{k=0}^{n-2} \alpha_k - \frac{1}{2} \sum_{k=0}^{n-2} \alpha_k^2 - \sum_{l=3}^{\infty} \sum_{k=0}^{n-2} \frac{1}{l} \alpha_k^l + \log(1 - \alpha_{n-1})$$

$$=: Z_{n,1} + Z_{n,2} + Z_{n,3} + Z_{n,4}.$$

Firstly, we shall prove

$$\frac{Z_{n,1}}{\sqrt{\log n}} \xrightarrow{d} N_{\mathbb{C}}(0, 1). \qquad (2.60)$$

It is equivalent to proving

$$\frac{1}{\sqrt{\log n}} \sum_{k=0}^{n-2} |\alpha_k| \cos \eta_k \xrightarrow{d} N\left(0, \frac{1}{2}\right) \qquad (2.61)$$

and

$$\frac{1}{\sqrt{\log n}} \sum_{k=0}^{n-2} |\alpha_k| \sin \eta_k \xrightarrow{d} N\left(0, \frac{1}{2}\right) \qquad (2.62)$$

where $\eta_k = arg\alpha_k$.

We only prove (2.61) since (2.62) is similar. In view of Lemma 2.12,

$$E|\alpha_k|^2 = \frac{1}{n-k}, \quad E|\alpha_k|^4 = \frac{2}{(n-k)(n-k+1)}$$

and

$$E \cos^2 \eta_k = \frac{1}{2}, \quad E \cos^4 \eta_k \le 1.$$

Since η_k is uniform, it is easy to check

$$\frac{1}{\log n}\sum_{k=0}^{n-2}E|\alpha_k|^2 E\cos^2\eta_k \to \frac{1}{2}$$

and

$$\sum_{k=0}^{n-2}E|\alpha_k|^4 \le \frac{\pi^2}{3}.$$

Hence (2.61) is now a direct consequence of the Lyapunov CLT.

Secondly, to deal with $Z_{n,2}$, note by Lemma 2.12

$$E\left|\sum_{k=0}^{n-2}\alpha_k^2\right|^2 = \sum_{k=0}^{n-2}E|\alpha_k|^4 \le \frac{\pi^2}{3}.$$

This, together with the Markov inequality, easily implies

$$\frac{1}{\sqrt{\log n}}Z_{n,2} \xrightarrow{P} 0. \tag{2.63}$$

Thirdly, it is easy to check $EZ_{n,3} = 0$ and

$$E|Z_{n,3}|^2 \le \sum_{l=3}^{\infty}\sum_{k=0}^{n-2}\frac{1}{l}E|\alpha_k|^{2l} \le \frac{2\pi^2}{3}.$$

So, it follows

$$\frac{1}{\sqrt{\log n}}Z_{n,3} \xrightarrow{P} 0. \tag{2.64}$$

Finally, for $Z_{n,4}$, note $\alpha_{n-1} = e^{i\eta}$ is uniformly distributed on $\mathbb{T}$, then almost surely α_{n-1} is not equal to 1, and so $\log(1 - \alpha_{n-1}) < \infty$. Hence it follows

$$\frac{1}{\sqrt{\log n}}Z_{n,4} \xrightarrow{P} 0. \tag{2.65}$$

Gathering (2.60), (2.63), (2.64) and (2.65) together implies (2.59). □

Turn to Theorem 2.12. Let $\alpha_k = \alpha_k(U_n, \mathbf{e}_1)$, the Verblunsky coefficients associated to $(U_n, \mathbf{e}_1)$, and construct the B_k and ψ_k as in (2.39) and (2.40), then

$$\{e^{i\theta} : B_{n-1}(e^{i\theta}) = e^{-i\eta}\} = \{e^{i\theta} : \psi_{n-1}(\theta) \in 2\pi\mathbb{Z} + \eta\}$$

is the eigenvalues of U_n. In particular, the number of angles lying in the arc $(a, b) \subseteq [0, 2\pi)$ is approximately $(\psi_{n-1}(b) - \psi_{n-1}(a))/2\pi$; indeed, it follows

$$\left|N_n(a, b) - \frac{\psi_{n-1}(b) - \psi_{n-1}(a)}{2\pi}\right| \le 1.$$

In this way, it suffices to show that asymptotically, $\psi_{n-1}(b)$ and $\psi_{n-1}(a)$ follow a joint normal law. This is what Killip and Nenciu (2008) employed in the study of general CβE.

Lemma 2.13. *Assume $a, b \in \mathbb{R}$ and $\alpha \sim \Theta_v$. Define*

$$\Upsilon(a, \alpha) = -2Im \log\left(1 - \alpha e^{ia}\right), \quad \tilde{\Upsilon}(a, \alpha) = 2Im\left(\alpha e^{ia}\right).$$

Then we have

$$E\Upsilon(a, \alpha) = E\tilde{\Upsilon}(a, \alpha) = 0, \tag{2.66}$$

$$E\tilde{\Upsilon}(a, \alpha)\tilde{\Upsilon}(b, \alpha) = \frac{4}{v+1}\cos(b - a), \tag{2.67}$$

$$E\tilde{\Upsilon}(a, \alpha)^4 = \frac{48}{(v+1)(v+3)}, \tag{2.68}$$

$$E|\Upsilon(a, \alpha) - \tilde{\Upsilon}(a, \alpha)|^2 \le \frac{16}{(v+1)(v+3)}, \tag{2.69}$$

$$E|\Upsilon(a, \alpha)|^2 \le \frac{8}{v+1}. \tag{2.70}$$

Proof. The fact that α follows a rotationally invariant law immediately implies

$$E\Upsilon(a, \alpha) = 0.$$

Specifically, for any $0 \le r < 1$,

$$\frac{1}{2\pi} \int_0^{2\pi} \log\left(1 - re^{i(\theta+a)}\right) d\theta = 0$$

by the mean value principle for harmonic functions. $E\tilde{\Upsilon}(a, \alpha) = 0$ is similar and simpler.

For (2.67), note

$$Im(\alpha e^{ia}) = |\alpha| \sin(a + arg\alpha)$$

and so it follows by Lemma 2.12

$$E\tilde{\Upsilon}(a, \alpha)\tilde{\Upsilon}(b, \alpha)$$
$$= 4E|\alpha|^2 \sin(a + arg\alpha)\sin(b + arg\alpha)$$
$$= \frac{4(v-1)}{2\pi} \int_0^1 \int_0^{2\pi} r^3(1 - r^2)^{(v-3)/2} \sin(\theta + a)\sin(\theta + b)drd\theta$$
$$= \frac{4}{v+1}\cos(b - a).$$

Similarly, we have

$$E\tilde{\Upsilon}(a,\alpha)^4 = \frac{16(v-1)}{2\pi} \int_0^1 \int_0^{2\pi} r^5 \left(1-r^2\right)^{(v-3)/2} \sin^4(\theta+a)dr d\theta$$

$$= \frac{48}{(v+1)(v+3)}.$$

Applying Plancherel's theorem to the power series formula for Υ gives

$$E\left|\Upsilon(a,\alpha) - \tilde{\Upsilon}(a,\alpha)\right|^2 = \sum_{l=2}^{\infty} \frac{2}{l^2} E|\alpha|^{2l}.$$

$$\leq 2\left(\frac{\pi^2}{6} - 1\right) E|\alpha|^4$$

(2.67) easily follows from Lemma 2.12.

Lastly, combining (2.67) and (2.69) implies (2.70). □

Lemma 2.14. *Assume that $a_k, \Theta_k, \gamma_k, \ k \geq 0$ are real valued sequences satisfying*

$$\Theta_{k+1} = \Theta_k + \delta + \gamma_k$$

with $0 < \delta < 2\pi$. Then we have

$$\left|1 - e^{i\delta}\right|\left|\sum_{k=1}^{n} a_k e^{i\Theta_k}\right|$$

$$\leq 2 \max_{1\leq k\leq n} |a_k| + \sum_{k=1}^{n} |a_k - a_{k+1}| + \sum_{k=1}^{n} |a_k \gamma_k|. \qquad (2.71)$$

Proof. Note

$$\left(e^{i\delta} - 1\right) \sum_{k=1}^{n} a_k e^{i\Theta_k}$$

$$= \sum_{k=1}^{n} a_k \left(e^{i(\Theta_k+\delta)} - e^{i\Theta_k}\right)$$

$$= \sum_{k=1}^{n} a_k \left(e^{i\Theta_{k+1}} - e^{i\Theta_k}\right) - \sum_{k=1}^{n} a_k \left(e^{i\Theta_{k+1}} - e^{i(\Theta_k+\delta)}\right).$$

Then (2.71) easily follows using summation by parts and the fact $\left|1-e^{i\gamma_k}\right| \leq |\gamma_k|$. □

Proof of Theorem 2.12. Start with 1-dimensional convergence. Note for any $0 < a < 2\pi$

$$\psi_{n-1}(a) - na = \sum_{k=1}^{n-1} \left(\psi_k(a) - \psi_{k-1}(a) - a \right)$$

$$= \sum_{k=0}^{n-2} \Upsilon\left(\psi_k(a), \alpha_k \right),$$

where $\Upsilon\left(\psi_k(a), \alpha_k \right)$ is defined as in (2.41). Since $\psi_k(a)$ depends only on $\alpha_0, \cdots, \alpha_{k-1}$, then it follows from Lemma 2.13

$$E\left(\Upsilon(\psi_k(a), \alpha_k) \big| \alpha_0, \cdots, \alpha_{k-1} \right) = 0.$$

Namely, $\Upsilon\left(\psi_k(a), \alpha_k \right)$, $0 \le k \le n-2$ is a martingale difference sequence. Define

$$\tilde{\Upsilon}\left(\psi_k(a), \alpha_k \right) = -2Im\left(\alpha_k e^{i\psi_k(a)} \right).$$

Similarly, $\tilde{\Upsilon}\left(\psi_k(a), \alpha_k \right)$, $0 \le k \le n-2$ is also a martingale difference sequence. Moreover, by Lemma 2.13 again

$$\sum_{k=0}^{n-2} E\left| \Upsilon(\psi_k(a), \alpha_k) - \tilde{\Upsilon}(\psi_k(a), \alpha_k) \right|^2 \le 3\pi^2.$$

Thus it suffices to prove

$$\frac{1}{\sqrt{2\log n}} \sum_{k=0}^{n-2} \tilde{\Upsilon}\left(\psi_k(a), \alpha_k \right) \xrightarrow{d} N(0,1).$$

It is in turn sufficient to verify

$$\frac{1}{2\log n} \sum_{k=0}^{n-2} E\left(\tilde{\Upsilon}(\psi_k(a), \alpha_k)^2 \big| \alpha_0, \cdots, \alpha_{k-1} \right) \xrightarrow{P} 1$$

and

$$\frac{1}{\log^2 n} \sum_{k=0}^{n-2} E\tilde{\Upsilon}\left(\psi_k(a), \alpha_k \right))^4 \to 0.$$

These directly follow from Lemma 2.12.

Next turn to 2-dimensional convergence. We need to verify

$$\frac{1}{\log n} \sum_{k=0}^{n-2} E\left(\tilde{\Upsilon}(\psi_k(a), \alpha_k) \tilde{\Upsilon}(\psi_k(b), \alpha_k) \big| \alpha_0, \cdots, \alpha_{k-1} \right) \xrightarrow{P} 0. \quad (2.72)$$

For distinct numbers a and b, we have by Lemma 2.12

$$E\big(\tilde{\Upsilon}(\psi_k(a),\alpha_k)\tilde{\Upsilon}(\psi_k(b),\alpha_k)\big|\alpha_0,\cdots,\alpha_{k-1}\big)$$
$$= \frac{2}{n-k}\cos\big(\psi_k(b)-\psi_k(a)\big).$$

Define

$$a_k = \frac{1}{n-k}, \quad \Theta_k = \psi_k(b)-\psi_k(a), \quad \delta = b-a$$

and

$$\gamma_k = \Upsilon\big(\psi_k(b),\alpha_k\big) - \Upsilon\big(\psi_k(a),\alpha_k\big),$$

then by (2.41)

$$\Theta_{k+1} = \Theta_k + \delta + \gamma_k.$$

Applying Lemma 2.14 and noting

$$E|\gamma_k| \le \frac{8}{(n-k)^{1/2}},$$

we have

$$\left|1 - e^{\mathrm{i}(b-a)}\right|\left|E\sum_{k=0}^{n-2}\frac{1}{n-k}e^{\mathrm{i}(\psi_k(b)-\psi_k(a))}\right| \le 9.$$

(2.72) is now valid. Thus by the martingale CLT

$$\frac{1}{\sqrt{2\log n}}\big(\psi_{n-1}(a)-na, \psi_{n-1}(b)-nb\big) \overset{d}{\longrightarrow} (Z_1, Z_2),$$

where Z_1, Z_2 are independent standard normal random variables. □

2.5 Circular β ensembles

The goal of this section is to extend the five diagonal matrix representation to the CβE.

Recall that the CβE represents a family of probability measures on n points of $\mathbb{T}$ with density function $p_{n,\beta}(e^{\mathrm{i}\theta_1},\cdots,e^{\mathrm{i}\theta_n})$ defined by (2.5). As observed in Section 2.1, $p_{n,2}$ describes the joint probability density of eigenvalues of a unitary matrix chosen from $\mathcal{U}_n$ according to Haar measure. Similarly, $p_{n,1}$ ($p_{n,4}$) describes the joint probability density of eigenvalues of an orthogonal (symplectic) matrix chosen from $\mathcal{O}_n$ ($\mathcal{S}_n$) according to Haar measure. However, no analog holds for general $\beta > 0$.

The following five diagonal matrix model discovered by Killip and Nenciu (2004) plays an important role in the study of CβE.

Theorem 2.17. *Assume that $\alpha_0, \alpha_1, \cdots, \alpha_{n-1}$ are independent complex random variables and $\alpha_k \sim \Theta_{\beta(n-k-1)+1}$-distributed for $\beta > 0$. Construct the CMV matrix C_n as in (2.43), then the eigenvalues of C_n obey the same law as $p_{n,\beta}$.*

The rest of this section is to prove the theorem. The proof is actually an ordinary use of the change of variables in standard probability theory. Let H_n^U be as in (2.46), then we have by Lemmas 2.7 and 2.8

$$\alpha_k(C_n, \mathbf{e}_1) = \alpha_k\left(H_n^U, \mathbf{e}_1\right) = \alpha_k.$$

So it suffices to prove the claim for H_n^U. Denote the ordered eigenvalues of H_n^U by $e^{i\theta_1}, \cdots, e^{i\theta_n}$. Then there must be a unitary matrix $V_n = (v_{ij})_{n \times n}$ such that

$$H_n^U = V_n \begin{pmatrix} e^{i\theta_1} & 0 & \cdots & 0 \\ 0 & e^{i\theta_2} & \cdots & 0 \\ \vdots & \vdots & \ddots & \vdots \\ 0 & 0 & \cdots & e^{i\theta_n} \end{pmatrix} V_n^*. \tag{2.73}$$

V_n may be chosen to consist of eigenvectors $\mathbf{v}_1, \mathbf{v}_2, \cdots, \mathbf{v}_n$. We also further require that $v_{11} := q_1 > 0$, $v_{12} := q_2 > 0$, $\cdots$, $v_{1n} := q_n > 0$, thus V_n is uniquely determined. It easily follows

$$q_1^2 + q_2^2 + \cdots + q_n^2 = 1 \tag{2.74}$$

because of orthonormality of V_n.

The following lemma gives an elegant identity between the eigenvalues and eigenvectors and the Verblunsky coefficients.

Lemma 2.15.

$$\prod_{l=1}^n q_l^2 \prod_{1 \le j < k \le n} \left| e^{i\theta_j} - e^{i\theta_k} \right|^2 = \prod_{l=0}^{n-2} \left(1 - |\alpha_l|^2\right)^{n-l-1}.$$

Proof. Define A and Q by

$$A = \begin{pmatrix} 1 & 1 & \cdots & 1 \\ e^{i\theta_1} & e^{i\theta_2} & \cdots & e^{i\theta_n} \\ \vdots & \vdots & \ddots & \vdots \\ e^{i(n-1)\theta_1} & e^{i(n-1)\theta_2} & \cdots & e^{i(n-1)\theta_n} \end{pmatrix}$$

and

$$Q = \begin{pmatrix} q_1^2 & 0 & \cdots & 0 \\ 0 & q_2^2 & \cdots & 0 \\ \vdots & \vdots & \ddots & \vdots \\ 0 & 0 & \cdots & q_n^2 \end{pmatrix}$$

then it follows

$$\prod_{l=1}^{n} q_l^2 \prod_{1 \le j < k \le n} \left| e^{i\theta_j} - e^{i\theta_k} \right|^2 = \det(AQA^*).$$

On the other hand, define

$$B = \begin{pmatrix} \Phi_0(e^{i\theta_1}) & \Phi_0(e^{i\theta_2}) & \cdots & \Phi_0(e^{i\theta_n}) \\ \Phi_1(e^{i\theta_1}) & \Phi_1(e^{i\theta_2}) & \cdots & \Phi_1(e^{i\theta_n}) \\ \vdots & \vdots & \ddots & \vdots \\ \Phi_{n-1}(e^{i\theta_1}) & \Phi_{n-1}(e^{i\theta_2}) & \cdots & \Phi_{n-1}(e^{i\theta_n}) \end{pmatrix},$$

where $\Phi_0, \Phi_1, \cdots, \Phi_{n-1}$ are monic orthogonal polynomials associated to the Verblunsky coefficients $\alpha_0, \alpha_1, \cdots, \alpha_{n-1}$. Then it is trivial to see

$$\det(A) = \det(B).$$

In addition, from the orthogonality property of the Φ_l, it follows

$$BQB^* = \begin{pmatrix} \|\Phi_0\|^2 & 0 & \cdots & 0 \\ 0 & \|\Phi_1\|^2 & \cdots & 0 \\ \vdots & \vdots & \ddots & 0 \\ 0 & 0 & \cdots & \|\Phi_{n-1}\|^2 \end{pmatrix},$$

where $\|\Phi_l\|^2 = \sum_{j=1}^{n} q_j^2 |\Phi_l(e^{i\theta_j})|^2$.

Hence according to (2.38),

$$\det(AQA^*) = \det(BQB^*) = \prod_{l=0}^{n-1} \|\Phi_l\|^2$$

$$= \prod_{l=0}^{n-2} \left(1 - |\alpha_l|^2\right)^{n-l-1}$$

just as required.   □

A key ingredient to the proof of Theorem 2.17 is to look for a proper change of variables and to compute explicitly the corresponding determinant of the Jacobian. For any $|t| < 1$, it follows from (2.73)

$$\left(I_n - tH_n^U\right)^{-1} = V_n \begin{pmatrix} (1 - te^{i\theta_1})^{-1} & \cdots & 0 \\ \vdots & \ddots & \vdots \\ 0 & \cdots & (1 - te^{i\theta_n})^{-1} \end{pmatrix} V_n^*. \qquad (2.75)$$

Applying the Taylor expansion of $(1 - x)^{-1}$, and equating powers of t of the $(1, 1)$ entries on both sides of (2.75), we can get the following system of equations

$$\bar{\alpha}_0 = \sum_{j=1}^{n} q_j^2 e^{i\theta_j}$$

$$* + \rho_0^2 \bar{\alpha}_1 = \sum_{j=1}^{n} q_j^2 e^{i2\theta_j}$$

$$* + \rho_0^2 \rho_1^2 \bar{\alpha}_2 = \sum_{j=1}^{n} q_j^2 e^{i3\theta_j} \qquad (2.76)$$

$$\vdots \qquad \qquad \vdots$$

$$* + \rho_0^2 \rho_1^2 \cdots \rho_{n-2}^2 \bar{\alpha}_{n-1} = \sum_{j=1}^{n} q_j^2 e^{in\theta_j}$$

where the $*$ denotes terms involving only variables already having appeared on the left hand side of the preceding equations.

In this way, we can naturally get a one-to-one mapping from $(\alpha_0, \alpha_1, \cdots, \alpha_{n-1})$ to $(e^{i\theta_1}, \cdots, e^{i\theta_n}, q_1, \cdots, q_{n-1})$. Recall that α_k, $0 \leq k \leq n - 2$ has an independent real and imaginary part, while α_{n-1}, $e^{i\theta_j}$ have unit modulus. We see that the number of variables is equal to $2n - 1$. In particular, let $\alpha_k = a_k + \mathbf{i}b_k$ and define J to be the determinant of the Jacobian matrix for the change of variables, namely

$$J = \frac{\bigwedge_{k=0}^{n-2} da_k \wedge db_k \wedge d\bar{\alpha}_{n-1}}{\bigwedge_{l=1}^{n-1} dq_l \bigwedge_{j=1}^{n} d\theta_j}$$

where $\wedge$ stands for the wedge product.

We shall compute explicitly the J following Forrester and Rains (2006) below. First, taking differentials on both sides of (2.74) immediately yields

$$q_n dq_n = - \sum_{j=1}^{n-1} q_j dq_j. \qquad (2.77)$$

Similarly, taking differentials on both sides of (2.76) gives

$$d\bar{\alpha}_0 = 2\sum_{j=1}^{n} e^{i\theta_j} q_j dq_j + i\sum_{j=1}^{n} q_j^2 e^{i\theta_j} d\theta_j$$

$$\rho_0^2 d\bar{\alpha}_1 = 2\sum_{j=1}^{n} e^{i2\theta_j} q_j dq_j + i2\sum_{j=1}^{n} q_j^2 e^{i2\theta_j} d\theta_j$$

$$\rho_0^2 \rho_1^2 d\bar{\alpha}_2 = 2\sum_{j=1}^{n} e^{i3\theta_j} q_j dq_j + i3\sum_{j=1}^{n} q_j^2 e^{i3\theta_j} d\theta_j \qquad (2.78)$$

$$\vdots \qquad\qquad \vdots$$

$$\rho_0^2 \rho_1^2 \cdots \rho_{n-2}^2 d\bar{\alpha}_{n-1} = 2\sum_{j=1}^{n} e^{in\theta_j} q_j dq_j + in\sum_{j=1}^{n} q_j^2 e^{in\theta_j} d\theta_j.$$

Forming the complex conjugates of all these equations but last, we get

$$d\alpha_0 = 2\sum_{j=1}^{n} e^{-i\theta_j} q_j dq_j - i\sum_{j=1}^{n} q_j^2 e^{-i\theta_j} d\theta_j$$

$$\rho_0^2 d\alpha_1 = 2\sum_{j=1}^{n} e^{-i2\theta_j} q_j dq_j - i2\sum_{j=1}^{n} q_j^2 e^{-i2\theta_j} d\theta_j$$

$$\rho_0^2 \rho_1^2 d\alpha_2 = 2\sum_{j=1}^{n} e^{-i3\theta_j} q_j dq_j - i3\sum_{j=1}^{n} q_j^2 e^{-i3\theta_j} d\theta_j \qquad (2.79)$$

$$\vdots \qquad\qquad \vdots$$

$$\rho_0^2 \rho_1^2 \cdots \rho_{n-3}^2 d\alpha_{n-2} = 2\sum_{j=1}^{n} e^{-i(n-1)\theta_j} q_j dq_j - i(n-1)\sum_{j=1}^{n} q_j^2 e^{-i(n-1)\theta_j} d\theta_j.$$

Now taking the wedge products of both sides of these $2n-1$ equations in (2.78) and (2.79), and using (2.77) shows

$$\rho_0^2 \rho_1^2 \cdots \rho_{n-2}^2 \prod_{l=0}^{n-2} \rho_l^{4(n-l-2)} \bigwedge_{l=0}^{n-2} d\bar{\alpha}_l \wedge d\alpha_l \bigwedge d\bar{\alpha}_{n-1}$$

$$= (2i)^{n-1} q_n^2 \prod_{l=1}^{n-1} q_l^3 D\left(e^{i\theta_1}, \cdots, e^{i\theta_n}\right) \bigwedge_{l=1}^{n-1} dq_l \bigwedge_{j=1}^{n} d\theta_j, \qquad (2.80)$$

where $D\left(e^{i\theta_1}, \cdots, e^{i\theta_n}\right)$ is defined by

$$D(x_1, \cdots, x_n) = \det\left(\begin{bmatrix} x_k^j - x_n^j \\ x_k^{-j} - x_n^{-j} \end{bmatrix}_{j,k=1,\cdots,n-1} \begin{bmatrix} jx_k^j \\ -jx_k^{-j} \end{bmatrix}_{\substack{j=1,\cdots,n-1 \\ k=1,\cdots,n}} \right).$$
$$\qquad\qquad\qquad\qquad [x_k^n - x_n^n]_{k=1,\cdots,n-1} \quad [nx_k^n]_{k=1,\cdots,n}$$

$$(2.81)$$

Lemma 2.16. *We have*

$$D(x_1, \cdots, x_n) = (-1)^{(n-1)(n-2)/2} \frac{\prod_{1 \le j < k \le n} (x_j - x_k)^4}{\prod_{j=1}^{n} x_j^{2n-3}}.$$

Proof. By inspection, the determinant $D(x_1, \cdots, x_n)$ is a symmetric function of $x_1, \cdots, x_n$ which is homogeneous of degree n. Upon multiplying columns 1 and n by x_1^{2n-3}, we see that $D(x_1, \cdots, x_n)$ becomes a polynomial in x_1, so it must be of the form

$$\frac{p(x_1, \cdots, x_n)}{\prod_{j=1}^{n} x_j^{2n-3}},$$

where $p(x_1, \cdots, x_n)$ is a symmetric polynomial of $x_1, \cdots, x_n$ of degree $2n(n-1)$.

We see immediately from (2.81) that $D(x_1, \cdots, x_n) = 0$ when $x_1 = x_2$. Furthermore, it is straightforward to verify that

$$\left(x_1 \frac{\partial}{\partial x_1}\right)^j D(x_1, \cdots, x_n) = 0, \quad j = 1, 2, 3$$

when $x_1 = x_2$. This is equivalent to saying

$$\frac{\partial^j}{\partial x_1^j} D(x_1, \cdots, x_n) = 0, \quad j = 1, 2, 3$$

when $x_1 = x_2$. The polynomial $p(x_1, \cdots, x_n)$ must thus contain as a factor $(x_1 - x_2)^4$, and so $\prod_{1 \le j < k \le n} (x_j - x_k)^4$ by symmetry. As this is of degree $2n(n-1)$, it follows that $p(x_1, \cdots, x_n)$ must in fact be proportional to $\prod_{1 \le j < k \le n} (x_j - x_k)^4$, which gives

$$D(x_1, \cdots, x_n) = c_n \frac{\prod_{1 \le j < k \le n} (x_j - x_k)^4}{\prod_{j=1}^{n} x_j^{2n-3}}$$

for some constant c_n.

To decide the c_n, let us look at the coefficient of the term $\prod_{j=1}^{n} x_j^{-(2n-3)} \prod_{j=1}^{n} x_j^{4(j-1)}$ in the determinant $D(x_1, \cdots, x_n)$. For sake of clarity, we consider two cases separately: n is either even or odd.

Assume $n = 2k$. Let us add $n - 1$ times the first column to the nth column. Then we see the coefficient of $x_1^{-(2n-3)}$ is given by a cofactor of the following 2×2 matrix

$$\begin{pmatrix} x_1^{-(n-2)} - x_n^{-(n-2)} & x_1^{-(n-2)} - (n-1)x_n^{-(n-2)} \\ x_1^{-(n-1)} - x_n^{-(n-1)} & -(n-1)x_n^{-(n-1)} \end{pmatrix}.$$

In the cofactor, we add $n-3$ times the first column to the $(n-1)$th column. Then we see the coefficient of $x_2^{-(2n-7)}$ is given by a cofactor of the following 2×2 matrix

$$\begin{pmatrix} x_2^{-(n-4)} - x_n^{-(n-4)} & x_2^{-(n-4)} - (n-1)x_n^{-(n-4)} \\ x_2^{-(n-3)} - x_n^{-(n-3)} & -(n-3)x_n^{-(n-3)} \end{pmatrix}.$$

Proceeding in this manner, we see the coefficient of x_{k-1}^{-5} is given by the determinant of the $n+1 \times n+1$ matrix

$$\begin{pmatrix} x_k - x_n & \cdots & x_{n-1} - x_n & x_k & \cdots & x_n \\ x_k^{-1} - x_n^{-1} & \cdots & x_{n-1}^{-1} - x_n^{-1} & -x_k^{-1} & \cdots & -x_n^{-1} \\ x_k^2 - x_n^2 & \cdots & x_{n-1}^2 - x_n^2 & 2x_k^2 & \cdots & 2x_n^2 \\ \vdots & \vdots & \vdots & \vdots & \vdots & \vdots \\ x_k^{n-1} - x_n^{n-1} & \cdots & x_{n-1}^{n-1} - x_n^{n-1} & (n-1)x_k^{n-1} & \cdots & (n-1)x_n^{n-1} \\ x_k^n - x_n^n & \cdots & x_{n-1}^n - x_n^n & nx_k^n & \cdots & nx_n^n \end{pmatrix}.$$

Interchange the top two rows to get

$$\begin{pmatrix} x_k^{-1} - x_n^{-1} & \cdots & x_{n-1}^{-1} - x_n^{-1} & -x_k^{-1} & \cdots & -x_n^{-1} \\ x_k - x_n & \cdots & x_{n-1} - x_n & x_k & \cdots & x_n \\ x_k^2 - x_n^2 & \cdots & x_{n-1}^2 - x_n^2 & 2x_k^2 & \cdots & 2x_n^2 \\ \vdots & \vdots & \vdots & \vdots & \vdots & \vdots \\ x_k^{n-1} - x_n^{n-1} & \cdots & x_{n-1}^{n-1} - x_n^{n-1} & (n-1)x_k^{n-1} & \cdots & (n-1)x_n^{n-1} \\ x_k^n - x_n^n & \cdots & x_{n-1}^n - x_n^n & nx_k^n & \cdots & nx_n^n \end{pmatrix}. \quad (2.82)$$

We postpone deciding the coefficient of x_k^{-1}, but we turn to the term x_n^{2n-1}. In the determinant of (2.82), we first subtract the kth column from columns $1, 2, \cdots, k-1$ to get

$$\begin{pmatrix} x_k^{-1} - x_{n-1}^{-1} & \cdots & x_{n-1}^{-1} - x_n^{-1} & -x_k^{-1} & \cdots & -x_n^{-1} \\ x_k - x_{n-1} & \cdots & x_{n-1} - x_n & x_k & \cdots & x_n \\ x_k^2 - x_{n-1}^2 & \cdots & x_{n-1}^2 - x_n^2 & 2x_k^2 & \cdots & 2x_n^2 \\ \vdots & \vdots & \vdots & \vdots & \vdots & \vdots \\ x_k^{n-1} - x_{n-1}^{n-1} & \cdots & x_{n-1}^{n-1} - x_n^{n-1} & (n-1)x_k^{n-1} & \cdots & (n-1)x_n^{n-1} \\ x_k^n - x_{n-1}^n & \cdots & x_{n-1}^n - x_n^n & nx_k^n & \cdots & nx_n^n \end{pmatrix}.$$

Then we add n times the kth column to $(n+1)$th column to see the coefficient of x_n^{2n-1} is given by the determinant of the $n-1 \times n-1$ matrix

$$\begin{pmatrix} x_k^{-1} - x_{n-1}^{-1} & \cdots & x_{n-2}^{-1} - x_{n-1}^{-1} & -x_k^{-1} & \cdots & -x_n^{-1} \\ x_k - x_{n-1} & \cdots & x_{n-2} - x_{n-1} & x_k & \cdots & x_n \\ x_k^2 - x_{n-1}^2 & \cdots & x_{n-2}^2 - x_{n-1}^2 & 2x_k^2 & \cdots & 2x_n^2 \\ \vdots & \vdots & \vdots & \vdots & \vdots & \vdots \\ x_k^{n-3} - x_{n-1}^{n-3} & \cdots & x_{n-2}^{n-3} - x_{n-1}^{n-3} & (n-3)x_k^{n-3} & \cdots & (n-3)x_{n-1}^{n-3} \\ x_k^{n-2} - x_{n-1}^{n-2} & \cdots & x_{n-2}^{n-2} - x_{n-1}^{n-2} & (n-2)x_k^{n-2} & \cdots & (n-2)x_{n-1}^{n-2} \end{pmatrix}.$$

Repeating this operation, we get the coefficient of x_{k+1}^3 is $-x_k^{-1}$. In summary, the coefficient of the $\prod_{j=1}^n x_j^{-(2n-3)} \prod_{j=1}^n x_j^{4(j-1)}$ is $(-1)^{k-1}$.

Assume $n = 2k + 1$. Then we can almost completely repeat the procedure above to see that the coefficient of the $\prod_{j=1}^n x_j^{-(2n-3)} \prod_{j=1}^n x_j^{4(j-1)}$ is $(-1)^k$.

Finally, note the coefficient of $\prod_{j=1}^n x_j^{4(j-1)}$ in $\prod_{1 \leq j < k \leq n} (x_j - x_k)^4$ is 1. So it follows

$$c_n = (-1)^{(n-1)(n-2)/2},$$

as desired. $\qquad \square$

Proceed to computing the determinant J. We have

Lemma 2.17.

$$|J| = \frac{\prod_{l=0}^{n-2} \left(1 - |\alpha_l|^2\right)}{q_n \prod_{l=1}^n q_l}.$$

Proof. Note

$$d\alpha_k = da_k + i db_k,$$
$$d\bar{\alpha}_k = da_k - i db_k.$$

It easily follows

$$d\bar{\alpha}_k \wedge d\alpha_k = \det \begin{pmatrix} 1 & -i \\ 1 & i \end{pmatrix} da_k \wedge db_k$$
$$= 2i\, da_k \wedge db_k. \tag{2.83}$$

Inserting (2.83) into (2.80) gives

$$J = \frac{q_n^2 \prod_{l=1}^{n-1} q_l^3}{\prod_{l=0}^{n-2} \rho_l^2 \prod_{l=0}^{n-2} \rho_l^{4(n-l-2)}} D\left(e^{i\theta_1}, \cdots, e^{i\theta_n}\right).$$

According to Lemmas 2.15 and 2.16, it immediately follows

$$|J| = \frac{q_n^2 \prod_{l=1}^{n-1} q_l^3}{\prod_{l=0}^{n-2} \rho_l^2 \prod_{l=0}^{n-2} \rho_l^{4(n-l-2)}} \prod_{1 \leq j < k \leq n} \left|e^{i\theta_j} - e^{i\theta_k}\right|^4$$
$$= \frac{\prod_{l=0}^{n-2} \left(1 - |\alpha_l|^2\right)}{q_n^2 \prod_{l=1}^n q_l}.$$

The proof is now complete. $\qquad \square$

Lemma 2.18. *Let*

$$\Delta_n = \{(q_1, q_2, \cdots, q_{n-1}) : q_i > 0, q_1^2 + \cdots + q_{n-1}^2 < 1\}.$$

Then

$$\int_{\Delta_n} \frac{1}{q_n} \prod_{j=1}^n q_j^{\beta-1} dq_1 \cdots dq_{n-1} = \frac{\Gamma(\frac{\beta}{2})^n}{2^{n-1}\Gamma(\frac{\beta n}{2})},$$

where $q_n = \sqrt{1 - q_1^2 - \cdots - q_{n-1}^2}$.

Proof. We only consider the case of $n \geq 2$ since the other case is trivial. Start with $n = 2$. Using the change of variable, we have

$$\int_{\Delta_2} \frac{1}{q_2} \prod_{j=1}^2 q_j^{\beta-1} dq_1 = \int_0^1 (1 - q_1^2)^{\beta/2-1} q_1^{\beta-1} dq_1$$

$$= \frac{1}{2} \int_0^1 (1 - q_1)^{\beta/2-1} q_1^{\beta/2-1} dq_1$$

$$= \frac{\Gamma(\frac{\beta}{2})^2}{2\Gamma(\beta)}.$$

Assume by induction that the claim is valid for some $n \geq 2$. It easily follows

$$\int_{\Delta_{n+1}} \frac{1}{q_{n+1}} \prod_{i=1}^{n+1} q_i^{\beta-1} dq_1 \cdots dq_n$$

$$= \int_{q_1^2 + \cdots + q_{n-1}^2 \leq 1 - q_n^2} (1 - q_n^2 - q_1^2 - \cdots - q_{n-1}^2)^{\beta/2-1} \prod_{i=1}^n q_i^{\beta-1} dq_1 \cdots dq_n$$

$$= \int_0^1 (1 - q_n^2)^{\beta/2-1} q_n^{\beta-1} dq_n \int_{q_1^2 + \cdots + q_{n-1}^2 \leq 1 - q_n^2} \prod_{i=1}^{n-1} q_i^{\beta-1}$$

$$\cdot \left(1 - \frac{q_1^2}{1 - q_n^2} - \cdots - \frac{q_{n-1}^2}{1 - q_n^2}\right)^{\beta/2-1} dq_1 \cdots dq_{n-1}. \tag{2.84}$$

Making a change of variable, the inner integral becomes

$$(1 - q_n^2)^{(n-1)\beta/2} \int_{\Delta_n} (1 - q_1^2 - \cdots - q_{n-1}^2)^{\beta/2-1} \prod_{i=1}^{n-1} q_i^{\beta-1} dq_1 \cdots dq_{n-1},$$

which is in turn equal to

$$(1 - q_n^2)^{(n-1)\beta/2} \frac{\Gamma(\frac{\beta}{2})^n}{2^{n-1}\Gamma(\frac{n\beta}{2})}$$

using the induction hypothesis. Substituting in (2.84) gives

$$\int_{\Delta_{n+1}} \frac{1}{q_{n+1}} \prod_{i=1}^{n+1} q_i^{\beta-1} dq_1 \cdots dq_n = \frac{\Gamma(\frac{\beta}{2})^n}{2^{n-1}\Gamma(\frac{n\beta}{2})} \int_0^1 (1-q_n^2)^{n\beta/2-1} q_n^{\beta-1} dq_n$$

$$= \frac{\Gamma(\frac{\beta}{2})^{n+1}}{2^n \Gamma(\frac{(n+1)\beta}{2})}.$$

We conclude by induction the proof. □

Proof of Theorem 2.17. As remarked above, (2.73) naturally induces a one-to-one mapping from $(e^{i\theta_1}, \cdots, e^{i\theta_n}, q_1, \cdots, q_{n-1})$ to $(a_0, b_0, \cdots, a_{n-2}, b_{n-2}, \alpha_{n-1})$. Let $f_{n,\beta}$ and $h_{n,\beta}$ be their respective joint probability density functions. Then it follows by Lemmas 2.17 and 2.15

$$f_{n,\beta}\big(e^{i\theta_1}, \cdots, e^{i\theta_n}, q_1, \cdots, q_{n-1}\big)$$

$$= h_{n,\beta}(a_0, b_0, \cdots, a_{n-2}, b_{n-2}, \alpha_{n-1})|J|$$

$$= \frac{\beta^{n-1}}{(2\pi)^n}(n-1)! \prod_{l=0}^{n-2} \big(1-|\alpha_l|^2\big)^{\beta(n-l-1)/2} \frac{1}{q_n \prod_{l=1}^n q_l}$$

$$= \frac{\beta^{n-1}}{(2\pi)^n}(n-1)! \prod_{1\le j<k\le n} \big|e^{i\theta_j} - e^{i\theta_k}\big|^\beta \frac{1}{q_n} \prod_{l=1}^n q_l^{\beta-1}.$$

This trivially implies that $(e^{i\theta_1}, \cdots, e^{i\theta_n})$ is independent of $(q_1, \cdots, q_{n-1})$. Integrating out the q_j over Δ_n, we get by Lemma 2.18

$$g_{n,\beta}\big(e^{i\theta_1}, \cdots, e^{i\theta_n}\big) := \int_{\Delta_n} f_{n,\beta}\big(e^{i\theta_1}, \cdots, e^{i\theta_n}, q_1, \cdots, q_{n-1}\big) dq_1 \cdots dq_{n-1}$$

$$= \frac{n!}{(2\pi)^n Z_{n,\beta}} \prod_{1\le j<k\le n} \big|e^{i\theta_j} - e^{i\theta_k}\big|^\beta.$$

Dividing by $n!$ to eliminate the ordering of eigenvalues, we conclude the proof as desired. □

Having a CMV matrix model, we can establish the following asymptotic normal fluctuations for the CβE.

Theorem 2.18. *Let* $e^{i\theta_1}, \cdots, e^{i\theta_n}$ *be chosen on the unit circle according to the CβE. Then as* $n \to \infty$
(i) for any θ_0 *with* $0 \le \theta_0 < 2\pi$,

$$\frac{1}{\sqrt{\frac{2}{\beta}\log n}} \sum_{j=1}^n \log\big(1 - e^{i(\theta_j - \theta_0)}\big) \xrightarrow{d} N_{\mathbb{C}}(0,1);$$

(ii) for any $0 < a < b < 2\pi$,

$$\frac{N_n(a,b) - \frac{n(b-a)}{2\pi}}{\frac{1}{\pi}\sqrt{\frac{2}{\beta}\log n}} \xrightarrow{d} N(0,1),$$

where $N_n(a,b)$ denotes the number of the angles θ_j lying in the arc between a and b.

Proof.   Left to the reader.   □

Chapter 3

Gaussian Unitary Ensemble

3.1 Introduction

Let $\mathcal{H}_n$ be the set of all $n \times n$ Hermitian matrices. To each matrix $H \in \mathcal{H}_n$ assign a probability measure as follows

$$P_n(H)dH = \frac{2^{n(n-1)/2}}{(2\pi)^{n^2/2}} e^{-\frac{1}{2}tr H^2} dH, \qquad (3.1)$$

where dH is the Lebesgue measure on the algebraically independent entries of H. P_n is clearly invariant under unitary transform, namely $P_n(UHU^*) = P_n(H)$ for every unitary matrix U, see Chapter 2 of Deift and Gioev (2009) for a proof. The probability space $(\mathcal{H}_n, P_n)$ is called Gaussian Unitary Ensemble (GUE). It is the most studied object in random matrix theory. As a matter of fact, the GUE is a prototype of a large number of matrix models and related problems.

Note that the GUE can be realized in the following way. Let $z_{ii}, 1 \leq i \leq n$ be a sequence of i.i.d real standard normal random variables, $z_{ij}, 1 \leq i < j \leq n$ an array of i.i.d. complex standard normal random variables independent of the z_{ii}'s. Then $A_n := (z_{ij})_{n \times n}$ where $z_{ji} = z_{ij}^*, i < j$ will induce a probability measure as given by (3.1) in $\mathcal{H}_n$.

A remarkable feature of the GUE is that the eigenvalues have an explicit nice probability density function. Let $\lambda_1, \cdots, \lambda_n$ be n real unordered eigenvalues of A_n, then they are almost surely distinct to each other and are absolutely continuous with respect to Lebesgue measure on $\mathbb{R}^n$. In particular, we have

Theorem 3.1. *Let $p_n(\mathbf{x})$ denote the joint probability density function of $\lambda = (\lambda_1, \cdots, \lambda_n)$, then*

$$p_n(\mathbf{x}) = \frac{1}{(2\pi)^{n/2} \prod_{k=1}^n k!} \prod_{1 \leq i < j \leq n} |x_j - x_i|^2 \prod_{k=1}^n e^{-x_k^2/2}, \qquad (3.2)$$

where $\mathbf{x} = (x_1, \cdots, x_n) \in \mathbb{R}^n$.

This theorem, due to Weyl, plays an important role in the study of GUE. Its proof can be found in the textbooks Anderson, Guionnet and Zeitouni (2010), Deift and Gioev (2009). (3.2) should be interpreted as follows. Let $f : \mathcal{H}_n \mapsto \mathbb{R}$ be an invariant function, i.e., $f(H) = f(UHU^*)$ for each $H \in \mathcal{H}_n$ and unitary matrix U. Then

$$Ef(H) = \int_{\mathbb{R}^n} f(\mathbf{x})p_n(\mathbf{x})d\mathbf{x}.$$

It is worthy to remark that there are two factors in the righthand side of (3.2). One is the product of n standard normal density functions, while the other is the square of Vandermonde determinant. The probability that two eigenvalues neighbor each other very closely is very small. Hence intuitively speaking, eigenvalues should locate more neatly than i.i.d. normal random points in the real line. It is the objective of this chapter that we shall take a closer look at the arrangement of eigenvalues from global behaviours.

In order to analyze the precise asymptotics of $p_n(\mathbf{x})$, we need to introduce Hermite orthogonal polynomials and the associated wave functions. Let $h_l(x), l \geq 0$ be a sequence of monic orthogonal polynomials with respect to the weight function $e^{-x^2/2}$ with $h_0(x) = 1$. Then

$$h_l(x) = (-1)^l e^{x^2/2} \frac{d^l}{dx^l} e^{-x^2/2}$$

$$= l! \sum_{i=0}^{[l/2]} (-1)^i \frac{x^{l-2i}}{2^i i!(l-2i)!}, \quad l \geq 1. \tag{3.3}$$

Define

$$\varphi_l(x) = (2\pi)^{-1/4}(l!)^{-1/2}h_l(x)e^{-x^2/4} \tag{3.4}$$

so that we have

$$\int_{-\infty}^{\infty} \varphi_l(x)\varphi_m(x)dx = \delta_{l,m}, \quad \forall l, m \geq 0.$$

Now a simple matrix manipulation directly yields

$$\prod_{1 \leq i < j \leq n} (x_j - x_i) = \det \begin{pmatrix} 1 & 1 & \cdots & 1 \\ x_1 & x_2 & \cdots & x_n \\ \vdots & \vdots & \ddots & \vdots \\ x_1^{n-1} & x_2^{n-1} & \cdots & x_n^{n-1} \end{pmatrix}$$

$$= \det \begin{pmatrix} h_0(x_1) & h_0(x_2) & \cdots & h_0(x_n) \\ h_1(x_1) & h_1(x_2) & \cdots & h_1(x_n) \\ \vdots & \vdots & \ddots & \vdots \\ h_{n-1}(x_1) & h_{n-1}(x_2) & \cdots & h_{n-1}(x_n) \end{pmatrix}. \tag{3.5}$$

Furthermore, substituting (3.5) into (3.2) and noting (3.4) immediately leads to the following determinantal expression for $p_n(\mathbf{x})$.

Proposition 3.1.

$$p_n(\mathbf{x}) = \frac{1}{n!} \det \left(K_n(x_i, x_j) \right)_{n \times n}, \tag{3.6}$$

where K_n is defined by

$$K_n(x, y) = \sum_{l=0}^{n-1} \varphi_l(x) \varphi_l(y). \tag{3.7}$$

Such a expression like (3.6) turns out to be very useful in the study of asymptotics of eigenvalues. In fact, the GUE is one of the first examples of so-called determinantal point processes (see Section 3.3 below for more details). A nice observation about the kernel K_n is the following:

$$\int_{-\infty}^{\infty} K_n(x, z) K_n(z, y) dz = K_n(x, y).$$

As an immediate consequence, we can easily obtain any k-dimensional marginal density. Let $p_{n,k}(x_1, \cdots, x_k)$ be the probability density function of $(\lambda_1, \cdots, \lambda_k)$, then it follows

$$p_{n,k}(x_1, \cdots, x_k) = \frac{(n-k)!}{n!} \det \left(K_n(x_i, x_j) \right)_{k \times k}. \tag{3.8}$$

In particular, we have

$$p_{n,1}(x) = \frac{1}{n} K_n(x, x).$$

We collect some basic properties of Hermite wave functions $\varphi_l(x)$ below. See Szegö (1975) for more information.

Lemma 3.1. *For $l \geq 1$, it follows*
(i) recurrence equation

$$x\varphi_l(x) = \sqrt{l+1}\varphi_{l+1}(x) + \sqrt{l}\varphi_{l-1}(x);$$

(ii) differential relations

$$\varphi_l'(x) = -\frac{x}{2}\varphi_l(x) + \sqrt{l}\varphi_{l-1}(x),$$

$$\varphi_l''(x) = -\left(l + \frac{1}{2} - \frac{x^2}{4}\right)\varphi_l(x);$$

(iii) Christoffel-Darboux identities

$$\sum_{m=0}^{l-1} \varphi_m(x)\varphi_m(y) = \sqrt{l} \cdot \frac{\varphi_l(x)\varphi_{l-1}(y) - \varphi_l(y)\varphi_{l-1}(x)}{x - y}, \quad x \neq y \quad (3.9)$$

$$\sum_{m=0}^{l-1} \varphi_m^2(x) = \sqrt{l} \cdot \left(\varphi_l'(x)\varphi_{l-1}(x) - \varphi_l(x)\varphi_{l-1}'(x)\right); \quad (3.10)$$

(iv) boundedness

$$\kappa := \sup_{l \geq 0} \|\varphi_l\|_\infty < \infty. \quad (3.11)$$

The next lemma, known as Plancherel-Rotach formulae, provides asymptotic behavior formulae for the Hermite orthogonal polynomials.

Lemma 3.2. *We have as* $n \to \infty$
(i) for $|x| < 2 - \delta$ *with* $\delta > 0$,

$$n^{1/4}\varphi_{n+k}(\sqrt{n}x) = \sqrt{\frac{2}{\pi}}\frac{1}{(4 - x^2)^{1/4}} \cos\left(n\alpha(\theta) + \frac{k+1}{2}\theta - \frac{\pi}{4}\right) + O(n^{-1}),$$

where $x = 2\cos\theta$, $\alpha(\theta) = \theta - \sin 2\theta/2$, $k = -1, 0, 1$ *and the convergence is uniform in* x. *The asymptotics is also valid for* $|x| < 2 - \delta_n$ *with* $\delta_n^{-1} = o(n^{2/3})$;
(ii) for $x = \pm 2 + \zeta n^{-2/3}$ *with* $\zeta \in \mathbb{R}$,

$$n^{1/12}\varphi_n(\sqrt{n}x) = 2^{1/4}\left(Ai(\zeta) + O(n^{-3/4})\right),$$

where $Ai(\zeta)$ *stands for the standard Airy function;*
(iii) for $|x| > 2 + \delta$,

$$n^{1/4}\varphi_n(\sqrt{n}x) = \frac{1}{\sqrt{\pi \sinh\theta}}e^{-(2n+1)\beta(\theta)/2}\left(1 + O(n^{-1})\right),$$

where $x = 2\cosh\theta$ *with* $\theta > 0$ *and* $\beta(\theta) = \sinh 2\theta/2 - \theta$.

As a direct application, we obtain a limit of the first marginal probability density after suitably scaled.

Proposition 3.2. *Define*

$$\bar{p}_n(x) = \sqrt{n}p_{n,1}(\sqrt{n}x).$$

(i) We have as $n \to \infty$

$$\bar{p}_n(x) \to \rho_{sc}(x)$$

uniformly on any closed interval of $(-2, 2)$, *where* ρ_{sc} *was defined by (1.15).*

(ii) Given $s > 0$, there exist positive constants c_1 and c_2 such that for each $n \geq 1$

$$\bar{p}_n\left(2 + \frac{s}{n^{3/2}}\right) \leq \frac{c_1}{n^{1/3}s}e^{-c_2 s^{3/2}}, \tag{3.12}$$

$$\bar{p}_n\left(-2 - \frac{s}{n^{3/2}}\right) \leq \frac{c_1}{n^{1/3}s}e^{-c_2 s^{3/2}}.$$

(iii) Given $|x| > 2 + \delta$ with $\delta > 0$, there exist positive constants c_3 and c_4 such that for each $n \geq 1$

$$\bar{p}_n(x) \leq c_3 e^{-c_4 n |x|^{3/2}}.$$

Proof.   To start with the proof of (i). Note it follows from (ii) and (iii) of Lemma 3.1

$$\begin{aligned}
\bar{p}_n(x) &= \frac{1}{\sqrt{n}} K_n\left(\sqrt{n}x, \sqrt{n}x\right) \\
&= \varphi_n'(\sqrt{n}x)\varphi_{n-1}(\sqrt{n}x) - \varphi_n(\sqrt{n}x)\varphi_{n-1}'(\sqrt{n}x) \\
&= \sqrt{n}\varphi_{n-1}(\sqrt{n}x)^2 - \sqrt{n-1}\varphi_n(\sqrt{n}x)\varphi_{n-2}(\sqrt{n}x). \tag{3.13}
\end{aligned}$$

Fix a $\delta > 0$. For each x such that $|x| \leq 2 + \delta$, we have by Lemma 3.1 (i)

$$\sqrt{n}\varphi_{n-1}(\sqrt{n}x)^2 = \frac{2}{\pi(4 - x^2)^{1/2}} \cos^2\left(n\alpha(\theta) - \frac{\pi}{4}\right) + O(n^{-1})$$

and

$$\begin{aligned}
&\sqrt{n}\varphi_n(\sqrt{n}x)\varphi_{n-2}(\sqrt{n}x) \\
&= \frac{2}{\pi(4 - x^2)^{1/2}} \cos\left(n\alpha(\theta) + \frac{\theta}{2} - \frac{\pi}{4}\right)\cos\left(n\alpha(\theta) - \frac{\theta}{2} - \frac{\pi}{4}\right) + O(n^{-1}),
\end{aligned}$$

where $x = 2\cos\theta$, $\alpha(\theta) = \theta - \sin 2\theta/2$ and the convergence is uniform in x.

Now a simple algebra yields

$$\bar{p}_n(x) = \frac{1}{2\pi}\sqrt{4 - x^2} + O(n^{-1}),$$

as desired.

Proceed to prove (ii). Note that it follows by Lemma 3.1

$$\begin{aligned}
\frac{d}{dx}K_n(x, x) &= \sqrt{n}\left(\varphi_n''(x)\varphi_{n-1}(x) - \varphi_n(x)\varphi_{n-1}''(x)\right) \\
&= -\sqrt{n}\varphi_n(x)\varphi_{n-1}(x).
\end{aligned}$$

Since $K_n(x, x)$ vanishes at ∞, we have

$$K_n(x, x) = \sqrt{n}\int_x^\infty \varphi_n(u)\varphi_{n-1}(u)du,$$

from which it follows

$$\bar{p}_n(x) = \sqrt{n} \int_x^\infty \varphi_n(\sqrt{n}u)\varphi_{n-1}(\sqrt{n}u)du.$$

By the Cauchy-Schwarz inequality,

$$\bar{p}_n(x) \le \sqrt{n}\left(\int_x^\infty \varphi_n(\sqrt{n}u)^2 du\right)^{1/2}\left(\int_x^\infty \varphi_{n-1}(\sqrt{n}u)du\right)^{1/2}.$$

Let $x = 2 + s/n^{2/3}$. We will below control

$$\sqrt{n}\int_{2+s/n^{2/3}}^\infty \varphi_n(\sqrt{n}u)^2 du = \frac{1}{n^{1/6}}\int_s^\infty \varphi_n\left(\sqrt{n}(2 + \frac{u}{n^{2/3}})\right)^2 du$$

from the above. Write

$$\cosh\theta = 1 + \frac{u}{2n^{2/3}}, \quad \beta(\theta) = \frac{1}{2}\sinh 2\theta - \theta.$$

Then it follows by using asymptotic formula

$$\frac{1}{n^{1/6}}\int_s^\infty \varphi_n\left(\sqrt{n}(2 + \frac{u}{n^{2/3}})\right)^2 du = (1 + o(1))\frac{1}{\pi}\int_{\theta_s}^\infty e^{-(2n+1)\beta(\theta)}d\theta,$$

where $\cosh\theta_s = 1 + s/2n^{2/3}$.

Since $\beta'(\theta) = 2(\sinh\theta)^2$ is increasing, we have for $s \to \infty$ and $n \to \infty$

$$\frac{1}{\pi}\int_{\theta_s}^\infty e^{-(2n+1)\beta(\theta)}d\theta \le \frac{1}{2\pi n\beta'(\theta_s)}e^{-(2n+1)\beta(\theta_s)}.$$

Note an elementary inequality

$$\frac{\theta^2}{2} \le \cosh\theta - 1 \le (\sinh\theta)^2,$$

from which one can readily derive

$$\beta'(\theta_s) = 2(\sinh\theta_s)^2 \ge 2(\cosh\theta_s - 1)$$
$$= \frac{s}{n^{2/3}}$$

and

$$\beta(\theta_s) = 2\int_0^{\theta_s}(\sinh x)^2 dx \ge \frac{2}{\theta_s}\left(\int_0^{\theta_s}\sinh x dx\right)^2$$
$$\ge \frac{s^{3/2}}{2n}.$$

Thus

$$\sqrt{n}\int_{2+s/n^{2/3}}^\infty \varphi_n(\sqrt{n}u)^2 du \le \frac{(1 + o(1))}{2\pi n^{1/3}s}e^{-(1+o(1))s^{3/2}}. \tag{3.14}$$

Similarly, the upper bound of (3.14) holds for the integral of $\varphi_{n-1}(\sqrt{n}u)^2$, and so (3.12) is proven.

Last, we turn to (iii). The proof is completely similar to (ii). Now we conclude the proposition. $\qquad\square$

Now we are ready to state and prove the celebrated Wigner semicircle law for the GUE. Define the empirical spectral distribution for normalized eigenvalues by

$$F_n(x) = \frac{1}{n} \sum_{k=1}^{n} 1_{(\lambda_k \leq \sqrt{n}x)}, \quad -\infty < x < \infty. \tag{3.15}$$

Proposition 3.2 gives the limit of the mean spectral density. In fact, we can further prove the following

Theorem 3.2. *We have as $n \to \infty$*

$$F_n \xrightarrow{d} \rho_{sc} \quad in \ P, \tag{3.16}$$

where ρ_{sc} was defined by (1.15).

Proof. The statement (3.16) means that for any bounded continuous function f,

$$\int_{-\infty}^{\infty} f(x) dF_n(x) \xrightarrow{P} \int_{-\infty}^{\infty} f(x) \rho_{sc}(x) dx, \quad n \to \infty. \tag{3.17}$$

Note that f in (3.17) can be replaced by any bounded Lipschitz function.

Let f be a bounded 1-Lipschtiz function, we will prove the following claims

$$E \frac{1}{n} \sum_{k=1}^{n} f\left(\frac{\lambda_k}{\sqrt{n}}\right) \to \int_{-\infty}^{\infty} f(x) \rho_{sc}(x) dx \tag{3.18}$$

and

$$Var\left(\frac{1}{n} \sum_{k=1}^{n} f\left(\frac{\lambda_k}{\sqrt{n}}\right)\right) \to 0. \tag{3.19}$$

First, we prove (3.18). Note

$$E f\left(\frac{\lambda_k}{\sqrt{n}}\right) = \int_{-\infty}^{\infty} f\left(\frac{x}{\sqrt{n}}\right) p_{n,1}(x) dx$$

$$= \int_{-\infty}^{\infty} f(x) \bar{p}_n(x) dx. \tag{3.20}$$

Fix a small $\delta > 0$ and let $\delta_n = s_n/n^{2/3}$ satisfy $\delta_n n^{2/3} \to \infty$ and $\delta_n n^{1/2} \to 0$. Write the integral on the righthand side of (3.20) as the sum of integrals $I_k, k = 1, 2, 3, 4$, over the sets $A_1 = \{x : |x| < 2-\delta\}$, $A_2 = \{x : 2-\delta \leq |x| < 2-\delta_n\}$, $A_3 = \{x : 2-\delta_n \leq |x| < 2+\delta_n\}$ and $A_4 = \{x : 2+\delta_n \leq |x| < \infty\}$.

We will below estimate each integral, separately. First, it clearly follows from Proposition 3.2 (i) that

$$I_1 = \int_{A_1} f(x)\bar{p}_n(x)dx \to \int_{A_1} f(x)\rho_{sc}(x)dx.$$

It remains to show that I_k, $k = 2, 3, 4$ are asymptotically as small as δ. Note f is bounded. Then I_k is bounded by the corresponding integral of $\bar{p}_n(x)$ over A_k.

Since $s_n \to \infty$, then according to Lemma 3.2 (i), we have for $x \in A_2$,

$$n^{1/4}\varphi_{n+k}(\sqrt{n}x) = \frac{1}{\sqrt{\pi \sin \theta}} \cos\left(n\alpha(\theta) + \frac{k+1}{2}\theta - \frac{\pi}{4}\right)(1 + o(1)),$$

where $x = 2\cos\theta$, $\alpha(\theta) = \sin\theta/2 - \theta$. Hence it follows

$$\int_{A_2} \bar{p}_n(x)dx = \int_{A_2} \left(n^{1/4}\varphi_{n-1}(\sqrt{n}x)\right)^2 dx$$

$$- \sqrt{\frac{n-1}{n}} \int_{A_2} n^{1/2}\varphi_n(\sqrt{n}x)\varphi_{n-2}(\sqrt{n}x)dx$$

$$= O(\delta).$$

To estimate the integral over A_3, we note (3.13) and use the bound in (3.11). Then

$$\int_{A_3} \bar{p}_n(x)dx \leq 2\|\bar{p}_n\|_\infty \delta_n$$

$$\leq 4\kappa^2 n^{1/2}\delta_n \to 0.$$

To estimate the integral over A_4, we use Proposition 3.2 (ii) to get

$$\int_{I_4} \bar{p}_n(x)dx \leq \int_{s_n}^\infty \left(\bar{p}_n\left(2 + \frac{s}{n^{2/3}}\right) + \bar{p}_n\left(-2 - \frac{s}{n^{2/3}}\right)\right)ds.$$

$$\to 0.$$

Combining the above four estimates together yields

$$\lim_{n\to\infty} Ef\left(\frac{\lambda_1}{\sqrt{n}}\right) = \int_{-2+\delta}^{2-\delta} f(x)\rho_{sc}(x)dx + O(\delta).$$

Letting $\delta \to 0$, we can conclude the proof of (3.18).

Next we turn to the proof of (3.19). Observe

$$Var\left(\sum_{k=1}^n f\left(\frac{\lambda_k}{\sqrt{n}}\right)\right)$$

$$= n\left[Ef\left(\frac{\lambda_1}{\sqrt{n}}\right)^2 - \left(Ef\left(\frac{\lambda_1}{\sqrt{n}}\right)\right)^2\right]$$

$$+ n(n-1)\left[Ef\left(\frac{\lambda_1}{\sqrt{n}}\right)f\left(\frac{\lambda_2}{\sqrt{n}}\right) - \left(Ef\left(\frac{\lambda_1}{\sqrt{n}}\right)\right)^2\right]. \tag{3.21}$$

Note

$$Ef\left(\frac{\lambda_1}{\sqrt{n}}\right)^2 = \int_{-\infty}^{\infty} f(x)^2 \bar{p}_n(x) dx$$

$$= \frac{1}{\sqrt{n}} \int_{-\infty}^{\infty} f(x)^2 K_n(\sqrt{n}x, \sqrt{n}x) dx$$

$$= \int_{-\infty}^{\infty} \int_{-\infty}^{\infty} f(x)^2 K_n(\sqrt{n}x, \sqrt{n}y)^2 dx dy \qquad (3.22)$$

and

$$Ef\left(\frac{\lambda_1}{\sqrt{n}}\right) f\left(\frac{\lambda_2}{\sqrt{n}}\right)$$

$$= \int_{-\infty}^{\infty} \int_{-\infty}^{\infty} f(x) f(y) n p_{n,2}(\sqrt{n}x, \sqrt{n}y) dx dy$$

$$= \frac{1}{n-1} \int_{-\infty}^{\infty} \int_{-\infty}^{\infty} f(x) f(y) \det\left(K_n(\sqrt{n}x, \sqrt{n}y)\right) dx dy$$

$$= \frac{1}{n-1} \left(\int_{-\infty}^{\infty} f(x) K_n(\sqrt{n}x, \sqrt{n}x) dx\right)^2$$

$$- \frac{1}{n-1} \int_{-\infty}^{\infty} \int_{-\infty}^{\infty} f(x) f(y) K_n(\sqrt{n}x, \sqrt{n}y)^2 dx dy. \qquad (3.23)$$

Substituting (3.22) and (3.23) into (3.21) yields

$$Var\left(\sum_{k=1}^{n} f\left(\frac{\lambda_k}{\sqrt{n}}\right)\right)$$

$$= \frac{1}{2} \int_{-\infty}^{\infty} \int_{-\infty}^{\infty} (f(x) - f(y))^2 K_n(\sqrt{n}x, \sqrt{n}y)^2 dx dy$$

$$= \frac{n}{2} \int_{-\infty}^{\infty} \int_{-\infty}^{\infty} \left(\frac{f(x) - f(y)}{x - y}\right)^2$$

$$\left(\varphi_n(\sqrt{n}x)\varphi_{n-1}(\sqrt{n}y) - \varphi_n(\sqrt{n}y)\varphi_{n-1}(\sqrt{n}x)\right)^2 dx dy.$$

Since f is 1-Lipschitz function, it follows by the orthogonality of φ_l

$$Var\left(\sum_{k=1}^{n} f\left(\frac{\lambda_k}{\sqrt{n}}\right)\right)$$

$$\leq \frac{n}{2} \int_{-\infty}^{\infty} \int_{-\infty}^{\infty} \left(\varphi_n(\sqrt{n}x)\varphi_{n-1}(\sqrt{n}y) - \varphi_n(\sqrt{n}y)\varphi_{n-1}(\sqrt{n}x)\right)^2 dx dy$$

$$= 1, \qquad (3.24)$$

which implies the claim (3.19).

Now we conclude the proof of Theorem 3.2. $\qquad\qquad\square$

3.2 Fluctuations of Stieltjes transforms

Let A_n be the standard GUE matrix as in the Introduction. Consider the normalized matrix $H_n = A_n/\sqrt{n}$ and denote by $\lambda_1, \lambda_2, \cdots, \lambda_n$ its eigenvalues. Define the Green function

$$G_n(z) = (G_{ij}(z))_{n \times n} = \frac{1}{H_n - z}$$

and its normalized trace

$$m_n(z) = \frac{1}{n} tr G_n(z) = \frac{1}{n} tr \frac{1}{H_n - z} = \frac{1}{n} \sum_{i=1}^{n} G_{ii}(z).$$

It obviously follows that

$$m_n(z) = \int_{-\infty}^{\infty} \frac{1}{x - z} dF_n(x) = s_{F_n}(z)$$

where F_n is defined by (3.15).

In this section we shall first estimate $Em_n(z)$ and $Var(m_n(z))$ and then prove a central limit theorem for $m_n(z)$. Start with some basic facts and lemmas about the Green function and trace of a matrix. We occasionally suppress the dependence of functions on z when the context is clear, for example we may write G_{ij} instead of $G_{ij}(z)$, and so on.

Lemma 3.3. *Let* $H_n = (H_{ij})_{n \times n}$ *be a Hermitian matrix,* $G_n(z) = (G_{ij}(z))_{n \times n}$ *its Green function. Then it follows*
(i) matrix identity

$$G_{ij} = -\frac{1}{z} + \frac{1}{z} \sum_{k=1}^{n} G_{ik} H_{ki};$$

(ii) for $z = a + i\eta$ *where* $\eta \neq 0$ *and* $k \geq 1$

$$\sup_{1 \leq i, j \leq n} |(G_n^k)_{ij}(z)| \leq \frac{1}{|\eta|^k}, \quad \frac{1}{n} |tr G_n^k| \leq \frac{1}{|\eta|^k};$$

(iii) differential relations

$$\frac{\partial G_{kl}}{\partial H_{ii}} = -G_{ki} G_{il}$$

and for $i \neq j$

$$\frac{\partial G_{kl}}{\partial Re H_{ij}} = -(G_{ki} G_{jl} + G_{kj} G_{il}), \quad \frac{\partial G_{kl}}{\partial Im H_{ij}} = -i(G_{ki} G_{jl} - G_{kj} G_{il}).$$

Proof. (i) trivially follows form the fact $G_n(H_n - z) = I_n$. To prove (ii), let $U = (u_{ij})_{n \times n}$ be a unitary matrix such that

$$H_n = U \begin{pmatrix} \lambda_1 & 0 & \cdots & 0 \\ 0 & \lambda_2 & \cdots & 0 \\ \vdots & \vdots & \ddots & \vdots \\ 0 & 0 & \cdots & \lambda_n \end{pmatrix} U^*. \tag{3.25}$$

Then

$$G_n^k = U \begin{pmatrix} (\lambda_1 - z)^{-k} & 0 & \cdots & 0 \\ 0 & (\lambda_2 - z)^{-k} & \cdots & 0 \\ \vdots & \vdots & \ddots & \vdots \\ 0 & 0 & \cdots & (\lambda_n - z)^{-k} \end{pmatrix} U^*.$$

Hence we have

$$(G_n^k)_{ij} = \sum_{l=1}^{n} u_{il}(\lambda_l - z)^{-k} u_{lj}^*,$$

from which it follows

$$\left| (G_n^k)_{ij} \right| \le \frac{1}{|\eta|^k} \sum_{l=1}^{n} |u_{il}||u_{lj}^*| \le \frac{1}{|\eta|^k}.$$

We conclude (ii).

Finally, (iii) easily follows from the Sherman-Morrison equation

$$\frac{1}{H_n + \delta A - z} - \frac{1}{H_n - z} = -\delta \frac{1}{H_n + \delta A - z} A \frac{1}{H_n - z}. \qquad \square$$

The next lemma collects some important properties that will be used below about Gaussian random variables.

Lemma 3.4. *Assume that* $g_1, g_2, \cdots, g_m$ *are independent centered normal random variables with* $E g_k^2 = \sigma_k^2$. *Denote* $\sigma^2 = \max_{1 \le k \le m} \sigma_k^2$.
(i) Stein equation: if $F : \mathbb{R}^m \mapsto \mathbb{C}$ *is a differentiable function, then*

$$E g_k F(g_1, \cdots, g_m) = \sigma_k^2 E \frac{\partial F}{\partial g_k}(g_1, \cdots, g_m).$$

(ii) Poincaré-Nash upper bound: if $F : \mathbb{R}^m \mapsto \mathbb{C}$ *is a differentiable function, then*

$$E \left| F(g_1, \cdots, g_m) - EF(g_1, \cdots, g_m) \right|^2 \le \sigma^2 E |\nabla F|^2$$

where ∇F *stands for the gradient of* F.
(iii) Concentration of measure inequality: if $F : \mathbb{R}^m \mapsto \mathbb{C}$ *is a Lipschitz function, then for any* $t > 0$

$$P\left(\left| F(g_1, \cdots, g_m) - EF(g_1, \cdots, g_m) \right| > t \right) \le e^{-t^2/2\sigma^2 \|F\|_{lip}}.$$

Now we can use the above two lemmas to get a rough estimate for $Em_n(z)$ and $Var(m_n(z))$.

Proposition 3.3. *For each z with $Imz \neq 0$, it follows*
(i)

$$Em_n(z) = m_{sc}(z) + O(n^{-2}), \qquad (3.26)$$

where $m_{sc}(z)$ denotes the Stieltjes transform of ρ_{sc}, namely

$$m_{sc}(z) = -\frac{z}{2} + \frac{1}{2}\sqrt{z^2 - 4};$$

(ii)

$$E|m_n(z) - Em_n(z)|^2 = O(n^{-2}). \qquad (3.27)$$

Proof. Start with the proof of (3.27). Note that m_n is a function of independent centered normal random variables $\{H_{ii}, 1 \leq i \leq n\}$ and $\{ReH_{ij}, ImH_{ij}, 1 \leq i < j \leq n\}$. We use the Poincaré-Nash upper bound to get

$$E|m_n - Em_n|^2 \leq \frac{1}{n}\Big[\sum_{i=1}^{n} E\Big|\frac{\partial m_n}{\partial H_{ii}}\Big|^2 + \sum_{i<j} E\Big|\frac{\partial m_n}{\partial ReH_{ij}}\Big|^2$$

$$+ \sum_{i<j} E\Big|\frac{\partial m_n}{\partial ImH_{ij}}\Big|^2\Big]. \qquad (3.28)$$

It easily follows from the differential relations in Lemma 3.3 (iii) that

$$\frac{\partial m_n}{\partial H_{ii}} = -\frac{1}{n}\sum_{l=1}^{n} G_{li}G_{il},$$

$$\frac{\partial m_n}{\partial ReH_{ij}} = -\frac{1}{n}\sum_{l=1}^{n}(G_{li}G_{jl} + G_{lj}G_{il}),$$

$$\frac{\partial m_n}{\partial ImH_{ij}} = -i\frac{1}{n}\sum_{l=1}^{n}(G_{li}G_{jl} - G_{lj}G_{il}).$$

In turn, according to Lemma 3.3 (ii), we have (3.27).

Proceed to the proof of (3.26). First, use the matrix identity to get

$$Em_n$$

$$= \frac{1}{n}\sum_{i=1}^{n} EG_{ii}$$

$$= \frac{1}{n}\sum_{i=1}^{n} E\Big(-\frac{1}{z} + \frac{1}{z}\sum_{k=1}^{n} G_{ik}H_{ki}\Big)$$

$$= -\frac{1}{z} + \frac{1}{zn}\sum_{i=1}^{n} EG_{ii}H_{ii} + \frac{1}{zn}\sum_{i\neq k}^{n} EG_{ik}\big(ReH_{ki} + iImH_{ki}\big). \quad (3.29)$$

Also, we have by Lemma 3.3 (iii)

$$EG_{ii}H_{ii} = \frac{1}{n}E\frac{\partial G_{ii}}{\partial H_{ii}} = -\frac{1}{n}G_{ii}^2 \tag{3.30}$$

$$EG_{ik}ReH_{ki} = \frac{1}{2n}E\frac{\partial G_{ik}}{\partial ReH_{ki}} = -\frac{1}{2n}(G_{ik}^2 + G_{ii}G_{kk}) \tag{3.31}$$

$$EG_{ik}ImH_{ki} = \frac{1}{2n}E\frac{\partial G_{ik}}{\partial ImH_{ki}} = -\frac{i}{2n}(G_{ik}^2 - G_{ii}G_{kk}). \tag{3.32}$$

Substituting (3.30)-(3.32) into (3.29) immediately yields

$$Em_n = -\frac{1}{z} - \frac{1}{z}Em_n^2. \tag{3.33}$$

According to (ii),

$$Em_n^2 - (Em_n)^2 = O(n^{-2}).$$

Hence it follows

$$Em_n = -\frac{1}{z} - \frac{1}{z}(Em_n)^2 + O(n^{-2}). \tag{3.34}$$

Recall that $m_{sc}(z)$ satisfies the equation

$$m_{sc} = -\frac{1}{z} - \frac{1}{z}m_{sc}^2.$$

It is now easy to see

$$Em_n(z) = m_{sc}(z) + O(n^{-2}),$$

as desired. □

Remark 3.1. (3.27) can be extended to any linear eigenvalue statistic (see (3.52) below) with differentiable test function. See Proposition 2.4 of Lytova and Pastur (2009).

As a direct consequence of Proposition 3.3, we obtain

Corollary 3.1. *For each z with $Imz \neq 0$,*

$$m_n(z) \xrightarrow{P} m_{sc}(z), \quad n \to \infty. \tag{3.35}$$

According to Theorem 1.14, the Stieltjes continuity theorem, (3.35) is in turn equivalent to saying that as $n \to \infty$

$$F_n \xrightarrow{d} \rho_{sc} \quad \text{in } P.$$

Thus we have derived the Wigner semicircle law using the Green function approach.

In the following we shall be devoted to the refinement of estimates of Em_n and $Var(m_n)$ given in Proposition 3.3. A basic tool is still Stein's equation. As above, we will repeatedly use the Stein equation to get the precise coefficients in terms of n^{-2}. Main results read as follows

Theorem 3.3. *For each z with $Imz \neq 0$, it follows*

(i)

$$Em_n(z) = m_{sc}(z) + \frac{1}{2(z^2 - 4)^{5/2}} \frac{1}{n^2} + o(n^{-2});$$

(ii)

$$E\big(m_n(z) - Em_n(z)\big)^2 = \frac{1}{(z^2 - 4)^2} \frac{1}{n^2} + o(n^{-2}), \tag{3.36}$$

$$Cov\big(m_n(z_1), m_n(z_2)\big)$$
$$= \frac{1}{2(z_1 - z_2)^2} \left(\frac{z_1 z_2 - 4}{\sqrt{z_1^2 - 4} \cdot \sqrt{z_2^2 - 4}} - 1 \right) \frac{1}{n^2} + o(n^{-2}). \tag{3.37}$$

Proof. One can directly get (i) from (ii) by noting the equation (3.33). We shall mainly focus on the computation of (3.36), since (3.37) is similar. To do this, note

$$Em_n^2 = \frac{1}{n} \sum_{i=1}^{n} Em_n G_{ii}$$

$$= \frac{1}{n} \sum_{i=1}^{n} Em_n \left(-\frac{1}{z} + \frac{1}{z} \sum_{k=1}^{n} G_{ki} H_{ik} \right)$$

$$= -\frac{1}{z} Em_n + \frac{1}{zn} \sum_{i=1}^{n} Em_n G_{ii} H_{ii}$$

$$+ \frac{1}{zn} \sum_{i \neq k} Em_n G_{ki} ReH_{ik} + \frac{i}{zn} \sum_{i \neq k} Em_n G_{ki} ImH_{ik}.$$

Using Lemma 3.3 (iii) and some simple algebra we get

$$Em_n^2 = -\frac{1}{z} Em_n - \frac{1}{z} Em_n^3 - \frac{1}{zn^3} \sum_{i,j,k} EG_{ij} G_{jk} G_{ki}. \tag{3.38}$$

Hence we have

$$Em_n^2 - (Em_n)^2 = -\frac{1}{z} Em_n - \frac{1}{z} Em_n^3 - (Em_n)^2 - \frac{1}{zn^3} \sum_{i,j,k=1}^{n} EG_{ij} G_{jk} G_{ki}$$

$$= -\frac{2}{z} Em_n \big(Em_n^2 - (Em_n)^2 \big) - \frac{1}{zn^3} \sum_{i,j,k=1}^{n} EG_{ij} G_{jk} G_{ki}$$

$$- \frac{1}{z} E\big(m_n - (Em_n) \big)^3.$$

Solving this equation further yields

$$Em_n^2 - (Em_n)^2 = -\frac{1}{z + 2Em_n} \frac{1}{n^3} \sum_{i,j,k=1}^{n} EG_{ij}G_{jk}G_{ki} + o(n^{-2}). \quad (3.39)$$

We remark that the summand in the righthand side of (3.39) is asymptotically as small as n^{-2} by Lemma 3.3 (ii). It remains to precisely estimate this summand below. For this, we first observe

$$\frac{1}{n} \sum_{i,k=1}^{n} EG_{ik}G_{ki} = \frac{1}{n} \sum_{i,k=1}^{n} EG_{ik}\left(-\frac{1}{z}\delta_{i,k} + \frac{1}{z}\sum_{l=1}^{n} G_{kl}H_{li}\right)$$

$$= -\frac{1}{z}Em_n + \frac{1}{zn} \sum_{i,j,k=1}^{n} EG_{ik}G_{kl}H_{li}$$

$$= -\frac{1}{z}Em_n - \frac{2}{z}Em_n\frac{1}{n} \sum_{i,k=1}^{n} G_{ik}G_{ki}$$

$$= -\frac{1}{z}Em_n - \frac{2}{z}Em_n\frac{1}{n} \sum_{i,k=1}^{n} EG_{ik}G_{ki} + O(n^{-1}),$$

and so we have

$$\frac{1}{n} \sum_{i,k=1}^{n} EG_{ik}G_{ki} = -\frac{Em_n}{z + 2Em_n}\left(1 + O(n^{-1})\right). \quad (3.40)$$

In the same spirit, we have

$$\frac{1}{n} \sum_{i,j,k=1}^{n} EG_{ij}G_{jk}G_{ki} = -\frac{1}{zn} \sum_{i,j=1}^{n} EG_{ij}G_{ji} + \frac{1}{zn} \sum_{i,j,k,l=1}^{n} EG_{ij}G_{jk}G_{kl}H_{li}$$

$$= -\frac{1}{zn} \sum_{i,j=1}^{n} EG_{ij}G_{ji} - \frac{1}{z}\left(\frac{1}{n} \sum_{i,j=1}^{n} EG_{ij}G_{ji}\right)^2$$

$$- \frac{2}{z}Em_n\frac{1}{n} \sum_{i,j,k=1}^{n} EG_{ij}G_{jk}G_{ki} + O(n^{-1}).$$

Solving this equation and noting (3.40) yields

$$\frac{1}{n} \sum_{i,j,k=1}^{n} EG_{ij}G_{jk}G_{ki} = \frac{1 + o(1)}{z + 2Em_n}\left[\frac{1}{n} \sum_{i,k=1}^{n} EG_{ik}G_{ki} + \left(\frac{1}{n} \sum_{i,j=1}^{n} EG_{ij}G_{ji}\right)^2\right]$$

$$= -\frac{zEm_n + (Em_n)^2}{(z + 2Em_n)^3}(1 + o(1))$$

$$= \frac{1 + o(1)}{(z + 2Em_n)^3},$$

where in the last step we used the fact $zEm_n + (Em_n)^2 = -1 + o(1)$.

Next, we turn to prove (3.37). Since the proof is very similar to (3.36), we only give some main steps. It follows by the matrix identity

$$
Em_n(z_1)m_n(z_2)
$$

$$
= \frac{1}{n}\sum_{i=1}^{n} Em_n(z_1)G_{ii}(z_2)
$$

$$
= \frac{1}{n}\sum_{i=1}^{n} Em_n(z_1)\left(-\frac{1}{z_2} + \frac{1}{z_2}\sum_{k=1}^{n} G_{ik}(z_2)H_{ki}\right)
$$

$$
= -\frac{1}{z_2}Em_n(z_1) + \frac{1}{z_2 n}\sum_{i,k=1}^{n} Em_n(z_1)G_{ik}(z_2)H_{ki}. \qquad (3.41)
$$

Applying the Stein equation to $m_n(z_2)$,

$$
\frac{1}{n}\sum_{i,k=1}^{n} Em_n(z_1)G_{ik}(z_2)H_{ki} = -Em_n(z_1)m_n(z_2)^2
$$

$$
-\frac{1}{n}\sum_{i,k,l}^{n} EG_{kl}(z_1)G_{li}(z_1)G_{ik}(z_2). \qquad (3.42)
$$

Substituting (3.42) into (3.41) yields

$$
Em_n(z_1)m_n(z_2) = -\frac{1}{z_2}Em_n(z_1) - \frac{1}{z_2}Em_n(z_1)m_n(z_2)^2
$$

$$
-\frac{1}{z_2 n}\sum_{i,k,l}^{n} EG_{kl}(z_1)G_{li}(z_1)G_{ik}(z_2).
$$

We now have

$$
Em_n(z_1)m_n(z_2) - Em_n(z_1)Em_n(z_2)
$$

$$
= -\frac{1}{z_2}Em_n(z_1) - \frac{1}{z_2}Em_n(z_1)m_n(z_2)^2 - Em_n(z_1)Em_n(z_2)
$$

$$
-\frac{1}{z_2 n}\sum_{i,k,l=1}^{n} EG_{kl}(z_1)G_{li}(z_1)G_{ik}(z_2).
$$

By virtue of Proposition 3.3 and (3.33), it follows

$$
Em_n(z_1)m_n(z_2) - Em_n(z_1)Em_n(z_2)
$$

$$
= -\frac{1}{z_2 + 2Em_n(z_2)}\frac{1}{n}\sum_{i,k,l=1}^{n} EG_{kl}(z_1)G_{li}(z_1)G_{ik}(z_2) + o(n^{-2}). \qquad (3.43)
$$

It remains to estimate the summand of the righthand side of (3.43). To this end, note

$$\frac{1}{n}\sum_{k,l=1}^{n} EG_{kl}(z_1)G_{lk}(z_2)$$

$$= \frac{1}{n}\sum_{k,l=1}^{n} EG_{kl}(z_1)\left(-\frac{1}{z_2}\delta_{k,l} + \frac{1}{z_2}\sum_{m=1}^{n} G_{lm}(z_2)H_{mk}\right)$$

$$= -\frac{1}{z_2}Em_n(z_1) + \frac{1}{z_2}\sum_{k,l,m=1}^{n} EG_{kl}(z_1)G_{lm}(z_2)H_{mk}$$

$$= -\frac{1}{z_2}Em_n(z_1) - \frac{1}{z_2}Em_n(z_1)\frac{1}{n}\sum_{k,l=1}^{n} G_{kl}(z_1)G_{lk}(z_2)$$

$$\quad -\frac{1}{z_2}Em_n(z_2)\frac{1}{n}\sum_{k,l=1}^{n} G_{kl}(z_1)G_{lk}(z_2)$$

$$= -\frac{1}{z_2}Em_n(z_1) - \frac{1}{z_2}Em_n(z_1)\frac{1}{n}\sum_{k,l=1}^{n} EG_{kl}(z_1)G_{lk}(z_2)$$

$$\quad -\frac{1}{z_2}Em_n(z_2)\frac{1}{n}\sum_{k,l=1}^{n} EG_{kl}(z_1)G_{lk}(z_2) + o(1)$$

which immediately gives

$$\frac{1}{n}\sum_{k,l=1}^{n} EG_{kl}(z_1)G_{lk}(z_2) = -\frac{Em_n(z_1)}{z_2 + Em_n(z_1) + Em_n(z_2)}(1 + o(1)).$$

Applying once again the matrix identity and Stein equation, we obtain

$$\frac{1}{n}\sum_{i,k,l=1}^{n} EG_{kl}(z_1)G_{li}(z_1)G_{ik}(z_2)$$

$$= -\frac{1}{z_2 n}\sum_{k,l=1}^{n} EG_{kl}(z_1)G_{lk}(z_1)$$

$$\quad -\frac{1}{z_2}E\left(\frac{1}{n}\sum_{k,l=1}^{n} G_{kl}(z_1)G_{lk}(z_1)\right)E\left(\frac{1}{n}\sum_{k,l=1}^{n} G_{kl}(z_1)G_{lk}(z_2)\right)$$

$$\quad -\frac{1}{z_2 n}Em_n(z_1)\frac{1}{n}\sum_{i,k,l=1}^{n} EG_{kl}(z_1)G_{li}(z_1)G_{ik}(z_2)$$

$$\quad -\frac{1}{z_2 n}Em_n(z_2)\frac{1}{n}\sum_{i,k,l=1}^{n} EG_{kl}(z_1)G_{li}(z_1)G_{ik}(z_2) + o(1).$$

In combination, we have

$$Em_n(z_1)m_n(z_2) - Em_n(z_1)Em_n(z_2)$$

$$= -\frac{1+o(1)}{z_2 + 2Em_n(z_2)} \frac{Em_n(z_1)}{z_2 + Em_n(z_2) + Em_n(z_1)}$$

$$\times \frac{1}{z_1 + 2Em_n(z_1)} \left(1 - \frac{Em_n(z_1)}{z_2 + Em_n(z_2) + Em_n(z_1)}\right) \frac{1}{n^2}. \quad (3.44)$$

To simplify the righthand side of (3.44), we observe the following asymptotic formulae

$$Em_n(z) = m_{sc}(z)(1 + o(1))$$

and

$$\frac{Em_n(z_1)}{z_2 + Em_n(z_2) + Em_n(z_1)} = \frac{Em_n(z_2)}{z_1 + Em_n(z_2) + Em_n(z_1)}(1 + o(1)).$$

Thus a simple calculus now easily yields

$$Em_n(z_1)m_n(z_2) - Em_n(z_1)Em_n(z_2)$$

$$= \frac{1+o(1)}{\sqrt{z_1^2 - 4}\sqrt{z_2^2 - 4}} \frac{2}{\sqrt{z_1^2 - 4} + \sqrt{z_2^2 - 4} + z_1 - z_2}$$

$$\times \frac{2}{\sqrt{z_1^2 - 4} + \sqrt{z_2^2 - 4} - (z_1 - z_2)} \frac{1}{n^2}$$

$$= \frac{1+o(1)}{\sqrt{z_1^2 - 4}\sqrt{z_2^2 - 4}} \frac{2}{z_1 z_2 - 4 + \sqrt{(z_1^2 - 4)(z_2^2 - 4)}} \frac{1}{n^2}$$

$$= \frac{1+o(1)}{2\sqrt{z_1^2 - 4}\sqrt{z_2^2 - 4}} \frac{z_1 z_2 - 4 - \sqrt{(z_1^2 - 4)(z_2^2 - 4)}}{(z_1 - z_2)^2} \frac{1}{n^2},$$

as desired.   □

We have so far proved a kind of law of large numbers for $m_n(z)$ and provided a precise estimate of $Em_n(z)$ and $Var(m_n(z))$. Having these, one may ask how $m_n(z)$ fluctuates around its average. In the rest of this section we will deal with such a issue. It turns out that $m_n(z)$ asymptotically follows a normal fluctuation. Moreover, we have

Theorem 3.4. *Define a random process by*

$$\zeta_n(z) = n(m_n(z) - Em_n(z)), \quad z \in \mathbb{C} \setminus \mathbb{R}.$$

Then there is a Gaussian process $\Xi = \{\Xi(z), z \in \mathbb{C} \setminus \mathbb{R}\}$ *with the covariance structure*

$$Cov\big(\Xi(z_1), \Xi(z_2)\big) = \frac{1}{2(z_1 - z_2)^2}\left(\frac{z_1 z_2 - 4}{\sqrt{z_1^2 - 4} \cdot \sqrt{z_1^2 - 4}} - 1\right)$$

such that

$$\zeta_n \Rightarrow \Xi, \quad n \to \infty.$$

Proof. We use a standard argument in the context of weak convergence of processes. That is, we will verify both finite dimensional distribution convergence and uniform tightness below.

Start by proving the uniform tightness. As in (3.28), we have

$$E|m_n(z_1) - m_n(z_2)|^2 \leq \frac{1}{n} E|\nabla(m_n(z_1) - m_n(z_2))|^2$$

$$= \frac{1}{n} E\Big[\sum_{i=1}^{n} \Big|\frac{\partial(m_n(z_1) - m_n(z_2))}{\partial H_{ii}}\Big|^2$$

$$+ \sum_{i<j} \Big|\frac{\partial(m_n(z_1) - m_n(z_2))}{\partial ReH_{ij}}\Big|^2$$

$$+ \sum_{i<j} \Big|\frac{\partial(m_n(z_1) - m_n(z_2))}{\partial ImH_{ij}}\Big|^2\Big]. \qquad (3.45)$$

Observe the eigendecomposition (3.25). Then it follows

$$\frac{\partial \lambda_k}{\partial H_{ii}} = u_{ik} u_{ik}^*,$$

$$\frac{\partial \lambda_k}{\partial ReH_{ij}} = u_{ik} u_{jk}^* + u_{ik}^* u_{jk}$$

$$= 2Re(u_{ik} u_{jk}^*),$$

$$\frac{\partial \lambda_k}{\partial ImH_{ij}} = \mathbf{i}(u_{ik}^* u_{jk} - u_{ik} u_{jk}^*)$$

$$= 2Im(u_{jk} u_{jk}^*).$$

Hence we have

$$\frac{\partial(m_n(z_1) - m_n(z_2))}{\partial H_{ii}}$$

$$= \sum_{k=1}^{n} \frac{\partial(m_n(z_1) - m_n(z_2))}{\partial \lambda_k} \cdot \frac{\partial \lambda_k}{\partial H_{ii}}$$

$$= \frac{1}{n} \sum_{k=1}^{n} \Big(\frac{1}{(\lambda_k - z_1)^2} - \frac{1}{(\lambda_k - z_2)^2}\Big) u_{ik} u_{ik}^*,$$

and so

$$\Big|\frac{\partial(m_n(z_1) - m_n(z_2))}{\partial H_{ii}}\Big|^2$$

$$\leq \frac{1}{n^2} \sum_{k,l=1}^{n} \Big(\frac{1}{(\lambda_k - z_1)^2} - \frac{1}{(\lambda_k - z_2)^2}\Big)$$

$$\times \Big(\frac{1}{(\lambda_l - \bar{z}_1)^2} - \frac{1}{(\lambda_l - \bar{z}_2)^2}\Big) u_{ik} u_{ik}^* u_{il} u_{il}^*. \qquad (3.46)$$

Similarly,

$$\left| \frac{\partial(m_n(z_1) - m_n(z_2))}{\partial Re H_{ij}} \right|^2$$

$$\leq \frac{4}{n^2} \sum_{k,l=1}^{n} \left(\frac{1}{(\lambda_k - z_1)^2} - \frac{1}{(\lambda_k - z_2)^2} \right)$$

$$\times \left(\frac{1}{(\lambda_l - \bar{z}_1)^2} - \frac{1}{(\lambda_l - \bar{z}_2)^2} \right) Re(u_{ik}u_{jk}^*) Re(u_{il}u_{jl}^*) \qquad (3.47)$$

and

$$\left| \frac{\partial(m_n(z_1) - m_n(z_2))}{\partial Im H_{ij}} \right|^2$$

$$\leq \frac{4}{n^2} \sum_{k,l=1}^{n} \left(\frac{1}{(\lambda_k - z_1)^2} - \frac{1}{(\lambda_k - z_2)^2} \right)$$

$$\times \left(\frac{1}{(\lambda_l - \bar{z}_1)^2} - \frac{1}{(\lambda_l - \bar{z}_2)^2} \right) Im(u_{ik}u_{jk}^*) Im(u_{il}u_{jl}^*). \qquad (3.48)$$

Substituting (3.46)-(3.48) into (3.45) yields

$$E|m_n(z_1) - m_n(z_2)|^2 \leq \frac{4}{n^3} \sum_{i,j,k,l=1}^{n} \left(\frac{1}{(\lambda_k - z_1)^2} - \frac{1}{(\lambda_k - z_2)^2} \right)$$

$$\times \left(\frac{1}{(\lambda_l - \bar{z}_1)^2} - \frac{1}{(\lambda_l - \bar{z}_2)^2} \right) u_{ik}u_{jk}^* u_{il}^* u_{jl}.$$

Note by orthonormality

$$\sum_{i=1}^{n} u_{ik}u_{il}^* = \delta_{k,l}, \quad \sum_{j=1}^{n} u_{jk}u_{jl}^* = \delta_{k,l}$$

$$\sum_{i,j,k,l=1}^{n} u_{ik}u_{jk}^* u_{il}^* u_{jl} = n.$$

We have

$$E|m_n(z_1) - m_n(z_2)|^2 \leq \frac{4}{n^3} \sum_{k=1}^{n} \left| \frac{1}{(\lambda_k - z_1)^2} - \frac{1}{(\lambda_k - z_2)^2} \right|^2$$

$$\leq \frac{4}{n^2 \eta^6} |z_1 - z_2|^2,$$

from which we can establish the uniform tightness for $m_n(z)$.

Proceed to proving finite dimensional distribution convergence. Fix $z_1, z_2, \cdots, z_q \in \mathbb{C} \setminus \mathbb{R}$. It is enough to prove

$$(\zeta_n(z_1), \cdots, \zeta_n(z_q)) \xrightarrow{d} (\Xi(z_1), \cdots, \Xi(z_q)), \quad n \to \infty.$$

Equivalently, for any $c_1, \cdots, c_q$

$$\sum_{l=1}^{q} c_l \zeta_n(z_l) \xrightarrow{d} \sum_{l=1}^{q} c_l \Xi(z_l), \quad n \to \infty.$$

Let

$$X_n = \sum_{l=1}^{q} c_l \zeta_n(z_l).$$

We shall prove that for any $t \in \mathbb{R}$,

$$EX_n e^{itX_n} - itEX_n^2 E e^{itX_n} \to 0.$$

This will be again done using the Stein equation. For simplicity and clarity, we only deal with the 1-dimensional case below. In particular, we shall prove

$$E\zeta_n e^{it\zeta_n} - itE\zeta_n^2 E e^{it\zeta_n} \to 0.$$

Namely,

$$nEm_n e^{it\zeta_n} - nEm_n E e^{it\zeta_n} - itn^2 \big(Em_n^2 - (Em_n)^2\big) E e^{it\zeta_n} \to 0. \quad (3.49)$$

Following the strategy in the proof of Proposition 3.3, it follows

$$Em_n e^{it\zeta_n} = -\frac{1}{z} E e^{it\zeta_n} - \frac{1}{z} Em_n^2 e^{it\zeta_n} - \frac{it}{zn} E \sum_{i,k,l=1}^{n} G_{ik}G_{kl}G_{li} e^{it\zeta_n}.$$

We have by virtue of (3.33)

$$n\big(Em_n e^{it\zeta_n} - Em_n E e^{it\zeta_n}\big) = -\frac{n}{z}\big(Em_n^2 e^{it\zeta_n} - Em_n^2 E e^{it\zeta_n}\big)$$

$$-\frac{it}{zn} E \sum_{i,k,l=1}^{n} G_{ik}G_{kl}G_{li} e^{it\zeta_n}. \quad (3.50)$$

Likewise, it follows

$$Em_n^2 e^{it\zeta_n} = -\frac{1}{z} Em_n e^{it\zeta_n} - \frac{1}{z} Em_n^3 e^{it\zeta_n} - \frac{1}{zn^3} E \sum_{i,k,l=1}^{n} G_{ik}G_{kl}G_{li} e^{it\zeta_n}$$

$$-\frac{it}{zn^3} Em_n \sum_{i,k,l=1}^{n} G_{ik}G_{kl}G_{li} e^{it\zeta_n}.$$

We have by virtue of (3.38)

$$Em_n^2 e^{it\zeta_n} - Em_n^2 E e^{it\zeta_n}$$

$$= -\frac{1}{z} E(m_n - Em_n) e^{it\zeta_n} - \frac{1}{z} E\big(m_n^3 - Em_n^3\big) e^{it\zeta_n}$$

$$-\frac{it}{zn^3}\Big(E \sum_{i,k,l=1}^{n} G_{ik}G_{kl}G_{li} e^{it\zeta_n} - E \sum_{i,k,l=1}^{n} G_{ik}G_{kl}G_{li} E e^{it\zeta_n}\Big)$$

$$-\frac{it}{zn^3} Em_n \sum_{i,k,l=1}^{n} G_{ik}G_{kl}G_{li} e^{it\zeta_n}.$$

Also, using some simple algebra and Proposition 3.3 yields

$$E\big(m_n^3 - Em_n^3\big)e^{it\zeta_n} = 3(Em_n)^2 E(m_n - Em_n)e^{it\zeta_n} + o(n^{-2}).$$

In turn, this implies

$$Em_n^2 e^{it\zeta_n} - Em_n^2 Ee^{it\zeta_n}$$

$$= -\frac{1 + 3(Em_n)^2}{z} E(m_n - Em_n)e^{it\zeta_n}$$

$$- \frac{itEm_n}{zn^3} E \sum_{i,k,l=1}^n G_{ik}G_{kl}G_{li}e^{it\zeta_n} + o(n^{-2}). \qquad (3.51)$$

Substituting (3.51) into (3.50) and solving the equation,

$$\Big(1 - \frac{1 + 3(Em_n)^2}{z^2}\Big)n(Em_n e^{it\zeta_n} - Em_n Ee^{it\zeta_n})$$

$$= it\Big(\frac{Em_n}{z^2} - \frac{1}{z}\Big)\frac{1}{n}E \sum_{i,k,l=1}^n G_{ik}G_{kl}G_{li}Ee^{it\zeta_n} + o(1).$$

Note by (3.34)

$$1 - \frac{1 + 3(Em_n)^2}{z^2} = -(z + 2Em_n)\Big(\frac{Em_n}{z^2} - \frac{1}{z}\Big) + o(1).$$

In combination with (3.39), it is now easy to see that (3.49) holds true, as desired. □

To conclude this section, let us turn to linear eigenvalue statistics. This is a very interesting and well studied object in the random matrix theory. Let $f : \mathbb{R} \to \mathbb{R}$ be a real valued measurable function. A linear eigenvalue statistic with test function f is defined by

$$T_n(f) = \frac{1}{n}\sum_{i=1}^n f(\lambda_i), \qquad (3.52)$$

where the λ_i's are eigenvalues of normalized GUE matrix H_n.

As shown in Theorem 3.2, if f is bounded and continuous, then

$$T_n(f) \xrightarrow{P} \int_{-2}^{2} f(x)\rho_{sc}(x)dx, \quad n \to \infty.$$

This is a certain weak law of large numbers for eigenvalues. From a probabilistic view, the next natural issue is to take a closer look at the fluctuation. Under what conditions could one have asymptotic normality? As a matter of fact, this is usually a crucial problem in the statistical inference theory.

As an immediate application of Theorem 3.4, we can easily derive a central limit theorem for a class of analytic functions.

Theorem 3.5. *Suppose that* $f : \mathbb{R} \to \mathbb{R}$ *is a bounded continuous function and analytic in a region including the real line. Then*

$$n\left(T_n(f) - \int_{-2}^{2} f(x)\rho_{sc}(x)dx\right) \xrightarrow{d} N(0, \sigma_f^2), \quad n \to \infty \qquad (3.53)$$

where the variance σ_f^2 *is given by*

$$\sigma_f^2 = \frac{1}{4\pi^2} \int_{-2}^{2}\int_{-2}^{2} \frac{4 - xy}{\sqrt{4-x^2}\sqrt{4-y^2}} \frac{(f(x) - f(y))^2}{(x-y)^2} dxdy. \qquad (3.54)$$

Proof. Without loss of generality, we may and do assume f is analytic in the region $\{z = x + i\eta : x \in \mathbb{R}, |\eta| \le 1\}$. According to the Cauchy integral formula,

$$f(x) = \frac{1}{2\pi i} \int_{|z|=1} \frac{f(z)}{x - z} dz,$$

which in turn implies

$$T_n(f) = \frac{1}{2\pi i} \int_{|z|=1} f(z) m_n(z) dz.$$

Hence it follows from Proposition 3.3

$$\begin{aligned}
ET_n(f) &= \frac{1}{2\pi i} \int_{|z|=1} f(z) Em_n(z) dz \\
&= \frac{1}{2\pi i} \int_{|z|=1} f(z) m_{sc}(z)\left(1 + O(n^{-2})\right) dz \\
&= \int_{-2}^{2} f(x)\rho_{sc}(x)dx + O(n^{-2}).
\end{aligned}$$

In addition, it also follows from Theorem 3.4

$$\begin{aligned}
n\left(T_n(f) - ET_n(f)\right) &= \frac{1}{2\pi i} \int_{|z|=1} f(z) n\left(m_n(z) - Em_n(z)\right) dz \\
&\xrightarrow{d} \frac{1}{2\pi i} \int_{|z|=1} f(z) \Xi(z) dz,
\end{aligned}$$

where the convergence is a standard application of continuous mapping theorem.

To get the variance, note the following integral identity

$$\begin{aligned}
Cov(\Xi(z_1), \Xi(z_2)) &= \frac{1}{2(z_1 - z_2)^2}\left(\frac{z_1 z_2 - 4}{\sqrt{z_1^2 - 4} \cdot \sqrt{z_1^2 - 4}} - 1\right) \\
&= \frac{1}{4\pi^2} \int_{-2}^{2}\int_{-2}^{2} \frac{xy - 4}{(x-y)^2\sqrt{x^2 - 4}\sqrt{x^2 - 4}} \\
&\quad \times \left(\frac{1}{z_1 - x} - \frac{1}{z_1 - y}\right)\left(\frac{1}{z_2 - x} - \frac{1}{z_2 - y}\right) dxdy.
\end{aligned}$$

Therefore we have

$$\sigma_f^2 = \frac{1}{(2\pi i)^2} \int_{|z_1|=1} \int_{|z_2|=1} f(z_1) f(z_2) Cov(\Xi(z_1), \Xi(z_2)) dz_1 dz_2$$

$$= \frac{1}{4\pi^2} \int_{-2}^{2} \int_{-2}^{2} \frac{xy - 4}{(x-y)^2 \sqrt{x^2 - 4} \sqrt{y^2 - 4}} dx dy \frac{1}{(2\pi i)^2} \int_{|z_1|=1} \int_{|z_2|=1}$$

$$f(z_1) f(z_2) \left(\frac{1}{z_1 - x} - \frac{1}{z_1 - y} \right) \left(\frac{1}{z_2 - x} - \frac{1}{z_2 - y} \right) dz_1 dz_2$$

$$= \frac{1}{4\pi^2} \int_{-2}^{2} \int_{-2}^{2} \frac{xy - 4}{\sqrt{x^2 - 4} \sqrt{y^2 - 4}} \frac{(f(x) - f(y))^2}{(x-y)^2} dx dy.$$

The proof is now complete.   □

It has been an interesting issue to study fluctuations of linear eigenvalue statistics for as wide as possible class of test functions. In Theorem 3.5, the analyticity hypothesis was only required to use the Cauchy integral formula. This condition can be replaced by other regularity properties. For instance, Lytova and Pastur (2009) proved that Theorem 3.5 is valid for a bounded continuous differentiable test function with bounded derivative. Johansson (1998) studied the global fluctuation of eigenvalues to manifest the regularity of eigenvalue distribution. In particular, assume that f : $\mathbb{R} \mapsto \mathbb{R}$ is not too large for large values of x:

(i) $f(x) \leq L(x^2 + 1)$ for some constant L and all $x \in \mathbb{R}$;

(ii) $|f'(x)| \leq q(x)$ for some polynomial $q(x)$ and all $x \in \mathbb{R}$;

(iii) For each x_0, there exists an $\alpha > 0$ such that $f(x)\psi_{x_0}(x) \in H^{2+\alpha}$, where $H^{2+\alpha}$ is standard Sobolev space and $\psi_{x_0}(x)$ is an infinitely differentiable function such that $|\psi_{x_0}(x)| \leq 1$ and

$$\psi_{x_0}(x) = \begin{cases} 1 & |x| \leq x_0, \\ 0 & |x| > x_0 + 1. \end{cases}$$

Then (3.53) is also valid with σ_f^2 given by.

$$\sigma_f^2 = \frac{1}{4\pi^2} \int_{-2}^{2} \int_{-2}^{2} \frac{f'(x) f(y) \sqrt{4 - x^2}}{(y - x) \sqrt{4 - y^2}} dx dy. \qquad (3.55)$$

Here we note that the righthand sides of (3.54) and (3.55) are equal.

3.3   Number of eigenvalues in an interval

In this section we are further concerned with such an interesting issue like how many eigenvalues locate in an interval. Consider the standard GUE matrix A_n, and denote by $\lambda_1, \lambda_2, \cdots, \lambda_n$ its eigenvalues. For $[a, b] \subseteq (-2, 2)$, define $N_n(a, b)$ to be the number of normalized eigenvalues lying in $[a, b]$. Namely,

$$N_n(a, b) = \#\{1 \le i \le n : \sqrt{n}\,a \le \lambda_i \le \sqrt{n}\,b\}.$$

According to the Wigner semicircle law, Theorem 3.2,

$$\frac{N_n(a, b)}{n} \xrightarrow{P} \int_a^b \rho_{sc}(x)dx, \quad n \to \infty.$$

In fact, using the asymptotic behaviors for Hermitian orthogonal polynomials as in Section 3.1, we can further have

Proposition 3.4.

$$EN_n(a, b) = n \int_a^b \rho_{sc}(x)dx + O(1) \tag{3.56}$$

and

$$Var\big(N_n(a, b)\big) = \frac{1}{2\pi^2} \log n\big(1 + o(1)\big). \tag{3.57}$$

Proof. (3.56) is trivial since the average spectral density function $\bar{p}_n(x)$ converges uniformly to $\rho_{sc}(x)$ in $[a, b]$.

To prove (3.57), note the following variance formula

$$
\begin{aligned}
Var(N_n(a, b)) &= \sqrt{n} \int_a^b K_n\big(\sqrt{n}x, \sqrt{n}x\big)dx \\
&\quad - n \int_a^b \int_a^b K_n\big(\sqrt{n}x, \sqrt{n}y\big)^2 dxdy \\
&= n \int_a^b \int_b^\infty K_n\big(\sqrt{n}x, \sqrt{n}y\big)^2 dxdy \\
&\quad + n \int_a^b \int_{-\infty}^a K_n\big(\sqrt{n}x, \sqrt{n}y\big)^2 dxdy \\
&=: I_1 + I_2.
\end{aligned}
$$

We shall estimate the integrals I_1 and I_2 below. The focus is upon I_1, since I_2 is completely similar. A change of variables easily gives

$$
\begin{aligned}
I_1 &= n \int_{b-a}^\infty dv \int_0^{b-a} K_n\big(\sqrt{n}(b - u), \sqrt{n}(b - u + v)\big)^2 du \\
&\quad + n \int_0^{b-a} dv \int_0^v K_n\big(\sqrt{n}(b - u), \sqrt{n}(b - u + v)\big)^2 du \\
&=: I_{1,1} + I_{1,2}.
\end{aligned}
$$

It is easy to control $I_{1,1}$ from above. In fact, when $v \geq b - a$

$$K_n(\sqrt{n}x, \sqrt{n}y)^2$$
$$\leq \frac{1}{(b-a)^2} \left(\varphi_n(\sqrt{n}x)\varphi_{n-1}(\sqrt{n}y) - \varphi_n(\sqrt{n}y)\varphi_{n-1}(\sqrt{n}x) \right)^2,$$

where $x = b - u, y = b - u + v$. Hence we have by the orthogonality of φ_n and φ_{n-1}

$$I_{1,1} \leq \frac{1}{(b-a)^2} \int_{-\infty}^{\infty} \int_{-\infty}^{\infty} n \big(\varphi_n(\sqrt{n}x)\varphi_{n-1}(\sqrt{n}y)$$
$$- \varphi_n(\sqrt{n}y)\varphi_{n-1}(\sqrt{n}x) \big)^2 dx dy$$
$$\leq \frac{2}{(b-a)^2}.$$

Turn to estimating $I_{1,2}$. Note

$$\lim_{y \to x} K_n(x, y) = \sqrt{n} \big(\varphi'_n(x)\varphi_{n-1}(x) - \varphi_n(x)\varphi'_{n-1}(x) \big)$$

and

$$\|\varphi_l\|_\infty \leq \kappa, \quad \|\varphi'_l\|_\infty \leq \sqrt{l}\kappa.$$

So,

$$n \int_0^{\frac{1}{n}} dv \int_0^v K_n\big(\sqrt{n}(b-u), \sqrt{n}(b-u+v) \big)^2 du = O(1).$$

For the integral over $(1/n, b - a)$, we use Lemma 3.2 to get

$$n^{1/2} \big(\varphi_n(\sqrt{n}x)\varphi_{n-1}(\sqrt{n}y) - \varphi_n(\sqrt{n}y)\varphi_{n-1}(\sqrt{n}x) \big)$$
$$= \frac{2}{\pi} \frac{1}{(4-x^2)^{1/4}(4-y^2)^{1/4}} \left[\cos\left(n\alpha(\theta_1) + \frac{\theta_1}{2} - \frac{\pi}{4} \right) \cos\left(n\alpha(\theta_2) - \frac{\pi}{4} \right) \right.$$
$$\left. - \cos\left(n\alpha(\theta_2) + \frac{\theta_2}{2} - \frac{\pi}{4} \right) \cos\left(n\alpha(\theta_1) - \frac{\pi}{4} \right) \right] + O(n^{-1})$$
$$= \frac{1}{2\pi} \frac{(4-xy)^{1/2}}{(4-x^2)^{1/4}(4-y^2)^{1/4}} + O(n^{-1}).$$

Thus with $x = b - u$ and $y = b - u + v$,

$$n \int_{\frac{1}{n}}^{b-a} dv \int_0^v K_n\big(\sqrt{n}(b-u), \sqrt{n}(b-u+v) \big)^2 du$$
$$= \frac{1}{4\pi^2} \int_{\frac{1}{n}}^{b-a} dv \int_0^v du \frac{4-xy}{v^2(4-x^2)^{1/2}(4-y^2)^{1/2}}$$
$$+ O(n^{-1}) \int_{\frac{1}{n}}^{b-a} dv \int_0^v \frac{1}{v^2} du.$$

Trivially, it follows

$$\int_{\frac{1}{n}}^{b-a} dv \int_0^v \frac{1}{v^2} du = O(\log n).$$

We also note

$$\sup_{x,y\in(a,b)} \frac{4-xy}{(4-x^2)^{1/2}(4-y^2)^{1/2}} \le C_{a,b}$$

for some positive constant $C_{a,b}$. Then for any $\varepsilon > 0$

$$\int_\varepsilon^{b-a} dv \int_0^v du \frac{4-xy}{v^2(4-x^2)^{1/2}(4-y^2)^{1/2}} = O(|\log\varepsilon|). \qquad (3.58)$$

On the other hand, it is easy to see

$$\frac{4-xy}{(4-x^2)^{1/2}(4-y^2)^{1/2}} = 1 + O(\varepsilon^2), \quad 0 < v < \varepsilon.$$

Hence we have

$$\int_{\frac{1}{n}}^\varepsilon dv \int_0^v du \frac{4-xy}{v^2(4-x^2)^{1/2}(4-y^2)^{1/2}} = (1+O(\varepsilon^2))(\log n + |\log\varepsilon|). \ (3.59)$$

Combining (3.58) and (3.59) together yields

$$\int_{\frac{1}{n}}^{b-a} dv \int_0^v du \frac{4-xy}{v^2(4-x^2)^{1/2}(4-y^2)^{1/2}}$$

$$= \int_{\frac{1}{n}}^\varepsilon dv \int_0^v du \frac{4-xy}{v^2(4-x^2)^{1/2}(4-y^2)^{1/2}}$$

$$+ \int_\varepsilon^{b-a} dv \int_0^v du \frac{4-xy}{v^2(4-x^2)^{1/2}(4-y^2)^{1/2}}$$

$$= (1+O(\varepsilon^2))(\log n + |\log\varepsilon|). \qquad \square$$

We remark that the linear eigenvalue statistic $\sum_{i=1}^n f(\lambda_i/\sqrt{n})$ has variance at most 1 whenever f is a 1-Lipschtiz test function, see (3.24). On the other hand, the counting function is not a 1-Lipschitz function. The Proposition 3.4 provides a $\log n$-like estimate for the size of variance of $N_n(a,b)$.

Having the Proposition, one would expect the asymptotic normal fluctuations for $N_n(a,b)$. Below is our main result of this section.

Theorem 3.6. *Under the above assumptions, as* $n \to \infty$

$$\frac{1}{\sqrt{\frac{1}{2\pi^2}\log n}}\left(N_n(a,b) - n\int_a^b \rho_{sc}(x)dx\right) \xrightarrow{d} N(0,1).$$

The rest of this section is devoted to the proof of Theorem 3.6. In fact, we shall prove the theorem under a more general setting. To do this, we need to introduce some basic definitions and properties about determinantal point processes. Recall a point process $\mathcal{X}$ on $\mathbb{R}$ is a random configuration such that any bounded set contains only finitely many points. The law of $\mathcal{X}$ is usually characterized by the family of integer-valued random variables $\{N_{\mathcal{X}}(A), A \in \mathcal{B}\}$, where $N_{\mathcal{X}}(A)$ denotes the number of $\mathcal{X}$ in A. Besides, the correlation function is becoming a very useful concept in describing the properties of point processes. The so-called correlation function was first introduced to study the point process by Macchi (1975). Given a point process $\mathcal{X}$, its k-point correlation function is defined by

$$\rho_k(x_1, \cdots, x_k) = \lim_{\delta \to 0} \frac{1}{(2\delta)^k} P\big((x_i - \delta, x_i + \delta) \cap \mathcal{X} \neq \emptyset, 1 \leq i \leq k\big),$$

where $x_1, \cdots, x_k \in \mathbb{R}$. Here we only considered the continuous case, the corresponding discrete case will be given in Chapter 4.

It turns out that the correlation functions is a powerful and nice tool in computing moments of $N_{\mathcal{X}}(A)$. In fact, it is easy to see

$$E\big(N_{\mathcal{X}}(A)\big)^{\downarrow k} = \int_{A^{\otimes k}} \rho_k(x_1, \cdots, x_k) dx_1 \cdots dx_k, \tag{3.60}$$

where $m^{\downarrow k} = m(m-1) \cdots (m - k + 1)$, and

$$E \prod_{i=1}^{k} N_{\mathcal{X}}(A_i) = \int_{A_1 \times \cdots \times A_k} \rho_k(x_1, \cdots, x_k) dx_1 \cdots dx_k.$$

A point process $\mathcal{X}$ is said to be determinantal if there exists a kernel function $K_{\mathcal{X}} : \mathbb{R} \times \mathbb{R} \mapsto \mathbb{R}$ such that

$$\rho_k(x_1, \cdots, x_k) = \det\big(K_{\mathcal{X}}(x_i, x_j)\big)_{k \times k}$$

for any $k \geq 1$ and $x_1, \cdots, x_k \in \mathbb{R}$.

The determinantal point processes have been attracting a lot of attention in the past two decades. More and more interesting examples have been found in the seemingly distinct problems. For instance, the GUE model A_n is a determinantal point process with $\mathcal{X} = \{\lambda_1, \lambda_2, \cdots, \lambda_n\}$ and kernel function $K_{\mathcal{X}} = K_n$ given by (3.7). Another well-known example is Poisson point process on $\mathbb{R}$. Let $\mathcal{P}$ be a Poisson point process with intensity function $\varrho(x)$, then $\mathcal{P}$ can be viewed as a determinantal process having $K_{\mathcal{P}}(x, y) = \varrho(x)\delta_{x,y}$. Note that Poisson point process is an independent point process, that is, a two-point correlation function is equal to the product of two one-point correlation functions. However, a general determinantal point process is a negatively associated process, since $\rho_2(x,y) \leq \rho_1(x)\rho_1(y)$.

Note that no claim is made about the existence or uniqueness of a determinantal point process for a given kernel K. To address these issues, we need to make some additional assumptions below. The kernel K is required to be symmetric and nonnegative definite, that is, $K(x,y) = K(y,x)$ for every $x,y \in \mathbb{R}$ and $\det(K(x_i,x_j))_{k \times k} \geq 0$ for any $x_1,x_2,\cdots,x_k \in \mathbb{R}$. we also further assume that K is locally square integrable on $\mathbb{R}^2$. This means that for any compact $D \subseteq \mathbb{R}$, we have

$$\int_{D^2} |K(x,y)|^2 dx dy < \infty.$$

Then we may use K as an integral kernel to define an associated integral operator as

$$\mathcal{K}f(x) = \int_{\mathbb{R}} K(x,y)f(y)dy < \infty$$

for functions $f \in L^2(\mathbb{R}, dx)$ with compact support.

For a compact set D, the restriction of $\mathcal{K}$ to D is the bounded linear operator $\mathcal{K}_D$ on $L^2(\mathbb{R})$ defined by

$$\mathcal{K}_D f(x) = \int_D K(x,y)f(y)dy, \quad x \in D.$$

Thus $\mathcal{K}_D$ is a self-adjoint compact operator. Let $q_n^D, n \geq 1$ be nonnegative eigenvalues of $\mathcal{K}_D$, the corresponding eigenfunctions ϕ_n^D forms a orthonormal basis on $L^2(D, dx)$. We say that $\mathcal{K}_D$ is of trace class if

$$\sum_{n=1}^{\infty} |q_n^D| < \infty.$$

If $\mathcal{K}_D$ is of trace class for every compact subset D, then we say that $\mathcal{K}$ is locally of trace class. The following two lemmas characterize the existence and uniqueness of a determinantal point process with a given kernel.

Lemma 3.5. *Let $\mathcal{X}$ be a determinantal point process with kernel $K_{\mathcal{X}}$. If $EN_{\mathcal{X}}(A) < \infty$, then $Ee^{tN_{\mathcal{X}}(A)} < \infty$ for any $t \in \mathbb{R}$. Consequently, for each compact set D, the distribution of $N_{\mathcal{X}}(D)$ is uniquely determined by $K_{\mathcal{X}}$.*

Proof. It easily follows

$$Ee^{tN_{\mathcal{X}}(A)} = E\big(1 + (e^t - 1)\big)^{N_{\mathcal{X}}(A)}$$

$$= \sum_{k=0}^{\infty} \frac{(e^t - 1)^k}{k!} E\big(N_{\mathcal{X}}(A)\big)^{\downarrow k}.$$

Also, by (3.60) and the Hadamard inequality for nonnegative definite matrix

$$E\big(N_{\mathcal{X}}(A)\big)^{\downarrow k} = \int_{A^{\otimes k}} \det\big(K_{\mathcal{X}}(x_i, x_j)\big) dx_1 \cdots dx_k$$

$$\leq \left(\int_A K_{\mathcal{X}}(x_1, x_1)) dx_1\right)^k = \big(EN_{\mathcal{X}}(A)\big)^k.$$

Therefore, we have

$$Ee^{tN_{\mathcal{X}}(A)} \leq \sum_{k=0}^{\infty} \frac{(e^t - 1)^k}{k!} \big(EN_{\mathcal{X}}(A)\big)^k < \infty.$$

For any compact set D, $EN_{\mathcal{X}}(D) < \infty$ since $\int_D K_{\mathcal{X}}(x, x) dx < \infty$, so $Ee^{tN_{\mathcal{X}}(D)} < \infty$ for all $t \in \mathbb{R}$.   □

Lemma 3.6. *Assume that K is a symmetric and nonnegative definite kernel function such that the integral operator $\mathcal{K}$ is a locally trace class. Then K defines a determinantal point process on $\mathbb{R}$ if and only if the spectrum of $\mathcal{K}$ is contained in [0,1].*

Proof.   See Theorem 4.5.5 of Soshnikov (2000).   □

Theorem 3.7. *Let $\mathcal{X}_n, n \geq 1$ be a sequence of determinantal point processes with kernel $K_{\mathcal{X}_n}$ on $\mathbb{R}$, let $I_n, n \geq 1$ be a sequence of intervals on $\mathbb{R}$. Assume that $K_{\mathcal{X}_n} \cdot \mathbf{1}_{I_n}$ define an integrable operator of locally trace class. Set $N_n = N_{\mathcal{X}_n}(I_n)$. If $Var(N_n) \to \infty$ as $n \to \infty$, then*

$$\frac{N_n - EN_n}{\sqrt{Var(N_n)}} \xrightarrow{d} N(0, 1).$$

The theorem was first proved by Costin and Lebowitz (1995) in the very special case. They only considered the Sine point process with kernel $K_{Sine}(x, y) = \sin(x - y)/(x - y)$, and Widom suggested it would hold for the GUE model. Later on Soshnikov (2002) extended it to general determinantal random points fields, including Bessel, Airy point processes.

The proof of the theorem is quite interesting. A basic strategy is to use the moment method, namely Theorem 1.8. Set

$$X_n = \frac{N_n - EN_n}{\sqrt{Var(N_n)}}.$$

Trivially, $\tau_1(X_n) = 0$, $\tau_2(X_n) = 1$ and

$$\tau_k(X_n) = \frac{\gamma_k(N_n)}{(Var N_n)^{k/2}}, \quad k \geq 3.$$

Then it suffices to show

$$\gamma_k(N_n) = o(\gamma_2(N_n)^{k/2}), \quad k \geq 3$$

provided $\gamma_2(N_n) = Var(N_n) \to \infty$.

Proof. For the sake of clarity, we write γ_k for $\gamma_k(N_n)$. A key ingredient is to express each γ_k in terms of correlation functions and so kernel functions of $\mathcal{X}_n$. Start with γ_3. It follows from (1.17)

$$\begin{aligned}
\gamma_3 &= EN_n^3 - 3EN_n^2 EN_n + 2(EN_n)^3 \\
&= E(N_n)_3 + 3E(N_n)_2 + EN_n - 3E(N_n)_2 EN_n \\
&\quad -3(EN_n)^2 + 2(EN_n)^3.
\end{aligned}$$

Also, we have by (3.60) and a simple algebra

$$\gamma_3 = 2 \int_{I_n^{\otimes 3}} K_n(x_1, x_2) K_n(x_2, x_3) K_n(x_3, x_1) dx_1 dx_2 dx_3$$

$$-3 \int_{I_n^{\otimes 2}} K_n(x_1, x_2) K_n(x_2, x_1) dx_1 dx_2 + \int_{I_n} K_n(x_1, x_1) dx_1.$$

To obtain a general equation for γ_k, we need to introduce k-point cluster function, namely

$$\alpha_k(x_1, \cdots, x_k) = \sum_G (-1)^{l-1}(l-1)! \prod_{j=1}^{l} \rho_{|G_j|}(\bar{x}(G_j)),$$

where $1 \leq l \leq k$, $G = (G_1, \cdots, G_l)$ is a partition of the set $\{1, 2, \cdots, k\}$, $|G_j|$ stands for the size of G_j, $\bar{x}(G_j) = \{x_i, i \in G_j\}$.

Using Möbius inversion formula, we can express the correlation functions in terms of Ursell functions as follows

$$\rho_k(x_1, \cdots, x_k) = \sum_G \prod_{j=1}^{l} \alpha_{|G_j|}(\bar{x}(G_j)).$$

Moreover, we have an elegant formula in the setting of determinantal point processes

$$\alpha_k(x_1, \cdots, x_k)$$
$$= (-1)^{k-1} \sum_\sigma K_{\mathcal{X}_n}(x_1, x_{\sigma(1)}) K_{\mathcal{X}_n}(x_{\sigma(1)}, x_{\sigma(2)}) \cdots K_{\mathcal{X}_n}(x_{\sigma(k)}, x_1), \quad (3.61)$$

where the sum is over all cyclic permutations $(\sigma(1), \cdots, \sigma(k))$ of $(1, 2, \cdots, k)$. Define

$$\beta_k = \int_{I_n^{\otimes k}} \alpha_k(x_1, \cdots, x_k) dx_1 \cdots dx_k, \quad k \geq 1.$$

Then it is not hard to see

$$\beta_k = \sum_G (-1)^{l-1}(l-1)! \prod_{j=1}^{l} \int_{I_n^{\otimes |G_j|}} \rho_{|G_j|}(\bar{x}(G_j)) d\bar{x}(G_j)$$

$$= \sum_G (-1)^{l-1}(l-1)! \prod_{j=1}^{l} E(N_n)^{\downarrow |G_j|}$$

$$= \sum_{\tau \mapsto k} \frac{k!}{\prod \tau_i! m_{\tau_i}!} (-1)^{l-1}(l-1)! \prod_{j=1}^{l} E(N_n)^{\downarrow \tau_j},$$

where $\tau = (\tau_1, \cdots, \tau_l) \mapsto k$ is an integer partition of k, m_{τ_i} stands for the multiplicity of τ_i in τ. We can derive from (1.17)

$$\sum_{k=1}^{\infty} \frac{\beta_k}{k!} (e^t - 1)^k = \log \sum_{k=0}^{\infty} \frac{E(N_n)^{\downarrow k}}{k!} (e^t - 1)^k$$

$$= \log E e^{t N_n} = \sum_{k=1}^{\infty} \frac{\gamma_k}{k!} t^k. \qquad (3.62)$$

Comparing the coefficients of the term t^k at both sides of (3.62), we obtain

$$\gamma_k = \sum_{l=1}^{k} \frac{\beta_l}{l!} \sum_{\tau_1 + \cdots + \tau_l = k} \frac{k!}{\tau_1! \cdots \tau_l!}.$$

Equivalently,

$$\gamma_k = \beta_k + \sum_{j=1}^{k-1} b_{k,j} \gamma_j \qquad (3.63)$$

where the coefficients $b_{k,j}$ are given by

$$b_{k,1} = (-1)^k (k-1)!, \quad b_{k,k} = -1, \quad k \geq 2$$

and

$$b_{k,j} = b_{k-1,j-1} - (k-1)b_{k-1,j}, \quad 2 \leq j \leq k-1.$$

Since it follows from (3.63)

$$\gamma_k = \beta_k + (-1)^k (k-1)! \gamma_1 + \sum_{j=2}^{k-1} b_{k,j} \gamma_j, \quad k \geq 3,$$

then it suffices to show

$$\beta_k + (-1)^k (k-1)! \gamma_1 = o(\gamma_2^{k/2}), \quad k \geq 3. \qquad (3.64)$$

To do this, use (3.61) to get

$$\beta_k = (-1)^{k-1} \sum_\sigma \int_{I_n^{\otimes k}} K_{\mathcal{X}_n}(x_1, x_{\sigma(1)}) \cdots K_{\mathcal{X}_n}(x_{\sigma(k)}, x_1) dx_1 \cdots dx_k$$

$$= (-1)^{k-1}(k-1)! \int_{I_n^{\otimes k}} K_{\mathcal{X}_n}(x_1, x_2) \cdots K_{\mathcal{X}_n}(x_k, x_1) dx_1 \cdots dx_k,$$

and so

$$\beta_k + (-1)^k (k-1)! \gamma_1 = (-1)^k (k-1)! \Big(\int_{I_n} K_{\mathcal{X}_n}(x_1, x_1) dx_1$$

$$- \int_{I_n^{\otimes k}} K_{\mathcal{X}_n}(x_1, x_2) \cdots K_{\mathcal{X}_n}(x_k, x_1) dx_1 \cdots dx_k \Big).$$

Define an integrable operator $\mathcal{K}_{I_n} : L^2(I_n, dx) \to L^2(I_n, dx)$ by

$$\mathcal{K}_{I_n} f(x) = \int_{I_n} f(y) K_{I_n}(x, y) dy, \quad x \in I_n.$$

Then it follows

$$\beta_k + (-1)^k (k-1)! \gamma_1 = (-1)^k (k-1)! \big(tr\mathcal{K}_{I_n} - tr\mathcal{K}_{I_n}^k \big)$$

$$= (-1)^k (k-1)! \sum_{l=2}^k tr\mathcal{K}_{I_n}^{l-2} \big(\mathcal{K}_{I_n} - \mathcal{K}_{I_n}^2 \big).$$

According to Lemma 3.6, we have

$$\big| \beta_k + (-1)^k (k-1)! \gamma_1 \big| \leq k! \big(tr\mathcal{K}_{I_n} - tr\mathcal{K}_{I_n}^2 \big)$$

$$= k! \gamma_2,$$

which gives (3.64). Now we conclude the proof of the theorem. $\qquad\square$

As the reader may see, Theorem 3.7 is of great universality for determinantal point processes in the sense that there is almost no requirement on the kernel function. However, the theorem itself does not tell what the expectation and variance of $N_{\mathcal{X}_n}(I_n)$ look like. To have numerical evaluation of expectation and variance, one usually needs to know more information about the kernel function. In the case of GUE, the kernel function is given by Hermite orthogonal polynomials so that we can give precise estimates of expectation and variance. This was already done in Proposition 3.4.

It is believed that Theorem 3.7 would have a wide range of applications. We only mention the work of Gustavsson (2005), in which he studied the kth greatest eigenvalue $\lambda_{(k)}$ of GUE model and used Theorem 3.7 to prove the $\lambda_{(k_n)}$ after properly scaled has a Gaussian fluctuation around its average as $k_n/n \to a \in (0, 1)$. He also dealt with the case of $k_n \to \infty$ and

$k_n/n \to 0$. These results are complement to the Tracy-Widom law for largest eigenvalues, see Section 4.1 and Figure 5.2.

In the end of this section, we shall provide a conceptual proof of Theorem 3.7. This is based on the following expression for the number of points lying in a set as a sum of independent Bernoulli random variables.

Let K be a kernel function such that the integral operator $\mathcal{K}$ is locally of trace class. Let $\mathcal{X}$ be a determinantal point process with K as its kernel. Let I be a bounded Borel set on $\mathbb{R}$, then $\mathcal{K} \cdot \mathbf{1}_I$ is locally trace class. Denote by $q_k, k \geq 1$ the eigenvalues of $\mathcal{K} \cdot \mathbf{1}_I$, the corresponding eigenfunctions ϕ_k form a orthonormal basis in $L^2(I)$. Define a new kernel function K^I by

$$K^I(x, y) = \sum_{k=1}^{\infty} q_k \phi_k(x) \phi_k(y),$$

which is a mixture of the q_k and ϕ_k.

It is evident that the point process $\mathcal{X} \cap I$ is determinantal. The following proposition implies that its kernel is given by K^I.

Proposition 3.5. *It holds almost everywhere with respect to Lebesgue measure*

$$K(x, y) = K^I(x, y).$$

Furthermore, assume that $\xi_k, k \geq 1$ is a sequence of independent Bernoulli random variables,

$$P(\xi_k = 1) = q_k, \quad P(\xi_k = 0) = 1 - q_k,$$

then we have

$$N_{\mathcal{X}}(I) \stackrel{d}{=} \sum_{k=1}^{\infty} \xi_k. \tag{3.65}$$

Proof. By assumption of trace class,

$$\int_I \left(\sum_{k=1}^{\infty} q_k \phi_k(x) \right)^2 dx = \sum_{k=1}^{\infty} q_k < \infty.$$

This shows that the series $\sum_{k=1}^{\infty} q_k^2 \phi_k(x)$ converges in $L^2(I)$ and also that it converges pointwise for every $x \in I \setminus I_0$ for some set I_0 of zero measure. By the Cauchy-Schwarz inequality,

$$\left(\sum_{k=n}^{\infty} q_k \phi_k(x) \phi_k(y) \right)^2 \leq \left(\sum_{k=n}^{\infty} q_k \phi_k(x)^2 \right) \left(\sum_{k=n}^{\infty} q_k \phi_k(y)^2 \right). \tag{3.66}$$

Hence the series $\sum_{k=1}^{\infty} q_k \phi_k(x) \phi_k(y)$ converges absolutely.

Let $f \in L^2(I)$. Write f in terms of the orthonormal basis $\{\phi_k\}$ to get for any $x \in I \setminus I_0$

$$f(x) = \sum_{k=1}^{\infty} \left(\int_I f(y)\phi_k(y)dy \right) \phi_k(x)$$

and so

$$\mathcal{K}f(x) = \sum_{k=1}^{\infty} \left(\int_I f(y)\phi_k(y)dy \right) \mathcal{K}\phi_k(x)$$

$$= \int_I f(y) \sum_{k=1}^{\infty} q_k \phi_k(y)\phi_k(x)dy$$

$$= \int_I f(y)K^I(x,y)dy.$$

This implies that we must have

$$K(x,y) = K^I(x,y) \quad \text{a.e.}$$

Turn to prove (3.65). We shall below prove

$$Ee^{tN_{\mathcal{X}}(I)} = Ee^{t\sum_{k=1}^{\infty} \xi_k}, \quad t \in \mathbb{R}.$$

First, it is easy to see

$$Ee^{t\sum_{k=1}^{\infty} \xi_k} = \prod_{k=1}^{\infty} Ee^{t\xi_k}$$

$$= \prod_{k=1}^{\infty} \left(1 + q_k(e^t - 1) \right)$$

$$= 1 + \sum_{k=1}^{\infty} \sum_{1 \le i_1 < \cdots < i_k < \infty} q_{i_1} \cdots q_{i_k}(e^t - 1)^k.$$

Second, to compute $Ee^{tN_{\mathcal{X}}(I)}$, we use the following formula

$$Ee^{tN_{\mathcal{X}}(I)} = \sum_{k=0}^{\infty} \frac{E(N_{\mathcal{X}}(I))_k}{k!}(e^t - 1)^k$$

$$= \sum_{k=0}^{\infty} \frac{(e^t - 1)^k}{k!} \int_{\mathbb{R}^k} \det \left(K^I(x_i, x_j) \right)_{k \times k} dx_1 \cdots dx_k. \quad (3.67)$$

For $k \geq 1$,

$$
\left(K^I(x_i, x_j)\right)_{k \times k} = \begin{pmatrix} q_1\phi_1(x_1) & q_2\phi_2(x_1) & \cdots & q_n\phi_n(x_1) & \cdots \\ q_1\phi_1(x_2) & q_2\phi_2(x_2) & \cdots & q_n\phi_n(x_2) & \cdots \\ \vdots & \vdots & \cdots & \vdots & \vdots \\ q_1\phi_1(x_k) & q_2\phi_2(x_k) & \cdots & q_n\phi_n(x_k) & \cdots \end{pmatrix}
$$

$$
\times \begin{pmatrix} \psi_1(x_1) & \psi_1(x_2) & \cdots & \psi_1(x_k) \\ \phi_2(x_1) & \phi_2(x_2) & \cdots & \phi_2(x_k) \\ \vdots & \vdots & \cdots & \vdots \\ \phi_n(x_1) & \phi_n(x_2) & \cdots & \phi_n(x_k) \\ \cdots & \cdots & \cdots & \cdots \end{pmatrix}
$$

$$
=: AB. \tag{3.68}
$$

Then according to the Cauchy-Binet formula

$$
\det\left(K^I(x_i, x_j)\right)_{k \times k} = \sum_{1 \leq i_1 < \cdots < i_k} \det(A_k B_k), \tag{3.69}
$$

where A_k is a $k \times k$ matrix consisting of row 1, $\cdots$, row k and column i_1, $\cdots$, column i_k from A, B_k is a $k \times k$ matrix consisting of column 1, $\cdots$, column k and row i_1, $\cdots$, row i_k from B.

Using the orthogonality of φ_i, we have

$$
\int_{\mathbb{R}^k} \det(A_k B_k) dx_1 \cdots dx_k = k! q_{i_1} \cdots q_{i_k}. \tag{3.70}
$$

Combining (3.67), (3.69) and (3.70) together yields

$$
E e^{t N_X(I)} = 1 + \sum_{k=1}^{\infty} \sum_{1 \leq i_1 < \cdots < i_k < \infty} q_{i_1} \cdots q_{i_k} (e^t - 1)^k.
$$

Thus we prove (3.65), and so conclude the proof. $\qquad \square$

Proof of Theorem 3.7. Having the identity (3.65) in law, a classic Lyapunov theorem (see (1.9)) can be used to establish the central limit theorem for N_n. Indeed, applying the Proposition 3.5, we get an array $\{\xi_{n,k}, n \geq 1, k \geq 1\}$ of independent Bernoulli random variables, so it suffices to show the central limit theorem holds for the sums $\sum_{k=1}^{\infty} \xi_{n,k}$. In turn, note

$$
\frac{\sum_{k=1}^{\infty} E|\xi_{n,k} - E\xi_{n,k}|^3}{(Var(\sum_{k=1}^{\infty} \xi_{n,k}))^{3/2}} \leq \frac{1}{(Var(N_n))^{1/2}} \to 0
$$

provided $Var(N_n) \to \infty$. Thus the Lyapunov condition is satisfied. $\qquad \square$

3.4 Logarithmic law

In this section we are concerned with the asymptotic behaviors of the logarithm of the determinant of the GUE matrix. Let $A_n = (z_{ij})_{n \times n}$ be the standard GUE matrix as given in the Introduction, denote its eigenvalues by $\lambda_1, \lambda_2, \cdots, \lambda_n$. Then we have

Theorem 3.8. *As* $n \to \infty$,

$$\frac{1}{\sqrt{\frac{1}{2} \log n}} \left(\log |\det A_n| - \frac{1}{2} \log n! + \frac{1}{4} \log n \right) \xrightarrow{d} N(0, 1). \quad (3.71)$$

The theorem is sometimes called the logarithmic law in literature. We remark that $\log |\det A_n| = \sum_{i=1}^{n} \log |\lambda_i|$ is a linear eigenvalue statistic with test function $f(x) = \log |x|$. However, the function $\log |x|$ is not so *nice* that one could not directly apply the results discussed in Section 3.2. The theorem was first proved by Girko in the 1970s using the martingale argument, see Girko (1979, 1990, 1998) and references therein for more details. Recently, Tao and Vu (2012) provided a new proof, which is based on a tridiagonal matrix representation due to Trotter (1984). We shall present their proof below. Before that, we want to give a parallel result about Ginibre model .

Let $M_n = (y_{ij})_{n \times n}$ be an $n \times n$ random matrix whose entries are *all* independent complex standard normal random variables. This is a rich and well-studied matrix model in the random matrix theory as well. Let $\nu_1, \nu_2, \cdots, \nu_n$ be its eigenvalues, then the joint probability density function is given by

$$\varrho_n(z_1, \cdots, z_n) = \frac{1}{n} \prod_{i<j} |z_i - z_j|^2 \prod_{i=1}^{n} e^{-|z_i|^2/2}, \quad z_i \in \mathbb{C}. \quad (3.72)$$

Define the bivariate empirical distribution function

$$F_n(x, y) = \frac{1}{n} \sum_{i=1}^{n} \mathbf{1}_{(Re\nu_i \le x, \, Im\nu_i \le y)}.$$

Then it follows

$$F_n \xrightarrow{d} \varrho_c \quad \text{in } P, \quad (3.73)$$

where ϱ_c stands for the uniform law in unit disk in the plane.

We leave the proofs of (3.72) and (3.73) to readers. Other more information can be found in Ginibre (1965) and Mehta (2004). As far as the determinant, a classic and interesting result is

Proposition 3.6. *As $n \to \infty$,*

$$\frac{1}{\sqrt{\frac{1}{4}\log n}}\left(\log|\det M_n| - \frac{1}{2}\log n! + \frac{1}{4}\log n\right) \xrightarrow{d} N(0,1). \quad (3.74)$$

Proof. First, observe the following identity in law

$$|\det M_n| \stackrel{d}{=} \frac{1}{2^{n/2}}\prod_{i=1}^{n}\chi_{2i}, \quad (3.75)$$

where the χ_i is a chi random variable with index i, and all chi random variables are independent.

Indeed, let $\mathbf{y}_1, \mathbf{y}_2, \cdots, \mathbf{y}_n$ denote row vectors of M_n. Then the absolute value of the determinant of M_n is equal to the volume of parallelnoid consisting of vectors $\mathbf{y}_1, \mathbf{y}_2, \cdots, \mathbf{y}_n$. In turn, the volume is equal to

$$|\mathbf{y}_1| \cdot |(I - P_1)\mathbf{y}_2| \cdots |(I - P_{n-1})\mathbf{y}_n|, \quad (3.76)$$

where P_i is an orthogonal projection onto the subspace spanned by vectors $\{\mathbf{y}_1, \mathbf{y}_2, \cdots, \mathbf{y}_i\}$, $1 \leq i \leq n - 1$. Note P_i is an idempotent projection with rank i, so $P_i\mathbf{y}_{i+1}$ is an i-variate complex standard normal random vector and is independent of $\{\mathbf{y}_1, \mathbf{y}_2, \cdots, \mathbf{y}_i\}$. Then letting $\chi_{2n} = \sqrt{2}|\mathbf{y}_1|$, $\chi_{2(n-i)} = \sqrt{2}|(I - P_i)\mathbf{y}_{i+1}|$, $1 \leq i \leq n - 1$ conclude the desired identity.

Second, note the χ_i has a density function

$$\frac{2^{1-i/2}}{\Gamma(i/2)}x^{i-1}e^{-x^2/2}, \quad x > 0$$

then it is easy to get

$$E\chi_i^k = 2^k \frac{\Gamma((i+k)/2)}{\Gamma(i/2)}$$

and the following asymptotic estimates

$$E\log\chi_i = \frac{1}{2}\log i - \frac{1}{2i} + O(i^{-2}), \quad Var(\log\chi_i) = \frac{1}{2i} + O(i^{-2}).$$

In addition, for each positive integer $k \geq 1$

$$E(\log\chi_i - E\log\chi_i)^{2k} = O(i^{-k}).$$

Lastly, note by (3.75)

$$\log|\det M_n| \stackrel{d}{=} -\frac{\log 2}{2}n + \sum_{i=1}^{n}\log\chi_{2i}.$$

(3.74) now directly follows from the classic Lyapunov CLT for sums of independent random variables. $\qquad\square$

The proof of Proposition 3.6 is simple and elegant. The hypothesis that all entries are independent plays an essential role. It is no longer true for A_n since it is Hermitian. We need to adopt a completely different method to prove Theorem 3.8. Start with a tridiagonal matrix representation of A_n.

Let $a_n, n \geq 1$ be a sequence of independent real standard normal random variables, $b_n, n \geq 1$ a sequence of independent random variables with each b_n distributed like χ_n. In addition, assume a_n's and b_n's are all independent. For each $n \geq 1$, construct a tridiagonal matrix

$$D_n = \begin{pmatrix} a_n & b_{n-1} & 0 & 0 & \cdots & 0 \\ b_{n-1} & a_{n-1} & b_{n-2} & 0 & \cdots & 0 \\ 0 & b_{n-2} & a_{n-2} & b_{n-3} & \cdots & 0 \\ 0 & \vdots & \ddots & \ddots & \ddots & \vdots \\ 0 & 0 & \cdots & b_2 & a_2 & b_1 \\ 0 & 0 & 0 & \cdots & b_1 & a_1 \end{pmatrix}. \tag{3.77}$$

Lemma 3.7. *The eigenvalues of D_n are distributed according to (3.2). In particular,*

$$\det A_n \overset{d}{=} \det D_n. \tag{3.78}$$

Proof. We shall obtain the D_n in (3.77) from A_n through a series of Householder transforms. Write

$$A_n = \begin{pmatrix} z_{11} & \mathbf{z}_1 \\ \mathbf{z}_1^* & A_{n,n-1} \end{pmatrix}$$

where $\mathbf{z}_1 = (z_{12}, \cdots, z_{1n})$. Let

$$w_1 = 0, \quad w_2 = -\frac{z_{21}}{|z_{21}|}\left(\frac{1}{2}\left(1 - \frac{|z_{21}|}{\alpha}\right)\right)^{1/2}$$

$$w_l = -\frac{z_{l1}}{(2\alpha(\alpha - |z_{21}|))^{1/2}}, \quad l \geq 3$$

where $\alpha > 0$ and

$$\alpha^2 = |z_{21}|^2 + |z_{31}|^2 + \cdots + |z_{n1}|^2.$$

Define the Householder transform by

$$V_n = I_n - 2\mathbf{w}_n\mathbf{w}_n^*$$

$$= \begin{pmatrix} 1 & 0 & \cdots & 0 \\ 0 & & & \\ \vdots & & V_{n,n-1} & \\ 0 & & & \end{pmatrix},$$

where $\mathbf{w}_n = (w_1, w_2, \cdots, w_n)^\tau$.

It is easy to check that V_n is a unitary matrix and

$$
V_n A_n V_n = \begin{pmatrix} z_{11} & (z_{12}, z_{13}, \cdots, z_{1n})V_{n,n-1} \\ V_{n,n-1}\begin{pmatrix} z_{12}^* \\ z_{13}^* \\ \vdots \\ z_{1n}^* \end{pmatrix} & V_{n,n-1}A_{n,n-1}V_{n,n-1} \end{pmatrix}
$$

$$
= \begin{pmatrix} z_{11} & \alpha\frac{z_{12}}{|z_{21}|} & 0 \\ \alpha\frac{z_{12}^*}{|z_{21}|} & V_{n,n-1}A_{n,n-1}V_{n,n-1} \\ 0 & \end{pmatrix}.
$$

To make the second entry in the first column nonnegative, we need to add one further configuration. Let R_n differ from the identity matrix by having $(2,2)$-entry $e^{-i\phi}$ with ϕ chosen appropriately and form $R_n V_n A_n V_n R_n^*$. Then we get the desired matrix

$$
\begin{pmatrix} z_{11} & \alpha & 0 \\ \alpha & \\ 0 & V_{n,n-1}A_{n,n-1}V_{n,n-1} \end{pmatrix},
$$

where $\alpha^2 = |z_{21}|^2 + |z_{31}|^2 + \cdots + |z_{n1}|^2$.

Define $a_n = z_{11}$, $b_{n-1} = \alpha$. $V_{n,n-1}$ is also a unitary matrix and is independent of $A_{n,n-1}$, so $V_{n,n-1}A_{n,n-1}V_{n,n-1}$ is an $n-1 \times n-1$ GUE matrix. Repeating the preceding procedure yields the desired matrix D_n. The proof is complete. □

According to (3.78), it suffices to prove (3.71) for $\log|\det D_n|$ below. Let $d_n = \det D_n$. It is easy to see the following recurrence relations

$$d_n = a_n d_{n-1} - b_{n-1}^2 d_{n-2} \tag{3.79}$$

$$d_{n-1} = a_{n-1}d_{n-2} - b_{n-2}^2 d_{n-3}. \tag{3.80}$$

Let $e_n = d_n/\sqrt{n!}$ and $c_n = (b_n^2 - n)/\sqrt{n}$. Note c_{n-k} is asymptotically normal as $n - k \to \infty$. So we deduce from (3.79) and (3.80)

$$e_n = \frac{a_n}{\sqrt{n}}e_{n-1} - \left(1 + \frac{c_{n-1}}{\sqrt{n}} - \frac{1}{2n}\right)e_{n-2} + \epsilon_1 \tag{3.81}$$

$$e_{n-1} = \frac{a_{n-1}}{\sqrt{n}}e_{n-2} - \left(1 + \frac{c_{n-2}}{\sqrt{n}} - \frac{1}{2n}\right)e_{n-3} + \epsilon_2, \tag{3.82}$$

where and in the sequel ϵ_1, ϵ_2 denote a small negligible quantity, whose value may be different from line to line.

In addition, substituting (3.82) into (3.81), we have

$$e_n = \left(-1 - \frac{c_{n-1}}{\sqrt{n}} + \frac{a_n a_{n-1}}{n} + \frac{1}{2n}\right)e_{n-2} - \left(\frac{a_n}{\sqrt{n}} + \frac{a_n c_{n-2}}{n}\right)e_{n-3} + \epsilon_1.$$

In terms of vectors, we have the following recurrence formula

$$\begin{pmatrix} e_n \\ e_{n-1} \end{pmatrix} = \left(-I_2 - \frac{1}{\sqrt{n}}S_{n,1} + \frac{1}{n}S_{n,2}\right)\begin{pmatrix} e_{n-2} \\ e_{n-3} \end{pmatrix} + \begin{pmatrix} \epsilon_1 \\ \epsilon_2 \end{pmatrix}, \qquad (3.83)$$

where

$$S_{n,1} = \begin{pmatrix} c_{n-1} & a_n \\ -a_{n-1} & c_{n-2} \end{pmatrix}$$

$$S_{n,2} = \frac{1}{2}I_2 + \begin{pmatrix} a_{n-1}a_n & -a_n c_{n-2} \\ 0 & 0 \end{pmatrix}.$$

Let $r_n^2 = e_{2n}^2 + e_{2n-1}^2$. It turns out that $\log r_n$ satisfies a CLT after properly scaled. This is stated as

Lemma 3.8. *As* $n \to \infty$,

$$\frac{\log r_n + \frac{1}{4}\log n}{\sqrt{\frac{1}{2}\log n}} \xrightarrow{d} N(0,1). \qquad (3.84)$$

Proof. Use (3.83) to get

$$r_n^2 = (e_{2n}, e_{2n-1})\begin{pmatrix} e_{2n} \\ e_{2n-1} \end{pmatrix}$$

$$= (e_{2n-2}, e_{2n-3})\Big(I_2 + \frac{1}{\sqrt{2n}}(S_{2n,1} + S_{2n,1}^\tau)$$

$$+ \frac{1}{2n}(S_{2n,1}^\tau S_{2n,1} - S_{2n,2} - S_{2n,2}^\tau)\Big)\begin{pmatrix} e_{2n-2} \\ e_{2n-3} \end{pmatrix} + \epsilon_0,$$

where ϵ_0 denotes a small negligible quantity, whose value may be different from line to line. Define

$$\xi_n = \frac{1}{r_{n-1}^2\sqrt{2n}}(e_{2n-2}, e_{2n-3})(S_{2n,1} + S_{2n,1}^\tau)\begin{pmatrix} e_{2n-2} \\ e_{2n-3} \end{pmatrix},$$

$$\eta_n = \frac{1}{r_{n-1}^2 2n}(e_{2n-2}, e_{2n-3})(S_{2n,1}^\tau S_{2n,1} - S_{2n,2} - S_{2n,2}^\tau)\begin{pmatrix} e_{2n-2} \\ e_{2n-3} \end{pmatrix}.$$

Then we have a recursive relation

$$r_n^2 = (1 + \xi_n + \eta_n + \epsilon_0)r_{n-1}^2.$$

Let $\mathcal{F}_n = \sigma\{a_1, \cdots, a_{2n}; b_1, \cdots, b_{2n-1}\}$. Then it follows for $n \geq 1$

$$E(\xi_n|\mathcal{F}_{n-1}) = 0, \quad E(\xi_n^2|\mathcal{F}_{n-1}) = \frac{2}{n}, \quad E(\xi_n^4|\mathcal{F}_{n-1}) = O(n^{-2}) \quad (3.85)$$

and

$$E(\eta_n|\mathcal{F}_{n-1}) = \frac{1}{2n}, \quad E(\eta_n^2|\mathcal{F}_{n-1}) = O(n^{-2}). \quad (3.86)$$

Using the Taylor expansion of $\log(1 + x)$ we obtain

$$\log r_n^2 = \log r_{n-1}^2 + \xi_n + \eta_n - \frac{\xi_n^2}{2} + \epsilon_0.$$

Let m_n be a sequence of integers such that $m_n/\log n \to 1$. Then

$$\log r_n^2 = \sum_{l=m_n+1}^{n} \xi_n + \sum_{l=m_n+1}^{n} \left(\eta_l - \frac{\xi_l^2}{2} + \epsilon_0\right) + \log r_{m_n}^2.$$

By the choice of m_n, we have

$$\frac{\log r_{m_n}^2}{\sqrt{\log n}} \xrightarrow{P} 0.$$

Also, by the Markov inequality, (3.85) and (3.86)

$$\frac{1}{\sqrt{\log n}} \sum_{l=m_n+1}^{n} \left(\eta_l - \frac{\xi_l^2}{2} + \frac{1}{2l} + \epsilon_0\right) \xrightarrow{P} 0.$$

Finally, by (3.85) and the martingale CLT we have

$$\frac{1}{\sqrt{\log n}} \sum_{l=m_n+1}^{n} \xi_l \xrightarrow{d} N(0, 1).$$

In combination, we have so far proven (3.84). $\qquad\square$

The above lemma describes asymptotically the magnitude of the vector (e_{2n}, e_{2n-1}). In order to obtain each component, we also need information about the phase of the vector.

Lemma 3.9. *Let $\theta_n \in (0, 2\pi)$ be such that $(e_{2n}, e_{2n-1}) = r_n(\cos\theta_n, \sin\theta_n)$. Then as $n \to \infty$,*

$$\theta_n \xrightarrow{d} \Theta, \quad (3.87)$$

where $\Theta \sim U(0, 2\pi)$.

Proof. Let us first look at the difference between θ_n and θ_{n-1}. Rewrite (3.83) as

$$r_n \begin{pmatrix} \cos\theta_n \\ \sin\theta_n \end{pmatrix} = r_{n-1}(-I_2 + D_n) \begin{pmatrix} \cos\theta_{n-1} \\ \sin\theta_{n-1} \end{pmatrix},$$

where

$$D_n = -\frac{1}{\sqrt{n}} S_{n,1} + \frac{1}{n} S_{n,2} + \epsilon$$

where ϵ is a negligible matrix. It then follows

$$\frac{r_n}{r_{n-1}} \begin{pmatrix} \cos\theta_n \\ \sin\theta_n \end{pmatrix} = - \begin{pmatrix} \cos\theta_{n-1} \\ \sin\theta_{n-1} \end{pmatrix} + D_n \begin{pmatrix} \cos\theta_{n-1} \\ \sin\theta_{n-1} \end{pmatrix}. \qquad (3.88)$$

Note $(\cos\theta_{n-1}, \sin\theta_{n-1})$ and $(-\sin\theta_{n-1}, \cos\theta_{n-1})$ form an orthonormal basis. It is easy to see

$$D_n \begin{pmatrix} \cos\theta_{n-1} \\ \sin\theta_{n-1} \end{pmatrix} = x_n \begin{pmatrix} \cos\theta_{n-1} \\ \sin\theta_{n-1} \end{pmatrix} + y_n \begin{pmatrix} -\sin\theta_{n-1} \\ \cos\theta_{n-1} \end{pmatrix}, \qquad (3.89)$$

where the coefficients x_n and y_n are given by

$$x_n = (\cos\theta_{n-1}, \sin\theta_{n-1}) D_n \begin{pmatrix} \cos\theta_{n-1} \\ \sin\theta_{n-1} \end{pmatrix}$$

and

$$y_n = (-\sin\theta_{n-1}, \cos\theta_{n-1}) D_n \begin{pmatrix} \cos\theta_{n-1} \\ \sin\theta_{n-1} \end{pmatrix}.$$

Substituting (3.89) back into (3.88) yields

$$\frac{r_n}{r_{n-1}} \begin{pmatrix} \cos\theta_n \\ \sin\theta_n \end{pmatrix} = (-1 + x_n) \begin{pmatrix} \cos\theta_{n-1} \\ \sin\theta_{n-1} \end{pmatrix} + y_n \begin{pmatrix} -\sin\theta_{n-1} \\ \cos\theta_{n-1} \end{pmatrix}.$$

Thus it is clear that

$$\tan(\theta_n - \theta_{n-1}) = \frac{y_n}{-1 + x_n},$$

which in turn leads to

$$\theta_n - \theta_{n-1} = \arctan \frac{y_n}{-1 + x_n}.$$

Next we estimate x_n and y_n. There is a constant $\varsigma > 0$ such that

$$x_n = O_P(n^{-\varsigma}), \quad y_n = O_P(n^{-\varsigma}).$$

In addition, for some $\iota > 1$

$$E(x_n | \theta_{n-1}) = O(n^{-\iota}), \quad E(y_n | \theta_{n-1}) = O(n^{-\iota})$$

$$E(y_n^2 | \theta_{n-1}) = \frac{1}{2n} + O(n^{-\iota}).$$

Using the Taylor expansions of $\arctan x$ and $(1+x)^{-1}$ we obtain

$$\theta_n - \theta_{n-1} = -(y_n - x_n y_n) + O(y_n^3),$$

and so

$$e^{ik\theta_n} = e^{ik\theta_{n-1}}\left(1 - iky_n + iky_n x_n - \frac{k^2}{2}y_n^2 + O(y_n^2(x_n + y_n))\right).$$

Hence we have

$$Ee^{ik\theta_n} = Ee^{ik\theta_{n-1}}\left(1 - iky_n + iky_n x_n - \frac{k^2}{2}y_n^2\right) + O(y_n^2(x_n + y_n)).$$

Moreover, using conditioning argument we get

$$Ee^{ik\theta_n} = Ee^{ik\theta_{n-1}}\left(1 - \frac{k^2}{4n}\right) + O(n^{-\iota}). \qquad (3.90)$$

Let $m_n \to \infty$ and $m_n/n \to 0$. Then repeatedly using (3.90) to get

$$Ee^{ik\theta_n} = \prod_{l=m_n+1}^{n}\left(1 - \frac{k^2}{4l}\right)Ee^{ik\theta_{m_n}} + O\left(\sum_{l=m_n+1}^{n} l^{-\iota}\right)$$
$$\to 0, \quad n \to \infty$$

where in the last limit we used $\iota > 1$ and the fact that $\prod_{l=m_n+1}^{n}\left(1 - \frac{k^2}{4l}\right) \to 0$ since $\sum_{l=m_n+1}^{n}\left(1 - \frac{k^2}{4l}\right) \to \infty$. Thus we complete the proof of (3.87). $\qquad\square$

Proof of Theorem 3.8. It easily follows from Lemma 3.9

$$P\left(|\cos\theta_n| < \frac{1}{\log n}\right) \to 0, \quad n \to \infty.$$

This in turn implies

$$\frac{\log|\cos\theta_n|}{\sqrt{\log n}} \xrightarrow{P} 0, \quad n \to \infty. \qquad (3.91)$$

On the other hand, we have

$$\log|e_{2n}| = \log r_n + \log|\cos\theta_n|.$$

Then according to Lemma 3.8 and (3.91),

$$\frac{\log|e_{2n}| + \frac{1}{4}\log n}{\sqrt{\frac{1}{2}\log n}} \xrightarrow{d} N(0,1).$$

The analog is valid for e_{2n-1}. Therefore we have proven Theorem 3.8. $\qquad\square$

To conclude this section, we shall mention the following variants of logarithmic law. Given $z \in \mathbb{R}$, we are interested in the asymptotic behaviors of the characteristic polynomials of A_n at z. We need to deal with two cases separately: either outside or inside the support of Wigner semicircle law.

Theorem 3.9. *If $z \in \mathbb{R} \setminus [-2, 2]$, then*

$$\log \left| \det(A_n - z\sqrt{n}) \right| - \frac{n}{2} \log n - n\mu_z \xrightarrow{d} N(0, \sigma_z^2),$$

where μ_z and σ_z^2 are given by

$$\mu_z = \frac{1}{2} \frac{z - \sqrt{z^2 - 4}}{z + \sqrt{z^2 - 4}} + \log \left| \frac{\sqrt{z^2 - 4} + z}{2} \right|,$$

$$\sigma_z^2 = \log \frac{|z| + \sqrt{z^2 - 1}}{2} - \frac{1}{2} \log(z^2 - 1).$$

Proof. This is a corollary to Theorem 3.5. Given $z \in \mathbb{R} \setminus [-2, 2]$, define $f_z(x) = \log|z - x|$. This is analytic outside a certain neighbourhood of $[-2, 2]$. In addition, a direct and lengthy computation shows

$$\mu_z = \int_{-2}^{2} f_z(x) \rho_{sc}(x) dx$$

$$= \frac{1}{2} \frac{z - \sqrt{z^2 - 4}}{z + \sqrt{z^2 - 4}} + \log \left| \frac{\sqrt{z^2 - 4} + z}{2} \right|$$

and

$$\sigma_z^2 = \frac{1}{4\pi^2} \int_{-2}^{2} \int_{-2}^{2} \frac{f_z'(x) f_z(y) \sqrt{4 - x^2}}{(y - x)\sqrt{4 - y^2}} dx dy$$

$$= \log \frac{|z| + \sqrt{z^2 - 1}}{2} - \frac{1}{2} \log(z^2 - 1). \qquad \square$$

Theorem 3.10. *If $z \in (-2, 2)$, then*

$$\frac{1}{\sqrt{\frac{1}{2} \log n}} \left(\log \left| \det(A_n - z\sqrt{n}) \right| - \frac{n}{2} \log n - \frac{n}{2} \left(\frac{z^2}{2} - 1 \right) \right)$$

$$\xrightarrow{d} N(0, 1). \tag{3.92}$$

To prove Theorem 3.10, we need a well-estimated result on the power of the characteristic polynomial for the GUE by Krasovsky (2007).

Proposition 3.7. *Fix $z \in (-2, 2)$. The following estimate holds*

$$E \left| \det(A_n - z\sqrt{n}) \right|^{2\alpha}$$

$$= C(\alpha) 2^{\alpha n} \left(1 - \frac{z^2}{4} \right)^{\alpha^2/2} \left(\frac{n}{2} \right)^{\alpha n + \alpha^2} e^{(z^2/2 - 1)\alpha n} (1 + \varepsilon_{\alpha, n}) \tag{3.93}$$

uniformly on any fixed compact set in the half plane $\operatorname{Re}\alpha > -1/2$. *Here*

$$C(\alpha) = \frac{1}{\Gamma(\alpha + \frac{1}{2})} \exp\left(2\int_0^\alpha \log\Gamma(s + \frac{1}{2})ds + \alpha^2\right)$$

$$= 2^{2\alpha^2} \frac{G(\alpha + 1)^2}{G(2\alpha + 1)}, \qquad (3.94)$$

where $G(\alpha)$ *is Barnes's G-function. The remainder term* $\varepsilon_{\alpha,n} = O(\log n/n)$ *is analytic in* α.

The proof is omitted. For α positive integers (3.93) has been found by Brézin and Hikami (2000), Forrester and Frankel (2004). For such α, $E|\det(A_n - z\sqrt{n})|^{2\alpha}$ can be reduced to the Hermite polynomials and their derivatives at the points z. However, it is not the case for noninteger α. In order to obtain (3.93), Krasovsky (2007) used Riemann-Hilbert problem approach to compute asymptotics of the determinant of a Hankel matrix whose support is supported on the real line and possesses power-like singularity.

Proof of Theorem 3.10. Start with computing expectation and variance of $\log|\det(A_n - z\sqrt{n})|$. For simplicity, write $M(\alpha)$ for $E|\det(A_n - z\sqrt{n})|^{2\alpha}$ below, and set

$$M(\alpha) = A(\alpha)B(\alpha),$$

where

$$A(\alpha) = 2^{\alpha n}\left(1 - \frac{z^2}{4}\right)^{\alpha^2/2}\left(\frac{n}{2}\right)^{\alpha n + \alpha^2} e^{(z^2/2 - 1)\alpha n}$$

and

$$B(\alpha) = C(\alpha)(1 + \varepsilon_{\alpha,n}).$$

It obviously follows

$$E\log|\det(A_n - z\sqrt{n})|^2 = M'(0)$$

and

$$E\left(\log|\det(A_n - z\sqrt{n})|^2\right)^2 = M''(0).$$

Thus we need only to evaluate $M'(0)$ and $M''(0)$. A direct calculation shows

$$M'(\alpha) = \left(n\log n + 2\alpha\log\frac{n}{2} + \left(\frac{z^2}{2} - 1\right)n + \frac{\alpha^2}{2}\left(1 - \frac{z^2}{4}\right)\right)A(\alpha)B(\alpha)$$
$$+ A(\alpha)B'(\alpha)$$

and

$$M''(\alpha) = 2\Big(n\log n + 2\alpha\log\frac{n}{2} + \Big(\frac{z^2}{2}-1\Big)n + \frac{\alpha^2}{2}\Big(1-\frac{z^2}{4}\Big)\Big)A(\alpha)B'(\alpha)$$

$$+\Big(n\log n + 2\alpha\log\frac{n}{2} + \Big(\frac{z^2}{2}-1\Big)n + \frac{\alpha^2}{2}\Big(1-\frac{z^2}{4}\Big)\Big)^2 A(\alpha)B(\alpha)$$

$$+\Big(2\log\frac{n}{2} + \frac{1}{2}\Big(1-\frac{z^2}{4}\Big)\Big)A(\alpha)B(\alpha) + A(\alpha)B''(\alpha).$$

It is easy to see $M(0) = A(0) = 1$, and so $B(0) = 1$. Furthermore, by the analyticity of $B(\alpha)$ for $Re\alpha > -1/2$ and using Cauchy's theorem, we have

$$B'(0) = C'(0) + O(n^{-1}\log n), \quad B''(0) = C''(0) + O(n^{-1}\log n).$$

Similarly, it follows from (3.94)

$$C'(\alpha) = C(\alpha)\Big(4\log 2\alpha + 2\frac{G'(\alpha+1)}{G(\alpha+1)} - 2\frac{G'(2\alpha+1)}{G(2\alpha+1)}\Big).$$

Note

$$\frac{G'(\alpha+1)}{G(\alpha+1)} = \frac{1}{2}\log(2\pi) + \frac{1}{2} - \alpha + \alpha\frac{\Gamma'(\alpha)}{\Gamma(\alpha)}$$

$$= \frac{1}{2}\log(2\pi) - \frac{1}{2} - (\gamma+1)\alpha + \frac{\pi^2}{6}\alpha + O(\alpha^3),$$

where γ is the Euler constant. So we have

$$C'(0) = 0, \quad C''(0) = 4\log 2 + 2(\gamma+1).$$

In combination, we obtain

$$M'(0) = \Big(n\log n + (\frac{z^2}{2}-1)n\Big) + B'(0)$$

and

$$M''(0) = \Big(n\log n + (\frac{z^2}{2}-1)n\Big)^2 + 2\Big(n\log n + (\frac{z^2}{2}-1)n\Big)B'(0)$$

$$+\Big(2\log\frac{n}{2} + \frac{1}{2}\Big(1-\frac{z^2}{4}\Big)\Big) + B''(0).$$

This in turn gives

$$E\log|\det(A_n - z\sqrt{n})| = \frac{1}{2}\Big(n\log n + (\frac{z^2}{2}-1)n\Big) + o(1) \qquad (3.95)$$

and

$$Var\big(\log|\det(A_n - z\sqrt{n})|\big)$$

$$= \frac{1}{2}\log n - \frac{z^2}{32} - \frac{1}{2}\log 2 + \frac{1}{2}\gamma + \frac{5}{8} + o(1). \qquad (3.96)$$

Next we turn to the proof of (3.92). Define for $t \in \mathbb{R}$

$$m_n(t) = E \exp\left(t\, \frac{\log |\det(A_n - z\sqrt{n})| - E \log |\det(A_n - z\sqrt{n})|}{\sqrt{Var(\log |\det(A_n - z\sqrt{n})|)}} \right).$$

It is sufficient to prove

$$m_n(t) \to e^{t^2/2}, \quad n \to \infty. \tag{3.97}$$

Indeed, using (3.95) and (3.96) we have

$$m_n(t) = \exp\left(-\frac{n \log n + (\frac{z^2}{2} - 1)n}{\sqrt{2 \log n}} t \right) M\left(\frac{t}{\sqrt{2 \log n}} \right)(1 + o(1))$$

$$= e^{t^2/2} C\left(\frac{t}{2 \log n} \right)(1 + o(1)).$$

As is known, the Barne's G-function is entire and $G(1) = 1$. It follows that $C\left(t/2 \log n \right) \to 1$ as $n \to \infty$, then we get (3.97) as desired. The proof is complete. $\qquad\qquad\qquad\qquad\qquad\qquad\qquad\qquad\qquad\qquad\qquad\qquad\square$

3.5   Hermite β ensembles

In the last section of this chapter, we will turn to the study of the Hermite β Ensemble (HβE), which is a natural extension of the GUE. By the HβE we mean an n-point process in the real line $\mathbb{R}$ with the following joint probability density function

$$p_{n,\beta}(x_1, \cdots, x_n) = Z_{n,\beta} \prod_{1 \leq i < j \leq n} |x_i - x_j|^\beta \prod_{j=1}^{n} e^{-x_j^2/2}, \tag{3.98}$$

where $x_1, \cdots, x_n \in \mathbb{R}$, $\beta > 0$ is a model parameter and

$$Z_{n,\beta} = \frac{1}{(2\pi)^{n/2}} \frac{\Gamma(\frac{\beta}{2})^n}{n! \prod_{j=1}^{n} \Gamma(\frac{\beta j}{2})}$$

by Selberg's integral. This model was first introduced by Dyson (1962) in the study of Coulomb lattice gas in the early sixties. The formula (3.98) can be rewritten as

$$p_{n,\beta}(x_1, \cdots, x_n) \propto e^{-\beta H_n(x_1, \cdots, x_n)},$$

where

$$H_n(x_1, \cdots, x_n) = \frac{1}{2\beta} \sum_{j=1}^{n} x_j^2 - \frac{1}{2} \sum_{i \neq j} \log |x_i - x_j|$$

is a Hamiltonian quantity, β may be viewed as inverse temperature. The quadratic function part means the points fall independently in the real line with normal law, while the extra logarithmic part indicates the points repel each other. The special cases $\beta = 1, 2, 4$ correspond to GOE, GUE and GSE respectively.

In the study of HβE, a remarkable contribution was made by Dumitriu and Edelman (2002), in which a tridiagonal matrix representation was discovered. Specifically speaking, let $a_n, n \geq 1$ be a sequence of independent normal random variables with mean 0 and variance 2. Let $b_n, n \geq 1$ be a sequence of independent chi random variables, each b_n having density function:

$$\frac{2^{1-\beta n/2}}{\Gamma(\frac{\beta n}{2})} x^{\beta n-1} e^{-x^2/2}, \quad x > 0.$$

In addition, all a_n's and b_n's are assumed to be independent. Define a tridiagonal matrix

$$D_{n,\beta} = \frac{1}{\sqrt{2}} \begin{pmatrix} a_n & b_{n-1} & 0 & 0 & \cdots & 0 \\ b_{n-1} & a_{n-1} & b_{n-2} & 0 & \cdots & 0 \\ 0 & b_{n-2} & a_{n-2} & b_{n-3} & \cdots & 0 \\ \vdots & \vdots & \ddots & \ddots & \ddots & \vdots \\ 0 & 0 & \cdots & b_2 & a_2 & b_1 \\ 0 & 0 & 0 & \cdots & b_1 & a_1 \end{pmatrix}.$$

Then we have

Theorem 3.11. *The eigenvalues of $D_{n,\beta}$ are distributed according to (3.98).*

As we see from Lemma 3.7, an explicit Householder transform can be used to produce $D_{n,2}$ from the GUE square matrix model. The general case will be below proved using eigendecomposition of a tridiagonal and the change of variables formula.

Given two sequences of real numbers $x_1, x_2, \cdots, x_n$ and $y_1, y_2, \cdots, y_{n-1}$, construct a tridiagonal matrix X_n as follows

$$X_n = \begin{pmatrix} x_n & y_{n-1} & 0 & 0 & \cdots & 0 \\ y_{n-1} & x_{n-1} & y_{n-2} & 0 & \cdots & 0 \\ 0 & y_{n-2} & x_{n-2} & y_{n-3} & \cdots & 0 \\ \vdots & \vdots & \ddots & \ddots & \ddots & \vdots \\ 0 & 0 & \cdots & y_2 & x_2 & y_1 \\ 0 & 0 & 0 & \cdots & y_1 & x_1 \end{pmatrix}.$$

Let $\lambda_1^{(n)}, \lambda_2^{(n)}, \cdots, \lambda_n^{(n)}$ be eigenvalues of X_n and assume that $\lambda_1^{(n)} > \lambda_2^{(n)} > \cdots > \lambda_n^{(n)}$. Write

$$X_n = Q\Lambda Q^\tau =: Q \begin{pmatrix} \lambda_1^{(n)} & 0 & \cdots & 0 \\ 0 & \lambda_2^{(n)} & \cdots & 0 \\ \vdots & \vdots & \ddots & \vdots \\ 0 & 0 & \cdots & \lambda_n^{(n)} \end{pmatrix} Q^\tau \qquad (3.99)$$

for eigendecomposition, where Q is eigenvector matrix such that $QQ^\tau = Q^\tau Q = I_n$ and the first row $\mathbf{q} = (q_1, q_2, \cdots, q_n)$ is strictly positive. Note that once $q_1, q_2, \cdots, q_n$ are specified, then other components of Q will be uniquely determined by eigenvalues and X_n. Conversely, starting from Λ and $\mathbf{q}$, one can reconstruct the matrix X_n.

Lemma 3.10.

$$\prod_{1 \leq i < j \leq n} \left(\lambda_i^{(n)} - \lambda_j^{(n)} \right) = \frac{\prod_{i=1}^{n-1} y_i^i}{\prod_{i=1}^{n} q_i}.$$

Proof. We similarly define X_k using $x_1, x_2, \cdots, x_k$ and $y_1, y_2, \cdots, y_{k-1}$ for $2 \leq k \leq n$. Let $P_k(\lambda)$ be the characteristic polynomial of X_k, and let also $\lambda_1^{(k)}, \lambda_2^{(k)}, \cdots, \lambda_k^{(k)}$ be the eigenvalues in decreasing order. Then it is easy to see the following recursive formula

$$P_n(\lambda) = (x_n - \lambda)P_{n-1}(\lambda) - y_{n-1}^2 P_{n-2}(\lambda). \qquad (3.100)$$

We can deduce from (3.100) that for any $1 \leq j \leq n-1$

$$\prod_{i=1}^{n} \left(\lambda_i^{(n)} - \lambda_j^{(n-1)} \right) = -y_{n-1}^2 \prod_{i=1}^{n-2} \left(\lambda_i^{(n-2)} - \lambda_j^{(n-1)} \right).$$

Hence it follows

$$\prod_{j=1}^{n-1} \prod_{i=1}^{n} \left(\lambda_i^{(n)} - \lambda_j^{(n-1)} \right) = (-1)^{n-1} y_{n-1}^{2(n-1)} \prod_{j=1}^{n-1} \prod_{i=1}^{n-2} \left(\lambda_i^{(n-2)} - \lambda_j^{(n-1)} \right)$$

$$= (-1)^{n(n-1)/2} \prod_{l=1}^{n-1} y_l^{2l}. \qquad (3.101)$$

On the other hand, note the following identity

$$\frac{P_{n-1}(\lambda)}{P_n(\lambda)} = \sum_{i=1}^{n} \frac{q_i^2}{\lambda_i^{(n)} - \lambda},$$

that is

$$P_{n-1}(\lambda) = \sum_{i=1}^{n} q_i^2 \prod_{l \neq i} \left(\lambda_l^{(n)} - \lambda\right).$$

This obviously implies for each $1 \leq j \leq n$

$$\prod_{i=1}^{n-1} \left(\lambda_i^{(n-1)} - \lambda_j^n\right) = q_j^2 \prod_{i \neq j} \left(\lambda_i^{(n)} - \lambda_j^{(n)}\right).$$

Hence it follows

$$\prod_{j=1}^{n} \prod_{i=1}^{n-1} \left(\lambda_i^{(n-1)} - \lambda_j^{(n)}\right)$$

$$= \prod_{j=1}^{n} q_j^2 \prod_{i \neq j} \left(\lambda_i^{(n)} - \lambda_j^{(n)}\right)$$

$$= (-1)^{n(n-1)/2} \prod_{j=1}^{n} q_j^2 \prod_{1 \leq i < j \leq n} \left(\lambda_i^{(n)} - \lambda_j^{(n)}\right)^2. \qquad (3.102)$$

Combining (3.101) and (3.102) together yields

$$\prod_{1 \leq i < j \leq n} \left(\lambda_i^{(n)} - \lambda_j^{(n)}\right)^2 = \frac{\prod_{i=1}^{n-1} y_i^{2i}}{\prod_{i=1}^{n} q_i^2}.$$

The proof is complete. $\qquad\qquad\qquad\qquad\qquad\qquad\qquad\qquad\qquad\qquad\square$

Consider the eigendecomposition (3.105), the $2n - 1$ variables

$$\mathbf{x} = (x_1, x_2, \cdots, x_n), \quad \mathbf{y} = (y_1, y_2, \cdots, y_{n-1})$$

can be put into a one-to-one correspondence with the $2n-1$ variables $(\mathbf{\Lambda}, \mathbf{q})$. Let J denote the determinant of the Jacobian for the change of variables from $(\mathbf{x}, \mathbf{y})$ to $(\mathbf{\Lambda}, \mathbf{q})$. Then we have

Lemma 3.11.

$$J = \frac{\prod_{i=1}^{n-1} y_i}{q_n \prod_{i=1}^{n} q_i}. \qquad (3.103)$$

Proof. Observe the following identity

$$\left(I_n - \lambda X_n\right)_{11}^{-1} = \sum_{i=1}^{n} \frac{q_i^2}{1 - \lambda\lambda_i^{(n)}}.$$

Use the Taylor expansion of $(1 - x)^{-1}$ to get

$$1 + \sum_{k=1}^{\infty} \lambda^k \left(X_n^k\right)_{11} = \sum_{k=0}^{\infty} \sum_{i=1}^{n} q_i^2 \lambda_i^{(n)k} \lambda^k.$$

Hence we have for each $k \geq 1$

$$\left(X_n^k\right)_{11} = \sum_{i=1}^{n} q_i^2 \lambda_i^{(n)k}.$$

In particular,

$$x_n = \sum_{i=1}^{n} q_i^2 \lambda_i^{(n)}$$

$$* + y_{n-1}^2 = \sum_{i=1}^{n} q_i^2 \lambda_i^{(n)2}$$

$$* + y_{n-1}^2 x_{n-1} = \sum_{i=1}^{n} q_i^2 \lambda_i^{(n)3} \qquad (3.104)$$

$$\cdots\cdots\cdots$$

$$* + y_{n-1}^2 \cdots y_1^2 x_1 = \sum_{i=1}^{n} q_i^2 \lambda_i^{(n)2n-1}$$

where $*$ stands for what already appeared in the preceding equation.

Taking differentials at both sides of equations in (3.104) and noting the fact $q_n^2 = 1 - \sum_{i=1}^{n-1} q_i^2$ yields

$$\mathbf{A}\begin{pmatrix} dx_n \\ dx_{n-1} \\ \vdots \\ dx_1 \\ dy_{n-1} \\ \vdots \\ dy_1 \end{pmatrix} = (\mathbf{B}_1, \mathbf{B}_2) \begin{pmatrix} d\lambda_1^{(n)} \\ d\lambda_2^{(n)} \\ \vdots \\ d\lambda_n^{(n)} \\ dq_1 \\ \vdots \\ dq_{n-1} \end{pmatrix},$$

where

$$\mathbf{A} = \operatorname{diag}\left(1, y_{n-1}^2, \cdots, \prod_{i=1}^{n-1} y_i^2, 2y_{n-1}, 2y_{n-2}y_{n-1}^2 \cdots, 2y_1 \prod_{l=2}^{n-1} y_l^2\right),$$

$$\mathbf{B}_1 = \left(2q_j(\lambda_j^{(n)i} - \lambda_n^{(n)i})\right)_{1 \leq i \leq 2n-1, 1 \leq j \leq n-1},$$

$$\mathbf{B}_2 = \left(iq_j^2 \lambda_j^{(n)i-1}\right)_{1 \leq i \leq 2n-1, 1 \leq j \leq n}.$$

Hence a direct computation gives

$$J = \det\left(\frac{\partial(\mathbf{x}, \mathbf{y})}{\partial(\lambda^{(n)}, \mathbf{q})}\right)$$

$$= \frac{1}{q_n} \frac{\prod_{i=1}^{n-1} y_i}{\prod_{i=1}^{n} q_i} \left(\frac{\prod_{i=1}^{n} q_i}{\prod_{i=1}^{n-1} y_i}\right)^4 \prod_{1 \leq i < j \leq n} \left(\lambda_i^{(n)} - \lambda_j^{(n)}\right)^4.$$

Now according to Lemma 3.10, (3.103) holds as desired. $\qquad\square$

Proof of Theorem 3.11. Denote by $\lambda_{1,\beta}, \lambda_{2,\beta}, \cdots, \lambda_{n,\beta}$ the eigenvalues of $D_{n,\beta}$. For clarity, we first assume that $\lambda_{1,\beta} > \lambda_{2,\beta} > \cdots > \lambda_{n,\beta}$, and write

$$D_{n,\beta} = Q\Lambda Q^\tau =: Q \begin{pmatrix} \lambda_{1,\beta} & 0 & \cdots & 0 \\ 0 & \lambda_{2,\beta} & \cdots & 0, \\ \vdots & \vdots & \ddots & \vdots \\ 0 & 0 & \cdots & \lambda_{n,\beta} \end{pmatrix} Q^\tau \qquad (3.105)$$

for eigendecomposition, where Q is eigenvector matrix such that $QQ^\tau = Q^\tau Q = I_n$ and the first row $\mathbf{q} = (q_1, q_2, \cdots, q_n)$ is strictly positive.

As remarked above, such an eigendecomposition is unique. Let T be a one-to-one correspondence between $D_{n,\beta}$ and (Λ, Q), then the determinant of its Jacobian is given by (3.103). Hence the joint probability density $p(\lambda, \mathbf{q})$ of $(\lambda_{1,\beta}, \lambda_{2,\beta}, \cdots, \lambda_{n,\beta})$ and $\mathbf{q} = (q_1, q_2, \cdots, q_{n-1})$ is equal to

$$\frac{1}{(2\pi)^{n/2}} \frac{2^{n-1}}{\prod_{i=1}^{n-1} \Gamma(\frac{\beta i}{2})} \prod_{i=1}^{n} e^{-\frac{1}{2}x_i^2} \prod_{i=1}^{n-1} y_i^{\beta i - 1} e^{-y_i^2} |J|$$

$$= \frac{1}{(2\pi)^{n/2}} \frac{2^{n-1}}{\prod_{i=1}^{n-1} \Gamma(\frac{\beta i}{2})} \prod_{i=1}^{n} e^{-\frac{1}{2}\lambda_i^2} \prod_{i=1}^{n-1} y_i^{\beta i - 1} \frac{\prod_{i=1}^{n-1} y_i}{q_n \prod_{i=1}^{n} q_i}$$

$$= \frac{1}{(2\pi)^{n/2}} \frac{2^{n-1}}{\prod_{i=1}^{n-1} \Gamma(\frac{\beta i}{2})} \prod_{i=1}^{n} e^{-\frac{1}{2}\lambda_i^2} \prod_{1 \le i < j \le n} (\lambda_i - \lambda_j)^\beta \frac{1}{q_n} \prod_{i=1}^{n} q_i^{\beta - 1}. \quad (3.106)$$

We see from (3.106) that $(\lambda_{1,\beta}, \lambda_{2,\beta}, \cdots, \lambda_{n,\beta})$ and $\mathbf{q} = (q_1, q_2, \cdots, q_{n-1})$ are independent and so the joint probability density $p_{n,\beta}(\lambda)$ of $(\lambda_{1,\beta}, \lambda_{2,\beta}, \cdots, \lambda_{n,\beta})$ can be obtained by integrating out the variable $\mathbf{q}$

$$p_{n,\beta}(\lambda_1, \cdots, \lambda_n)$$

$$= \frac{1}{(2\pi)^{n/2}} \frac{2^{n-1}}{\prod_{i=1}^{n-1} \Gamma(\frac{\beta i}{2})} \prod_{i=1}^{n} e^{-\lambda_i^2/2} \prod_{1 \le i < j \le n} (\lambda_i - \lambda_j)^\beta$$

$$\cdot \int_{\mathbf{q}: q_i > 0, \sum_{i=1}^{n} q_i^2 = 1} \frac{1}{q_n} \prod_{i=1}^{n} q_i^{\beta - 1} dq_1 \cdots dq_n$$

$$= \frac{1}{(2\pi)^{n/2}} \frac{\Gamma(\frac{\beta}{2})^n}{\prod_{i=1}^{n} \Gamma(\frac{\beta i}{2})} \prod_{1 \le i < j \le n} (\lambda_i - \lambda_j)^\beta \prod_{i=1}^{n} e^{-\lambda_i^2/2}, \quad (3.107)$$

where we used Lemma 2.18 to compute the integral in the second equation.

Finally, to obtain the joint probability density of unordered eigenvalue, we only need to multiply (3.107) by the factor $1/n!$. The proof is now concluded. □

The HβE is a rich and well studied model in random matrix theory. It possesses many nice properties similar to the GUE. In particular, there have been a lot of new advances in the study of asymptotic behaviours of point statistics since the tridiagonal matrix model was discovered. We will below quickly review some results related to limit laws without proofs. The interested reader is referred to original papers for more information.

To enable eigenvalues asymptotically fall in the interval $(-2\sqrt{n}, 2\sqrt{n})$, we consider $H_{n,\beta} =: \sqrt{\frac{2}{\beta}} D_{n,\beta}$. Denote by $\lambda_{1,\beta}, \lambda_{2,\beta}, \cdots, \lambda_{n,\beta}$ the eigenvalues of $H_{n,\beta}$, the corresponding empirical distribution function is

$$F_{n,\beta}(x) = \frac{1}{n} \sum_{i=1}^{n} \mathbf{1}_{(\lambda_{i,\beta} \leq \sqrt{n}x)}.$$

Dumitriu and Edelman (2002) used moment methods to prove the Wigner semicircle law as follows

$$F_{n,\beta} \xrightarrow{d} \rho_{sc} \quad \text{in } P.$$

In particular, for each bounded continuous function f,

$$\frac{1}{n} \sum_{i=1}^{n} f\left(\frac{\lambda_{i,\beta}}{\sqrt{n}}\right) \xrightarrow{P} \int_{-2}^{2} f(x)\rho_{sc}dx.$$

Moreover, if f satisfy a certain regularity condition, then according to Johansson (1998), the central limit theorem holds. Namely,

$$\sum_{i=1}^{n} f\left(\frac{\lambda_{i,\beta}}{\sqrt{n}}\right) - n \int_{-2}^{2} f(x)\rho_{sc}(x)dx \xrightarrow{d} N(0, \sigma_{\beta,f}^2),$$

where $\sigma_{\beta,f}^2$ is given by

$$\sigma_{\beta,f}^2 = \left(\frac{2}{\beta} - 1\right)\left(\frac{1}{4}(f(2) + f(-2)) - \int_{-2}^{2} f(x)\rho_{sc}'(x)dx\right)$$
$$- \frac{1}{2\pi^2\beta} \int_{-2}^{2} \int_{-2}^{2} \frac{f'(x)f(y)\rho_{sc}(x)}{(x-y)\rho_{sc}(y)} dxdy.$$

Following the line of the proof in Theorem 3.8, one could also prove the logarithmic law

$$\frac{1}{\sqrt{\frac{\beta}{4} \log n}}\left(\log|\det H_{n,\beta}| - \frac{1}{2}\log n! + \frac{1}{4}\log n\right) \xrightarrow{d} N(0,1).$$

As for the counting functions of eigenvalue point process, it is worthy mentioning the following two results both at the edge and inside the bulk. Let u_n be a sequence of real numbers. Define for $x \in \mathbb{R}$

$$N_{n,\beta}(x) = \sharp\{1 \leq i \leq n : n^{1/6}(\lambda_{i,\beta} - u_n) \text{ fall between 0 and } x\}.$$

Based on variational analysis, Ramírez, Rider and Virág (2011) proved that under the assumption $n^{1/6}(2\sqrt{n} - u_n) \to a \in \mathbb{R}$,

$$N_{n,\beta}(x) \xrightarrow{d} N_{Airy_\beta}(x), \quad x \in \mathbb{R},$$

where $Airy_\beta$ is defined as -1 times the point process of eigenvalues of the stochastic with parameter β, and $N_{Airy_\beta}(x)$ is the number of points between 0 and x.

In the same spirit, Valkó and Virág (2009) considered the eigenvalues around any location away from the spectral edge. Let u_n be a sequence of real numbers so that $n^{1/6}(2\sqrt{n} - |u_n|) \to \infty$. Define for $x \in \mathbb{R}$

$$N_{n,\beta}(x) = \sharp\left\{ 1 \leq i \leq n : \sqrt{4n - u_n^2}(\lambda_{i,\beta} - u_n) \text{ fall between } 0 \text{ and } x\right\},$$

then

$$N_{n,\beta}(x) \xrightarrow{d} N_{Sine_\beta}(x), \quad x \in \mathbb{R},$$

where $Sine_\beta$ is a translation invariant point process given by the Brownian carousel.

As the reader may see, the point process from the HβE is no longer determinantal except in special cases. Thus Theorem 3.7, the Costin-Lebowitz-Soshnikov theorem, is not applicable. However, we can follow the strategy of Valkó and Virág (2009) to prove the central limit theorem for the number of points of the HβE lying in the right side of the origin, see and (2010).

Theorem 3.12. *Let $N_n(0, \infty)$ be the number of points of the HβE lying in the right side of the origin. Then it follows*

$$\frac{1}{\sqrt{\frac{1}{\beta\pi^2}\log n}}\left(N_n(0, \infty) - \frac{n}{2}\right) \xrightarrow{d} N(0, 1). \tag{3.108}$$

We remark that the number $N_n(0, \infty)$, sometimes called the index, is a key object of interest to physicists. Cavagna, Garrahan and Giardina (2000) calculated the distribution of the index for GOE by means of the replica method and obtained Gaussian distribution with asymptotic variance like $\log n/\pi^2$. Majumdar, Nadal, Scardicchio and Vivo (2009) further computed analytically the probability distribution of the number $N_n[0, \infty)$ of positive points for HβE using the partition function and saddle point analysis. They computed the variance $\log n/\beta\pi^2 + O(1)$, which agrees with the corresponding variance in (3.108), while they thought the distribution is not strictly Gaussian due to an unusual logarithmic singularity in the rate function.

The rest part of the section will prove Theorem 3.12. The proof relies largely on the new phase evolution of eigenvectors invented by Valkó and

Virág (2009). Let $H_{n,\beta} = \sqrt{2/\beta}D_{n,\beta}$. We only need to consider the number of positive eigenvalues of $H_{n,\beta}$. A key idea is to derive from the tridiagonal matrix model a recurrence relation for a real number Λ to be an eigenvalue, which yields an evolution relation for eigenvectors. Specifically, let $s_j = \sqrt{n-j-1/2}$. Define

$$O_n = \begin{pmatrix} d_{11} & 0 & 0 & \cdots & 0 \\ 0 & d_{22} & 0 & \cdots & 0 \\ \vdots & \vdots & \vdots & \ddots & \vdots \\ 0 & 0 & 0 & \cdots & d_{nn} \end{pmatrix},$$

where

$$d_{11} = 1, \quad d_{ii} = \frac{b_{n-1-i}}{s_{i-1}}d_{i-1,i-1}, \quad 2 \le i \le n.$$

Let

$$X_i = \frac{a_{n-i}}{\sqrt{\beta}}, \qquad 0 \le i \le n-1$$

and

$$Y_i = \frac{b_{n-1-i}^2}{\beta s_{i+1}} - s_i, \qquad 0 \le i \le n-2.$$

Then

$$O_n^{-1}H_{n,\beta}O_n = \begin{pmatrix} X_0 & s_0 + Y_0 & 0 & \cdots & 0 \\ s_1 & X_1 & s_1 + Y_1 & \cdots & 0 \\ 0 & s_2 & X_2 & \cdots & 0 \\ \vdots & \vdots & \vdots & \ddots & \vdots \\ 0 & 0 & 0 & \cdots & X_{n-1} \end{pmatrix}$$

obviously have the same eigenvalues as $H_{n,\beta}$. However, there is a significant difference between these two matrices. The rows between $O_n^{-1}H_{n,\beta}O_n$ are independent of each other, while $H_{n,\beta}$ is symmetric so that the rows are not independent.

Assume that Λ is an eigenvalue of $O_n^{-1}H_{n,\beta}O_n$, then by definition there exists a nonzero eigenvector $\mathbf{v} = (v_1, v_2, \cdots, v_n)^\tau$ such that

$$O_n^{-1}H_{n,\beta}O_n\mathbf{v} = \Lambda\mathbf{v}.$$

Without loss of generality, we can assume $v_1 = 1$. Thus, Λ is an eigenvalue if and only there exists an eigenvector $\mathbf{v}^\tau = (1, v_2, \cdots, v_n)$ such that

$$\begin{pmatrix} X_0 & s_0 + Y_0 & 0 & \cdots & 0 \\ s_1 & X_1 & s_1 + Y_1 & \cdots & 0 \\ 0 & s_2 & X_2 & \cdots & 0 \\ \vdots & \vdots & \vdots & \ddots & \vdots \\ 0 & 0 & 0 & \cdots & X_{n-1} \end{pmatrix} \begin{pmatrix} 1 \\ v_2 \\ v_3 \\ \vdots \\ v_n \end{pmatrix} = \Lambda \begin{pmatrix} 1 \\ v_2 \\ v_3 \\ \vdots \\ v_n \end{pmatrix}.$$

It can be equivalently rewritten as

$$
\begin{pmatrix}
1 & X_0 & s_0 + Y_0 & 0 & \cdots & 0 & 0 \\
0 & s_1 & X_1 & s_1 + Y_1 & \cdots & 0 & 0 \\
0 & 0 & s_2 & X_2 & \cdots & 0 & 0 \\
\cdots & \vdots & \vdots & \vdots & \ddots & \vdots & \vdots \\
0 & 0 & 0 & 0 & \cdots & X_{n-1} & 1
\end{pmatrix}
\begin{pmatrix}
0 \\ 1 \\ v_2 \\ v_3 \\ \vdots \\ v_n \\ 0
\end{pmatrix}
= \Lambda
\begin{pmatrix}
1 \\ v_2 \\ v_3 \\ \vdots \\ v_n
\end{pmatrix}.
$$

Let $v_0 = 0, v_{n+1} = 0$ and define $r_l = v_{l+1}/v_l$, $0 \le l \le n$. Thus we have the following necessary and sufficient condition for Λ to be an eigenvalue in terms of evolution:

$$
\infty = r_0, \quad r_n = 0
$$

and

$$
r_{l+1} = \frac{1}{1 + \frac{Y_l}{s_l}} \left(-\frac{1}{r_l} + \frac{\Lambda - X_l}{s_l} \right), \quad 0 \le l \le n - 2. \tag{3.109}
$$

Since the (X_l, Y_l)'s are independent, then $r_0, r_1, \cdots, r_{n-1}, r_n$ forms a Markov chain with ∞ as initial state and 0 as destination state, and the next state r_{l+1} given a present state r_l will be attained through a random fractional linear transform.

Next we turn to the description of the phase evolution. Let $\mathbb{H}$ denote the upper half plane, $\mathbb{U}$ the Poincaré disk model, define the bijection

$$
\mathbf{U} : \bar{\mathbb{H}} \to \bar{\mathbb{U}}, \quad z \to \frac{i - z}{i + z},
$$

which is also a bijection of the boundary. As r moves on the boundary $\partial \mathbb{H} = \mathbb{R} \cup \{\infty\}$, its image under $\mathbf{U}$ will move along $\partial \mathbb{U}$.

In order to follow the number of times this image circles $\mathbb{U}$, we need to extend the action from $\partial \mathbb{U}$ to its universal cover, $\mathbb{R}' = \mathbb{R}$, where the prime is used to distinguish this from $\partial \mathbb{H}$. For an action $\mathbf{T}$ on $\mathbb{R}'$, the three actions are denoted by

$$
\bar{\mathbb{H}} \to \bar{\mathbb{H}} : z \to z_{\bullet}\mathbf{T}, \quad \bar{\mathbb{U}} \to \bar{\mathbb{U}} : z \to z_{\circ}\mathbf{T}, \quad \mathbb{R}' \to \mathbb{R}' : z \to z_{*}\mathbf{T}.
$$

Let $\mathbf{Q}(\alpha)$ denote the rotation by α in $\mathbb{U}$ about 0, i.e.,

$$
\varphi_{*}\mathbf{Q}(\alpha) = \varphi + \alpha.
$$

For $a, b \in \mathbb{R}$, let $\mathbf{A}(a, b)$ be the affine map $z \to a(z + b)$ in $\mathbb{H}$. Furthermore, define

$$
\mathbf{W}_l = \mathbf{A} \left(\frac{1}{1 + \frac{Y_l}{s_l}}, -\frac{X_l}{s_l} \right)
$$

and

$$\mathbf{R}_{l,\Lambda} = \mathbf{Q}(\pi)\mathbf{A}\left(1, \frac{\Lambda}{s_l}\right)\mathbf{W}_l, \quad 0 \le l \le n-1.$$

With these notations, the evolution of r in (3.109) becomes

$$r_{l+1} = r_l \bullet \mathbf{R}_{l,\Lambda}, \quad 0 \le l \le n-1$$

and Λ is an eigenvalue if and only if

$$\infty \bullet \mathbf{R}_{0,\Lambda} \cdots \mathbf{R}_{n-1,\Lambda} = 0.$$

For $0 \le l \le n$ define

$$\hat{\varphi}_{l,\Lambda} = \pi_* \mathbf{R}_{0,\Lambda} \cdots \mathbf{R}_{l-1,\Lambda}, \quad \hat{\varphi}_{l,\Lambda}^{\odot} = 0_* \mathbf{R}_{n-1,\Lambda}^{-1} \cdots \mathbf{R}_{l,\Lambda}^{-1},$$

then

$$\hat{\varphi}_{l,\Lambda} = \hat{\varphi}_{l,\Lambda}^{\odot} \mod 2\pi.$$

The following lemma summarizes nice properties about $\hat{\varphi}$ and $\hat{\varphi}^{\odot}$, whose proof can be found in Valkó and Virág (2012).

Lemma 3.12. *With the above notations, we have*
(i) $r_{l,\Lambda} \bullet \mathbf{U} = e^{\mathbf{i}\hat{\varphi}_{l,\Lambda}}$;
(ii) $\hat{\varphi}_{0,\Lambda} = \pi, \quad \hat{\varphi}_{n,\Lambda}^{\odot} = 0$;
(iii) for each $0 < l \le n$, $\hat{\varphi}_{l,\Lambda}$ is an analytic and strictly increasing in Λ. For $0 \le l < n$, $\hat{\varphi}_{l,\Lambda}^{\odot}$ is analytic and strictly decreasing in Λ;
(iv) for any $0 \le l \le n$, Λ is an eigenvalue of $H_{n,\beta}$ if and only if $\hat{\varphi}_{l,\Lambda} - \hat{\varphi}_{l,\Lambda}^{\odot} \in 2\pi\mathbb{Z}$.

Fix $-2 < x < 2$ and $n_0 = n(1 - x^2/4) - 1/2$. Let $\Lambda = x\sqrt{n} + \lambda/2\sqrt{n_0}$ and recycle the notation $r_{l,\lambda}$, $\hat{\varphi}_{l,\lambda}$, $\hat{\varphi}_{l,\lambda}^{\odot}$ for the quantities $r_{l,\Lambda}$, $\hat{\varphi}_{l,\Lambda}$, $\hat{\varphi}_{l,\Lambda}^{\odot}$.

Note that there is a macroscopic term $\mathbf{Q}(\pi)\mathbf{A}(1, \Lambda/s_l)$ in the evolution operator $\mathbf{R}_{l,\Lambda}$. So the phase function $\varphi_{l,\Lambda}$ exhibits fast oscillation in l. Let

$$\mathbf{J}_l = \mathbf{Q}(\pi)\mathbf{A}\left(1, \frac{x\sqrt{n}}{s_l}\right)$$

and

$$\rho_l = \sqrt{\frac{nx^2/4}{nx^2/4 + n_0 - l}} + \mathbf{i}\sqrt{\frac{n_0 - l}{nx^2/4 + n_0 - l}}.$$

Thus $\mathbf{J}_l$ is a rotation since $\rho_l \bullet \mathbf{J}_l = \rho_l$. We separate $\mathbf{J}_l$ from the evolution operator $\mathbf{R}$ to get

$$\mathbf{R}_{l,\lambda} = \mathbf{J}_l \mathbf{L}_{l,\lambda} \mathbf{W}_l, \quad \mathbf{L}_{l,\lambda} = \mathbf{A}\left(1, \frac{\lambda}{2\sqrt{n_0}s_l}\right).$$

Note that for any finite λ, $\mathbf{L}_{l,\lambda}$ and $\mathbf{W}_l$ become infinitesimal in the $n \to \infty$ limit while $\mathbf{J}_l$ does not. Let

$$\mathbf{T}_l = \mathbf{A}\left(\frac{1}{Im(\rho_l)}, -Re(\rho_l)\right),$$

then

$$\mathbf{J}_l = \mathbf{Q}(-2\arg(\rho_l))^{\mathbf{T}_l^{-1}}$$

where $A^B = B^{-1}AB$. Define

$$\mathbf{Q}_l = \mathbf{Q}(2\arg(\rho_0))\ldots\mathbf{Q}(2\arg(\rho_l))$$

and

$$\varphi_{l,\lambda} = \hat{\varphi}_{l,\lambda*}\mathbf{T}_l\mathbf{Q}_{l-1}, \quad \varphi_{l,\lambda}^{\odot} = \hat{\varphi}_{l,\lambda*}^{\odot}\mathbf{T}_l\mathbf{Q}_{l-1}.$$

The following lemma is a variant of Lemma 3.12.

Lemma 3.13. *With the above notations, we have for $1 \le l \le n-1$*
(i') $\varphi_{0,\lambda} = \pi$;
(ii') $\varphi_{l,\lambda}$ and $-\varphi_{l,\lambda}^{\odot}$ are analytic and strictly increasing in λ and are also independent;
(iii') with $\mathbf{S}_{l,\lambda} = \mathbf{T}_l^{-1}\mathbf{L}_\lambda\mathbf{W}_\lambda\mathbf{T}_{l+1}$ and $\eta_l = \rho_0^2\rho_1^2\cdots\rho_l^2$, we have

$$\Delta\varphi_{l,\lambda} := \varphi_{l+1,\lambda} - \varphi_{l,\lambda} = ash(\mathbf{S}_{l,\lambda}, -1, e^{i\varphi_{l,\lambda}}\bar{\eta}_l);$$

(iv') $\hat{\varphi}_{l,\lambda} = \varphi_{l,\lambda*}\mathbf{Q}_{l-1}^{-1}\mathbf{T}_l^{-1}$;
(v') for any $\lambda < \lambda'$ we have a.s.

$$N_n\left(x\sqrt{n} + \frac{\lambda}{2\sqrt{n_0}}, \quad x\sqrt{n} + \frac{\lambda'}{2\sqrt{n_0}}\right)$$
$$= \#((\varphi_{l,\lambda} - \varphi_{l,\lambda}^{\odot}, \quad \varphi_{l,\lambda'} - \varphi_{l,\lambda'}^{\odot}] \cap 2\pi\mathbb{Z}). \tag{3.110}$$

The difference $\Delta\varphi_{l,\lambda}$ can be estimated as follows. Let

$$Z_{l,\lambda} = \mathbf{i}_\bullet\mathbf{S}_{l,\lambda}^{-1} - \mathbf{i}$$
$$= \mathbf{i}_\bullet\mathbf{T}_{l+1}^{-1}(\mathbf{L}_\lambda\mathbf{W}_\lambda)^{-1}\mathbf{T}_l - \mathbf{i}$$
$$= v_{l,\lambda} + V_l,$$

where

$$v_{l,\lambda} = -\frac{\lambda}{2\sqrt{n_0}\sqrt{n_0-l}} + \frac{\rho_{l+1}-\rho_l}{Im(\rho_l)}, \quad V_l = \frac{X_l + \rho_{l+1}Y_l}{\sqrt{n_0-l}}.$$

Then

$$\Delta\varphi_{l,\lambda} = ash(\mathbf{S}_{l,\lambda}, -1, z\bar{\eta})$$

$$= Re\left[-(1+\bar{z}\eta)Z - \frac{\mathbf{i}(1+\bar{z}\eta)^2}{4}Z^2\right] + O(Z^3)$$

$$= -ReZ + \frac{ImZ^2}{4} + \eta \text{ terms} + O(Z^3),$$

where we used $Z = Z_{l,\lambda}$, $\eta = \eta_l$ and $z = e^{\mathbf{i}\varphi_{l,\lambda}}$.

Lemma 3.14. *Assume* $\lambda = \lambda_n = o(\sqrt{n})$. *For* $l \le n_0$, *we have*

$$E\big(\Delta\varphi_{l,\lambda}|\varphi_{l,\lambda} = x\big) = \frac{1}{n_0}b_n + \frac{1}{n_0}osc_1 + O\big((n_0 - l)^{-3/2}\big), \quad (3.111)$$

$$E\big((\Delta\varphi_{l,\lambda})^2|\varphi_{l,\lambda} = x\big) = \frac{1}{n_0}a_n + \frac{1}{n_0}osc_2 + O\big((n_0 - l)^{-3/2}\big), \quad (3.112)$$

$$E\big(|\Delta\varphi_{l,\lambda}|^d|\varphi_{l,\lambda}\big) = O\big((n_0 - l)^{-3/2}\big), \quad d > 2, \quad (3.113)$$

where

$$b_n = \frac{\sqrt{n_0}\lambda}{2\sqrt{n_0 - l}} - \frac{Re\frac{d\rho}{dt}}{Im\rho} + \frac{n_0 Im(\rho^2)}{2\beta\sqrt{n_0 - l}},$$

$$a_n = \frac{2n_0}{\beta(n_0 - l)} + \frac{n_0(3 + Re\rho^2)}{\beta(n_0 - l)}.$$

The oscillatory terms are

$$osc_1 = Re\Big((-v_\lambda - \mathbf{i}\frac{q}{2})e^{-\mathbf{i}x}\eta_l\Big) + \frac{1}{4}Re\big(\mathbf{i}e^{-2\mathbf{i}x}\eta_l^2 q\big),$$

$$osc_2 = p_n Re(e^{-\mathbf{i}x}\eta_l) + Re\Big[q_n(e^{-\mathbf{i}x}\eta_l + \frac{1}{2}e^{-\mathbf{i}2x}\eta_l^2)\Big],$$

where

$$p_n = \frac{4n_0}{\beta(n_0 - l)}, \qquad q_n = \frac{2n_0(1 + \rho_l)}{\beta(n_0 - l)}.$$

Lemma 3.15. *We have for* $0 < l \le n_0$
(i)

$$\hat{\phi}_{l,\infty} = \pi, \qquad \hat{\phi}_{l,\infty}^{\odot} = -2(n - l)\pi;$$

(ii)

$$\phi_{l,\infty} = (l + 1)\pi, \qquad \phi_{l,\infty}^{\odot} = -2n\pi + 3l\pi;$$

(iii)

$$\phi_{l,0}^{\odot} = \hat{\phi}_{l,0}^{\odot} + l\pi, \qquad \hat{\phi}_{l,0}^{\odot} = 0_*\mathbf{R}_{n-1}^{-1}\cdots\mathbf{R}_l^{-1}.$$

Proof. First, prove (i). Recall

$$\hat{\varphi}_{l,\infty} = \pi_* \mathbf{R}_{0,\infty} \cdots \mathbf{R}_{l-1,\infty}, \quad \hat{\varphi}^{\ominus}_{l,\infty} = 0_* \mathbf{R}^{-1}_{n-1,\infty} \cdots \mathbf{R}^{-1}_{l,\infty},$$

where $\mathbf{R}_{l,\infty} = \mathbf{Q}(\pi) \mathbf{A}(1, \infty) \mathbf{W}_l$.

Note the affine transformation $\mathbf{A}(1, \infty) \mathbf{W}_l$ maps any z to ∞, and the image of ∞ under the Möbius transform $\mathbf{U}$ is -1, which in turn corresponds to $\pi \in \mathbb{R}'$. Thus it easily follows $\hat{\phi}_{l,\infty} = \pi$.

As for $\hat{\phi}^{\ominus}_{l,\infty}$, note

$$\hat{\phi}^{\ominus}_{l,\infty} = \hat{\phi}^{\ominus}_{l+1,\infty *} \mathbf{R}^{-1}_{l,\infty},$$

where $\mathbf{R}^{-1}_{l,\infty} = \mathbf{W}^{-1}_l \mathbf{A}(1, -\infty) \mathbf{Q}(-\pi)$.

By the angular shift formula, we have

$$\hat{\phi}^{\ominus}_{l,\infty} = \hat{\phi}^{\ominus}_{l+1,\infty *} \mathbf{W}^{-1}_l \mathbf{A}(1, -\infty) - \pi$$

$$= \hat{\phi}^{\ominus}_{l+1,\infty} + ash\left(\mathbf{W}^{-1}_l \mathbf{A}(1, -\infty), -1, e^{\mathbf{i}\hat{\phi}^{\ominus}_{l+1,\infty}}\right) - \pi$$

$$= \hat{\phi}^{\ominus}_{l+1,\infty} + \arg_{[0,2\pi)}\left(\frac{e^{\mathbf{i}\hat{\phi}^{\ominus}_{l+1,\infty}} \circ \mathbf{W}^{-1}_l \mathbf{A}(1, -\infty)}{-1 \circ \mathbf{W}^{-1}_l \mathbf{A}(1, -\infty)}\right)$$

$$- \arg_{[0,2\pi)}\left(\frac{e^{\mathbf{i}\hat{\phi}^{\ominus}_{l+1,\infty}}}{-1}\right) - \pi$$

$$= \hat{\phi}^{\ominus}_{l+1,\infty} + \arg_{[0,2\pi)}(1) - \arg_{[0,2\pi)}(-e^{\mathbf{i}\hat{\phi}^{\ominus}_{l+1,\infty}}) - \pi,$$

from which and the fact $\hat{\phi}^{\ominus}_{n,\infty} = 0$ we can easily derive

$$\hat{\phi}^{\ominus}_{n-1,\infty} = -2\pi, \quad \hat{\phi}^{\ominus}_{n-2,\infty} = -4\pi, \cdots, \hat{\phi}^{\ominus}_{l,\infty} = -2(n-l)\pi.$$

Next we turn to the proof of (ii) and (iii). Since $x = 0$, then $\rho_l = \mathbf{i}$, and so $\mathbf{T}_l$ is the identity transform and $\mathbf{Q}_{l-1} = \mathbf{Q}(l\pi)$ for each $0 < l \leq n_0$. Thus we have by the fact that $\mathbf{Q}$ is a rotation,

$$\varphi_{l,\lambda} = \hat{\varphi}_{l,\lambda *} \mathbf{T}_l \mathbf{Q}_{l-1} \tag{3.114}$$

$$= \hat{\varphi}_{l,\lambda *} \mathbf{Q}(l\pi)$$

$$= \hat{\varphi}_{l,\lambda} + l\pi,$$

where $0 \leq \lambda \leq \infty$. Similarly, $\varphi^{\ominus}_{l,\lambda} = \hat{\varphi}^{\ominus}_{l,\lambda} + l\pi$. $\square$

Proof of Theorem 3.12. Let $x = 0$ and $n_0 = n - 1/2$. Taking $l = \lfloor n_0 \rfloor = n - 1$, we have by (3.110)

$$N_n(0, \infty) = \sharp\left(\left(\varphi_{n-1,0} - \varphi^{\ominus}_{n-1,0}, \quad \varphi_{n-1,\infty} - \varphi^{\ominus}_{n-1,\infty}\right) \cap 2\pi\mathbb{Z}\right),$$

from which it readily follows

$$\left| N_n(0,\infty) - \frac{1}{2\pi}\left(\varphi_{n-1,\infty} - \varphi^{\odot}_{n-1,\infty} - (\varphi_{n-1,0} - \varphi^{\odot}_{n-1,0})\right) \right| \leq 1.$$

Applying Lemma 3.15 to $l = n - 1$ immediately yields

$$\varphi_{n-1,\infty} - \varphi^{\odot}_{n-1,\infty} = 3\pi$$

and

$$\varphi^{\odot}_{n-1,0} = 0_* \mathbf{R}^{-1}_{n-1,0} + (n-1)\pi.$$

Also, it follows

$$\frac{0_* \mathbf{R}^{-1}_{n-1,0}}{\sqrt{\log n}} \xrightarrow{P} 0.$$

In combination, we need only to prove

$$\frac{\varphi_{n-1,0}}{\sqrt{\log n}} \xrightarrow{d} N\left(0, \frac{4}{\beta}\right).$$

To this end, we shall use the following CLT for Markov chain. Recall that $\pi = \varphi_{0,0}, \varphi_{1,0}, \cdots, \varphi_{n-1,0}$ forms a Markov chain. Let

$$z_{l+1} = \Delta\varphi_{l,0} - E(\Delta\varphi_{l,0}|\varphi_{l,0}).$$

Then $z_1, z_2, \cdots, z_{n-1}$ forms a martingale difference sequence. The martingale CLT implies: if the following conditions are satisfied:
(i)

$$B_n := \sum_{l=1}^{n-1} E z_l^2 \to \infty, \tag{3.115}$$

(ii)

$$\frac{1}{B_n} \sum_{l=1}^{n-1} E\left(z_l^2|\varphi_{l-1,0}\right) \xrightarrow{P} 1, \tag{3.116}$$

(iii) for any $\varepsilon > 0$

$$\frac{1}{B_n} \sum_{l=1}^{n-1} E\left(z_l^2 \mathbf{1}_{(|z_l|>\varepsilon\sigma_n)}|\varphi_{l-1,0}\right) \xrightarrow{P} 0, \tag{3.117}$$

then we have

$$\frac{1}{\sqrt{B_n}} \sum_{l=1}^{n-1} z_l \xrightarrow{d} N(0,1).$$

We will next verify conditions (3.115) - (3.117) by asymptotic estimates (3.111) - (3.113) for the increments $E(\Delta\varphi_{l,0}|\varphi_{l,0})$. Start with B_n. Note

$$E(\Delta\varphi_{l,0}|\varphi_{l,0}) = O\big((n_0 - l)^{-3/2}\big)$$

$$\begin{aligned}
E(\Delta\varphi_{l,0})^2 &= E\big(E((\Delta\varphi_{l,0})^2|\varphi_{l,0})\big) \\
&= \frac{4}{\beta(n_0 - l)} + \frac{4}{n_0 - l} E\mathrm{Re}\big((-1)^{l+1}e^{-i\varphi_{l,0}}\big) \\
&\quad + O\big((n_0 - l)^{-3/2}\big).
\end{aligned}$$

Hence a direct computation yields

$$\begin{aligned}
B_n &= \sum_{l=1}^{n-1} E(\Delta\varphi_{l,0})^2 - E(E(\Delta\varphi_{l,0}|\varphi_{l,0}))^2 \\
&= \frac{4}{\beta}\log n + O(1) \to \infty
\end{aligned}$$

and

$$\frac{1}{B_n}\sum_{l=1}^{n-1} E\big(z_l^2|\varphi_{l-1,0}\big) - 1$$

$$= \frac{1}{B_n}\sum_{l=1}^{n-1}\big(E(z_l^2|\varphi_{l-1,0}) - Ez_l^2\big)$$

$$= \frac{1}{B_n}\sum_{l=1}^{n-1}\big(E((\Delta\varphi_{l-1,0})^2|\varphi_{l-1,0}) - E(\Delta\varphi_{l-1,0})^2\big)$$

$$+ \frac{1}{B_n}\sum_{l=1}^{n-1} E\big(E(\Delta\varphi_{l-1,0}|\varphi_{l-1,0})\big)^2 - \big(E(\Delta\varphi_{l-1,0}|\varphi_{l-1,0})\big)^2$$

$$\xrightarrow{P} 0.$$

It also follows from (3.113) that

$$\sum_{l=1}^{n-1} E|\Delta\varphi_{l-1,0}|^3 = O(1),$$

which in turn immediately implies the Lindeberg condition (3.117). Thus we have completed the proof of the theorem. $\square$

Chapter 4

Random Uniform Partitions

4.1 Introduction

The theory of partitions is one of the very few branches of mathematics that can be appreciated by anyone who is endowed with little more than a lively interested in the subject. Its applications are found wherever discrete objects are to be counted or classified, whether in the molecular and the atomic studies of matter, in the theory of numbers, or in combinatorial problems from all sources.

Let n be a natural number. A partition of n is a finite nonincreasing sequence of positive integers $\lambda_1 \geq \lambda_2 \geq \cdots \geq \lambda_l > 0$ such that $\sum_{j=1}^{l} \lambda_j = n$. Set

$$r_k = \#\{1 \leq j \leq l : \lambda_j = k\}.$$

Trivially,

$$\sum_{k=1}^{\infty} r_k = l, \quad \sum_{k=1}^{\infty} k r_k = n.$$

As we remarked in Section 2.2, there is a close connection between partitions and permutations.

The set of all partitions of n are denoted by $\mathcal{P}_n$, and the set of all partitions by $\mathcal{P}$, i.e., $\mathcal{P} = \cup_{n=0}^{\infty} \mathcal{P}_n$. Here by convention the empty sequence forms the only partition of zero. Among the most important and fundamental is the question of enumerating various set of partitions. Let $p(n)$ be the number of partitions of n. Trivially, $p(0) = 0$, and $p(n)$ increases quite rapidly with n. In fact, $p(10) = 42$, $p(20) = 627$, $p(50) = 204226$, $p(100) = 190569292$, $p(200) = 3972999029388$.

The study of $p(n)$ dates back to Euler as early as in the 1750s, who proved many beautiful and significant partition theorems, and so laid the foundations of the theory of partitions. Many of the other great mathematicians have contributed to the development of the theory. The reader is referred to Andrews (1976), which is a first thorough survey of this field with many informative historic notes.

It turns out that generating function is a powerful tool for studying $p(n)$. Define the generating function of the $p(n)$ by

$$\mathcal{F}(z) = \sum_{n=0}^{\infty} p(n) z^n. \tag{4.1}$$

Euler started the analytic theory of partitions by providing the explicit formula

$$\mathcal{F}(z) = \prod_{k=1}^{\infty} \frac{1}{1 - z^k}. \tag{4.2}$$

We remark that on the one hand, for many problems it suffices to consider $\mathcal{F}(z)$ as a formal power series in z; on the other hand, much asymptotic work requires that $\mathcal{F}(z)$ be an analytic function of the complex variable z.

The asymptotic theory starts 150 years after Euler, with the first celebrated letters of Ramanujan to Hardy in 1913. In a celebrated series of memoirs published in 1917 and 1918, Hardy and Ramanujan found (and was perfected by Radamacher) very precise estimates for $p(n)$. In particular, we have

Theorem 4.1.

$$p(n) = \frac{1}{4\sqrt{3}\, n} e^{2c\sqrt{n}} \big(1 + o(1)\big), \tag{4.3}$$

where and in the sequel $c = \pi/\sqrt{6}$.

The complete proof of Theorem 4.1 can be found in §2.7 of the book Postnikov (1988). Instead, we prefer to give a rough sketch of the proof, without justifying anything. First, using the Cauchy integral formula for the coefficients of a power series, we obtain from (4.1) and (4.2)

$$p(n) = \frac{1}{2\pi} \int_{-\pi}^{\pi} r^{-n} e^{-in\theta} \mathcal{F}\big(re^{i\theta}\big) d\theta.$$

Choose $\theta_n > 0$ and split the integral expression for $p(n)$ into two parts:

$$p(n) = \frac{1}{2\pi} \bigg(\int_{|\theta| \le \theta_n} + \int_{|\theta| > \theta_n} \bigg) r^{-n} e^{-in\theta} \mathcal{F}\big(re^{i\theta}\big) d\theta.$$

If $\theta_n = n^{-3/4+\varepsilon}$, $\varepsilon > 0$, then it holds

$$p(n) \approx \frac{1}{2\pi} \int_{|\theta| \le \theta_n} r^{-n} e^{-in\theta} \mathcal{F}(re^{i\theta}) d\theta.$$

By Taylor's formula with two terms, for $|\theta| \le \theta_n$ we have

$$\log \mathcal{F}(re^{i\theta}) \approx \log \mathcal{F}(r) + i\theta \big(\log \mathcal{F}(r)\big)' - \frac{\theta^2}{2} \big(\log \mathcal{F}(r)\big)'',$$

and so

$$p(n) \approx \frac{\mathcal{F}(r)}{2\pi r^n} \int_{|\theta| \le \theta_n} e^{-i\theta(n - (\log \mathcal{F}(r))')} e^{-\theta^2 (\log \mathcal{F}(r))''/2} d\theta.$$

Up to this point, r has been a free parameter. We now choose r so that $n + \big(\log \mathcal{F}(r)\big)' = 0$, i.e., we must choose $r = e^{-v}$ to satisfy the equation

$$n = \sum_{k=1}^{\infty} \frac{k}{e^{kv} - 1}.$$

Note

$$\sum_{k=1}^{\infty} \frac{k}{e^{kv} - 1} = \frac{1}{v^2} \sum_{k=1}^{\infty} v \frac{kv}{e^{kv} - 1}$$

$$\approx \frac{1}{v^2} \int_0^{\infty} \frac{x}{e^x - 1} dx$$

$$= \frac{c^2}{v^2}.$$

Thus we must take v so that $n \approx c^2/v^2$, i.e., $v \approx c/\sqrt{n}$. For such a choice,

$$p(n) \approx \frac{\mathcal{F}(r)}{2\pi r^n} \int_{|\theta| \le \theta_n} e^{-\theta^2 (\log \mathcal{F}(r))''/2} d\theta$$

$$\approx \frac{\mathcal{F}(r)}{2\pi r^n} \int_{-\infty}^{\infty} e^{-\theta^2 (\log \mathcal{F}(r))''/2} d\theta$$

$$= \frac{\mathcal{F}(r)}{r^n \sqrt{2\pi \log(\mathcal{F}(r))''}}, \tag{4.4}$$

using the classical normal integral. To evaluate the value of (4.4), we need the following lemma, see Postnikov (1988).

Lemma 4.1. *Assume* $Re z > 0$ *and* $z \to 0$, *staying within some angle lying in the right half-plane. Then*

$$\log \mathcal{F}(e^{-z}) = \frac{c^2}{z} + \frac{1}{2} \log \frac{z}{2\pi} + O(|z|). \tag{4.5}$$

As a consequence, it easily follows with $r = e^{-c/\sqrt{n}}$

$$\log \mathcal{F}(r) \approx c\sqrt{n} + \frac{1}{4}\log\frac{1}{24n} \qquad (4.6)$$

and

$$(\log \mathcal{F}(r))'' \approx \frac{2}{c}n^{3/2}. \qquad (4.7)$$

Substituting (4.6) and (4.7) into (4.4) yields the desired result (4.3).

Another effective elementary device for studying partitions is graphical representation. To each partition λ is associated its Young diagram (shape), which can be formally defined as the set of point $(i, j) \in \mathbb{Z}^2$ such that $1 \le j \le \lambda_i$. In drawing such diagrams, by convention, the first coordinate i (the row index) increases as one goes downwards, the second coordinate j (the column index) increases as one goes from the left to the right and these points are left justified. More often it is convenient to replace the nodes by unit squares, see Figure 2.1.

Such a representation is extremely useful when we consider applications of partitions to plane partitions or Young tableaux. Sometimes we prefer the representation to be upside down in consistency with Descartes coordinate geometry.

The conjugate of a partition λ is the partition λ' whose diagram is the transpose of the diagram λ, i.e., the diagram obtained by reflection in the main diagonal. Hence the λ'_i is the number of squares in the ith column of λ, or equivalently,

$$\lambda'_i = \sum_{k=i}^{\infty} r_k. \qquad (4.8)$$

In particular, $\lambda'_1 = l(\lambda)$ and $\lambda_1 = l(\lambda')$. Obviously, $\lambda'' = \lambda$.

We have so far defined the set $\mathcal{P}_n$ of partitions of n and known how to count its size $p(n)$. Now we want to equip a probability measure on this set. As we will see, this set bears many various natural measures. The first natural measure is certainly uniform, i.e., choose at random a partition with equal probability. Let $P_{u,n}$ be the uniform measure defined by

$$P_{u,n}(\lambda) = \frac{1}{p(n)}, \quad \lambda \in \mathcal{P}_n, \qquad (4.9)$$

where the subscript u stands for Uniform. The primary goal of this chapter is to study the asymptotic behaviours of a typical partition under $(\mathcal{P}_n, P_{u,n})$ as its size $n \to \infty$. The first remarkable feature is that a typical Young diagram properly scaled has a limit shape. To be precise, define

$$\varphi_\lambda(t) = \sum_{k \ge t} r_k, \quad t \ge 0. \qquad (4.10)$$

In particular, $\varphi_\lambda(i) = \lambda'_i$, and $\varphi_\lambda(t)$ is a nonincreasing step function such that $\int_0^\infty \varphi_\lambda(t)dt = n$.

Theorem 4.2. *Under* $(\mathcal{P}_n, P_{u,n})$ *we have as* $n \to \infty$

$$\sup_{a \le t \le b} \left| \frac{1}{\sqrt{n}} \varphi_\lambda(\sqrt{n}t) - \Psi(t) \right| \xrightarrow{P} 0 \qquad (4.11)$$

where $0 < a < b < \infty$ *and*

$$\Psi(t) = \int_t^\infty \frac{e^{-cu}}{1 - e^{-cu}} du = -\frac{1}{c} \log(1 - e^{-ct}). \qquad (4.12)$$

We remark that the curve $\Psi(t)$ was first conjectured by Temperley (1952), who studied the number of ways in which a given amount of energy can be shared out among the different possible states of an assembly. The rigorous argument was given by Vershik (1994, 1996). In fact, Vershik and his school has been recognized to be the first group who started a systematic study of limit shapes of various random geometric objects. We also note that the curve $\Psi(t)$ has two asymptotic lines: $s = 0$ and $t = 0$, see Figure 4.1.

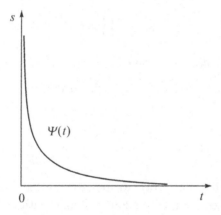

Fig. 4.1   Temperley-Vershik curve

From the probabilistic viewpoint, Theorem 4.2 is a kind of weak law of large numbers. Next, it is natural to ask what the fluctuation is of a typical Young diagram around the limit shape. That is the problem of the second order fluctuation. It turns out that we need to deal with two cases separately: at the edge and in the bulk. Let us first treat the edge case of $\varphi_\lambda(k)$, $k \ge 0$.

Note that λ and λ' have the same likelihood, then it follows by duality

$$\varphi_\lambda(k) = \lambda'_k \overset{d}{=} \lambda_k, \quad k \ge 1.$$

Hence it is sufficient to study the asymptotic distribution of the λ_k's, the largest parts of λ. Let us begin with the following deep and interesting result, due to Erdös and Lehner (1941).

Theorem 4.3. *As $n \to \infty$, we have for each $x \in \mathbb{R}$*

$$P_{u,n}\left(\frac{c}{\sqrt{n}}\lambda_1 - \log\frac{\sqrt{n}}{c} \le x\right) \longrightarrow e^{-e^{-x}}. \tag{4.13}$$

Note that the limit distribution in the right hand side of (4.13) is the famous Gumbel distribution, which appears widely in the study of extremal statistics for independent random variables.

Besides, one can further consider the jointly asymptotic distributions of the first m largest parts. Fristedt (1993) obtained the following

Theorem 4.4. *As $n \to \infty$, we have for $x_1 > x_2 > \cdots > x_m$*

$$\lim_{n\to\infty} P_{u,n}\left(\frac{c}{\sqrt{n}}\lambda_i - \log\frac{\sqrt{n}}{c} \le x_i, \quad 1 \le i \le m\right)$$
$$= \int_{-\infty}^{x_1} \cdots \int_{-\infty}^{x_m} p_0(x_1) \prod_{i=2}^{m} p(x_{i-1}, x_i) dx_1 \cdots dx_m, \tag{4.14}$$

where p_0 and p are defined as follows

$$p_0(x) = e^{-e^{-x}-x}, \quad x \in \mathbb{R}$$

and

$$p(x,y) = \begin{cases} e^{e^{-x}-e^{-y}-y}, & x > y, \\ 0, & x \le y. \end{cases}$$

To understand the limit (4.14), we remark the following nice fact. Let $\eta_1, \eta_2, \cdots$ be a sequence of random variables with (4.14) as their joint distribution functions, then for $x_1 > x_2 > \cdots > x_m$

$$P\left(\eta_m = x_m | \eta_{m-1} = x_{m-1}, \cdots, \eta_1 = x_1\right) = p(x_{m-1}, x_m).$$

Hence the η_k's form a Markov chain with $p(x,y)$ as the transition density.

Next, let us turn to the bulk case, i.e., treat $\varphi_\lambda(\sqrt{n}t)$ where $0 < t < \infty$. Define

$$X_n(t) = \frac{1}{n^{1/4}}\left(\varphi_\lambda(\sqrt{n}t) - \sqrt{n}\Psi(t)\right), \quad t > 0.$$

An interesting result is the following central limit theorem due to Pittel (1997).

Theorem 4.5. *Under* $(\mathcal{P}_n, P_{u,n})$ *we have as* $n \to \infty$

$$X_n \Rightarrow X \tag{4.15}$$

in terms of finite dimensional distributions. Here $X(t), t > 0$ *is a centered Gaussian process with the covariance structure*

$$Cov\big(X(t_1), X(t_2)\big) = \sigma_{t_2}^2 - s_{t_1} s_{t_2}, \quad t_1 \le t_2 \tag{4.16}$$

where

$$\sigma_t^2 = \int_t^\infty \frac{e^{-cu}}{(1 - e^{-cu})^2} du = \frac{e^{-ct}}{c(1 - e^{-ct})}$$

and

$$\begin{aligned}
s_t &= \int_t^\infty \frac{u e^{-cu}}{(1 - e^{-cu})^2} du \\
&= \frac{t e^{-ct}}{(1 - e^{-ct})} - \frac{1}{c^2} \log\big(1 - e^{-ct}\big).
\end{aligned}$$

What happens if $t = t_n$ tends to infinity? It turns out that a similar central limit theorem holds when t_n grows slowly. In particular, we have

Theorem 4.6. *Assume* $t_n, n \ge 1$ *is a sequence of positive numbers such that*

$$t_n \to \infty, \quad t_n - \frac{1}{2c} \log n \to -\infty. \tag{4.17}$$

Let

$$X_n(t_n) = \frac{e^{ct_n/2}}{n^{1/4}} \big(\varphi_\lambda(\sqrt{n} t_n) - \sqrt{n} \Psi(t_n)\big), \quad t > 0$$

then under $(\mathcal{P}_n, P_{u,n})$, *as* $n \to \infty$

$$X_n(t_n) \xrightarrow{d} N(0, 1). \tag{4.18}$$

Note that $e^{ct_n/2}/n^{1/4}$ goes to zero under the assumption (4.17). We have so far seen many interesting probability limit theorems for random uniform partitions. In the next two sections we shall provide rigorous proofs. A basic strategy is as follows. First, we will in Section 4.2 construct a larger probability space $(\mathcal{P}, Q_q)$ where $0 < q < 1$ is a model parameter, under which the r_k's are independent geometric random variables. Thus we can directly apply the classical limit theory to the partial sums $\sum_k r_k$. Second, we will in Section 4.3 transfer to the desired space $(\mathcal{P}_n, P_{u,n})$ using the fact $P_{u,n}$ is essentially the restriction of Q_q to $\mathcal{P}_n$. It is there that we develop a conditioning argument, which is consistent with the so-called transition between grand and small ensembles in physics literatures.

Having the convergence of finite dimensional distributions, one might expect a weak convergence of the processes $(X_n(t), t > 0)$. To this end, it is required to check the uniform tightness condition, i.e., for any $\varepsilon > 0$

$$\lim_{\delta \to 0} \lim_{n \to \infty} P_{u,n}\left(\sup_{|t-s| \leq \delta} |X_n(t) - X_n(s)| > \varepsilon \right) = 0.$$

However, we are not able to find a good way to verify such a condition. Instead, we shall in Section 4.4 state and prove a weaker stochastic equicontinuity condition: for any $\varepsilon > 0$

$$\lim_{\delta \to 0} \lim_{n \to \infty} \sup_{|t-s| \leq \delta} P_{u,n}\big(|X_n(t) - X_n(s)| > \varepsilon\big) = 0.$$

This together with Theorem 1.16 immediately implies a functional central limit theorem holds for a certain class of integral statistics of $X_n(t)$. We shall also give two examples at the end of Section 4.4. To conclude this chapter, we shall briefly discuss a generalized multiplicative random partitions induced by a family of analytic functions.

Throughout the chapter, $c_1, c_2, \cdots$ denote positive numeric constants, whose exact values are not of importance.

4.2   Grand ensembles

In this section we shall study unrestricted random partitions with multiplicative measures. Let $0 < q < 1$, define the multiplicative measure by

$$Q_q(\lambda) = \frac{q^{|\lambda|}}{\mathcal{F}(q)}, \quad \lambda \in \mathcal{P}, \tag{4.19}$$

where $|\lambda|$ denotes the size of the partition λ.

Note by (4.19) and (4.1)

$$\sum_{\lambda \in \mathcal{P}} Q_q(\lambda) = \frac{1}{\mathcal{F}(q)} \sum_{\lambda \in \mathcal{P}} q^{|\lambda|}$$

$$= \frac{1}{\mathcal{F}(q)} \sum_{n=0}^{\infty} \sum_{\lambda \in \mathcal{P}_n} q^n = 1.$$

Thus we can induce a probability space $(\mathcal{P}, Q_q)$, which is called a grand ensemble with parameter q. Surprisingly, this Q_q has an elegant property as follows.

Lemma 4.2. *Under* $(\mathcal{P}, Q_q)$, $r_1, r_2, \cdots$ *is a sequence of independent geometric random variables. In particular, we have*

$$Q_q(\lambda \in \mathcal{P} : r_k = j) = (1 - q^k)q^{jk}, \quad j = 0, 1, 2, \cdots.$$

Proof. The proof is easy. Indeed, note $\lambda = (1^{r_1}, 2^{r_2}, \cdots)$, so $|\lambda| = \sum_{k=1}^{\infty} k r_k$. Thus we have

$$Q_q(\lambda) = \prod_{k=1}^{\infty} q^{k r_k}(1 - q^k),$$

as desired. $\qquad\qquad\qquad\qquad\qquad\qquad\qquad\qquad\qquad\qquad\qquad\qquad\square$

This lemma will play a fundamental important role in the study of random uniform partitions. It will enable us to apply the classical limit theorems for sums of independent random variables. Denote by E_q expectation with respect to Q_q. As a direct consequence, we have

$$E_q r_k = \frac{q^k}{1 - q^k}, \quad Var_q(r_k) = \frac{q^k}{(1 - q^k)^2}$$

and

$$E_q z^{r_k} = \frac{1 - q^k}{1 - zq^k}.$$

Under $(\mathcal{P}, Q_q)$, the size $|\lambda|$ is itself a random variable. Let $q_n = e^{-c/\sqrt{n}}$, then it is easy to see

$$\begin{aligned}
\mu_n := E_{q_n}|\lambda| &= \sum_{k=1}^{\infty} k E_{q_n} r_k \\
&= \sum_{k=1}^{\infty} \frac{k q_n^k}{1 - q_n^k} = \sum_{k=1}^{\infty} \frac{k e^{-ck/\sqrt{n}}}{1 - e^{-ck/\sqrt{n}}} \\
&= n \int_0^{\infty} \frac{u e^{-cu}}{1 - e^{-cu}} du + O(\sqrt{n}) \\
&= n + O(\sqrt{n}), \qquad\qquad\qquad\qquad\qquad (4.20)
\end{aligned}$$

where in the last step we used the fact

$$\int_0^{\infty} \frac{u e^{-cu}}{1 - e^{-cu}} du = 1.$$

Similarly,

$$\sigma_n^2 := Var_{q_n}(|\lambda|) = \sum_{k=1}^{\infty} k^2 Var_{q_n}(r_k)$$

$$= \sum_{k=1}^{\infty} \frac{k^2 q_n^k}{(1 - q_n^k)^2} = \sum_{k=1}^{\infty} \frac{k^2 e^{-ck/\sqrt{n}}}{(1 - e^{-ck/\sqrt{n}})^2}$$

$$= n^{3/2} \int_0^{\infty} \frac{u^2 e^{-cu}}{(1 - e^{-cu})^2} du + O(n)$$

$$= \frac{2}{c} n^{3/2} + O(n). \tag{4.21}$$

Fristedt (1993) obtained the following refinement.

Theorem 4.7. *Under* $(\mathcal{P}, Q_{q_n})$, $|\lambda|$ *normally concentrates around* n. *Namely,*

$$\frac{|\lambda| - n}{n^{3/4}} \xrightarrow{d} N\left(0, \frac{2}{c}\right).$$

Moreover, we have the local limit theorem

$$Q_{q_n}(|\lambda| = n) = \frac{1}{(96)^{1/4} n^{3/4}} (1 + o(1)). \tag{4.22}$$

Proof. By virtue of (4.20) and (4.21), it suffices to prove

$$\frac{|\lambda| - \mu_n}{\sigma_n} \xrightarrow{d} N(0, 1). \tag{4.23}$$

In turn, this will be proved using characteristic functions below. Let

$$f_n(x) = E_{q_n} \exp\left(\frac{ix|\lambda|}{\sigma_n}\right).$$

Then it follows from Lemma 4.2

$$f_n(x) = E_{q_n} \exp\left(\frac{ix}{\sigma_n} \sum_{k=1}^{\infty} k r_k\right)$$

$$= \prod_{k=1}^{\infty} \frac{1 - q_n^k}{1 - q_n^k e^{ikx/\sigma_n}}.$$

Observe the following elementary Taylor formulas

$$\log(1 + z) = z + \frac{z^2}{2} + O(|z|^3), \quad |z| \to 0$$

and

$$e^{ix} = 1 + ix - \frac{x^2}{2} + O(|x|^3), \quad |x| \to 0.$$

We have

$$\log\left(1 - \frac{q_n^k(e^{ikx/\sigma_n} - 1)}{1 - q_n^k}\right) = -\frac{q_n^k(e^{ikx/\sigma_n} - 1)}{1 - q_n^k} + \frac{1}{2}\frac{q_n^{2k}(e^{ikx/\sigma_n} - 1)^2}{(1 - q_n^k)^2}$$

$$+ O\left(\frac{q_n^{3k}|e^{ikx/\sigma_n} - 1|^3}{(1 - q_n^k)^3}\right)$$

$$= -\frac{ix}{\sigma_n} \cdot \frac{kq_n^k}{1 - q_n^k} + \frac{x^2}{2\sigma_n^2} \cdot \frac{k^2 q_n^k}{(1 - q_n^k)^2}$$

$$+ O\left(\frac{k^3 q_n^{3k}}{(1 - q_n^k)^3}\right).$$

Taking summation over k yields

$$\sum_{k=1}^{\infty} \log\left(1 - \frac{q_n^k(e^{ikx/\sigma_n} - 1)}{1 - q_n^k}\right) = -\frac{ix}{\sigma_n} \sum_{k=1}^{\infty} \frac{kq_n^k}{1 - q_n^k} + \frac{x^2}{2\sigma_n^2} \sum_{k=1}^{\infty} \frac{k^2 q_n^k}{(1 - q_n^k)^2}$$

$$+ O\left(\frac{1}{\sigma_n^3} \sum_{k=1}^{\infty} \frac{k^3 q_n^{3k}}{(1 - q_n^k)^3}\right). \tag{4.24}$$

It follows by (4.21),

$$\frac{1}{\sigma_n^3} \sum_{k=1}^{\infty} \frac{k^3 q_n^{3k}}{(1 - q_n^k)^3} = O\left(n^{-1/4}\right), \tag{4.25}$$

which implies that the above Taylor expansions are reasonable.

Therefore we see from (4.24) and (4.25) that

$$\log f_n(x) = -\sum_{k=1}^{\infty} \log \frac{1 - q_n^k e^{ikx/\sigma_n}}{1 - q_n^k}$$

$$= -\sum_{k=1}^{\infty} \log\left(1 - \frac{q_n^k(e^{ikx/\sigma_n} - 1)}{1 - q_n^k}\right)$$

$$= \frac{ix\mu_n}{\sigma_n} - \frac{x^2}{2} + o(1).$$

We now conclude the desired assertion (4.23).

Next we turn to the proof of (4.22). To this end, we use the inverse formula for lattice random variables to get

$$Q_{q_n}(|\lambda| = n) = Q_{q_n}\left(\frac{|\lambda|}{\sigma_n} = \frac{n}{\sigma_n}\right)$$

$$= \frac{1}{2\pi\sigma_n} \int_{-\pi\sigma_n}^{\pi\sigma_n} e^{-ixn/\sigma_n} f_n(x)dx. \tag{4.26}$$

Let $\rho(n) = \pi\sigma_n^{1/3}$. Then for $|x| < \rho(n)$,

$$\log|f_n(x)| = -\frac{1}{2}\sum_{k=1}^{\infty}\log\left(1 + \frac{2q_n^k(1 - \cos kx/\sigma_n)}{(1 - q_n^k)^2}\right)$$

$$\leq -\frac{1}{2}\sum_{\sigma_n^{2/3}/2 < k \leq \sigma_n^{2/3}}\log\left(1 + \frac{c_1 x^2}{\sigma_n^{2/3}}\right)$$

$$\leq -c_2 x^2\sum_{\sigma_n^{2/3}/2 < k \leq \sigma_n^{2/3}}\sigma_n^{-2/3}$$

$$\leq -c_3 x^2.$$

Thus we get

$$|f_n(x)| \leq e^{-c_3 x^2}, \quad |x| \leq \rho(n).$$

For $|x| > \rho(n)$, let $S_x = \left\{k : k \leq \sigma_n^{2/3}, \cos kx/\sigma_n \leq 0\right\}$. Then

$$\log|f_n(x)| \leq -\frac{1}{2}\sum_{k \in S_x}\log\left(1 + \frac{2q_n^k(1 - \cos kx/\sigma_n)}{(1 - q_n^k)^2}\right)$$

$$\leq -c_4\sigma_n^{2/3},$$

which implies

$$\sup_{\rho(n) < |x| \leq \pi\sigma_n}|f_n(x)| \leq e^{-c_4\sigma_n^{2/3}} = o\left(\sigma_n^{-1}\right). \tag{4.27}$$

Next we shall estimate the integral in the right hand side of (4.26). Split the interval $(-\pi\sigma_n, \pi\sigma_n)$ into two disjoint subsets: $\{|x| \leq \rho(n)\}$ and $\{\rho(n) < |x| \leq \pi\sigma_n\}$, and evaluate the integral value over each one. Since $e^{-\mathrm{i}xn/\sigma_n}f_n(x) \to e^{-x^2/2}$, then by the control convergence theorem,

$$\int_{|x| \leq \rho(n)} e^{-\mathrm{i}xn/\sigma_n}f_n(x)dx \longrightarrow \int_{-\infty}^{\infty} e^{-x^2/2}dx = \sqrt{2\pi}.$$

Also, we have by (4.27)

$$\int_{\rho(n) < |x| \leq \pi\sigma_n} e^{-\mathrm{i}xn/\sigma_n}f_n(x)dx = o(1).$$

In combination, we get the desired assertion. $\qquad\square$

Theorem 4.8. *Under* $(\mathcal{P}, Q_{q_n})$ *we have as* $n \to \infty$

$$\sup_{a \leq t \leq b}\left|\frac{1}{\sqrt{n}}\varphi_\lambda(\sqrt{n}t) - \Psi(t)\right| \xrightarrow{P} 0 \tag{4.28}$$

where $0 < a < b < \infty$.

Proof. We first prove the convergence in (4.28) for each fixed $t > 0$. Indeed, it follows by (4.20)

$$E_{q_n} \frac{1}{\sqrt{n}} \varphi_\lambda(\sqrt{n}t) = \frac{1}{\sqrt{n}} \sum_{k \geq \sqrt{n}t} E_{q_n} r_k = \frac{1}{\sqrt{n}} \sum_{k \geq \sqrt{n}t} \frac{q_n^k}{1 - q_n^k}$$
$$= \int_t^\infty \frac{e^{-cu}}{1 - e^{-cu}} du + o(1) = \Psi(t) + o(1).$$

Similarly, by (4.21)

$$Var_{q_n} \left(\frac{1}{\sqrt{n}} \varphi_\lambda(\sqrt{n}t) \right) = O(n^{-1/2}).$$

Therefore according to the Markov inequality, we immediately have

$$\left| \frac{1}{\sqrt{n}} \varphi_\lambda(\sqrt{n}t) - \Psi(t) \right| \xrightarrow{P} 0. \tag{4.29}$$

Turn to the uniform convergence. Fix $0 < a < b < \infty$. For any $\varepsilon > 0$, there is an $m \geq 0$ and $a = t_0 < t_1 < \cdots < t_m < t_{m+1} = b$ such that

$$\max_{0 \leq i \leq m} |\Psi(t_i) - \Psi(t_{i+1})| \leq \varepsilon.$$

Also, by virtue of the monotonicity of φ_λ, we have

$$\sup_{a \leq t \leq b} \left| \frac{1}{\sqrt{n}} \varphi_\lambda(\sqrt{n}t) - \Psi(t) \right| \leq 3 \max_{0 \leq i \leq m} \left| \frac{1}{\sqrt{n}} \varphi_\lambda(\sqrt{n}t_i) - \Psi(t_i) \right|$$
$$+ \max_{0 \leq i \leq m} |\Psi(t_i) - \Psi(t_{i+1})|. \tag{4.30}$$

Hence it follows from (4.29) and (4.30)

$$Q_{q_n} \left(\sup_{a \leq t \leq b} \left| \frac{1}{\sqrt{n}} \varphi_\lambda(\sqrt{n}t) - \Psi(t) \right| > 4\varepsilon \right)$$
$$\leq m \max_{0 \leq i \leq m} Q_{q_n} \left(\left| \frac{1}{\sqrt{n}} \varphi_\lambda(\sqrt{n}t_i) - \Psi(t_i) \right| > \varepsilon \right)$$
$$\to 0, \quad n \to \infty.$$

The proof is complete. □

Theorem 4.9. *For any $x \in \mathbb{R}$, we have as $\to \infty$*

$$Q_{q_n} \left(\frac{c}{\sqrt{n}} \lambda_1 - \log \frac{\sqrt{n}}{c} \leq x \right) \longrightarrow e^{-e^{-x}}. \tag{4.31}$$

Proof.   Let $A_{n,x} = \sqrt{n}(\log \sqrt{n}/c + x)/c$. Since λ_1 is the largest part of λ, then it is easy to see

$$Q_{q_n}\left(\frac{c}{\sqrt{n}}\lambda_1 - \log \frac{\sqrt{n}}{c} \leq x\right) = Q_{q_n}(r_k = 0, \forall k \geq A_{n,x}).$$

It follows by Lemma 4.2

$$\prod_{k \geq A_{n,x}} Q_{q_n}(r_k = 0) = \prod_{k \geq A_{n,x}} (1 - q_n^k). \tag{4.32}$$

For each $x \in \mathbb{R}$, $q_n^k \to 0$ whenever $k \geq A_{n,x}$. Hence we have as $n \to \infty$,

$$\sum_{k \geq A_{n,x}} \log(1 - q_n^k) = -(1 + o(1)) \sum_{k \geq A_{n,x}} q_n^k$$

$$= -(1 + o(1)) \frac{q_n^{\lceil A_{n,x} \rceil}}{1 - q_n}$$

$$\to -e^{-x},$$

which together with (4.32) implies (4.31). $\qquad\square$

Theorem 4.10. *For $x_1 > \cdots > x_m$*

$$\lim_{n \to \infty} Q_{q_n}\left(\frac{c}{\sqrt{n}}\lambda_i - \log \frac{\sqrt{n}}{c} \leq x_i, \quad 1 \leq i \leq m\right)$$

$$= \int_{-\infty}^{x_1} \cdots \int_{-\infty}^{x_m} p_0(x_1) \prod_{i=2}^{m} p(x_{i-1}, x_i) dx_1 \cdots dx_m,$$

where p_0 and p are as in Theorem 4.4.

Proof.   For simplicity of notations, we only prove the statement in the case $m = 2$. Let $A_{n,x} = \sqrt{n}(\log \sqrt{n}/c + x)/c$. Then it is easy to see

$$Q_{q_n}\left(\frac{c}{\sqrt{n}}\lambda_i - \log \frac{\sqrt{n}}{c} \leq x_i, \quad i = 1, 2\right)$$

$$= Q_{q_n}\left(\bigcap_{k > A_{n,x_2}} \{r_k = 0\}\right)$$

$$+ \sum_{A_{n,x_2} < j \leq A_{n,x_1}} Q_{q_n}\left(\{r_j = 1\} \bigcap_{k > A_{n,x_2}, k \neq j} \{r_k = 0\}\right). \tag{4.33}$$

By Lemma 4.2, it follows for each $A_{n,x_2} < j \leq A_{n,x_1}$

$$Q_{q_n}\left(\{r_j = 1\} \bigcap_{k > A_{n,x_2}, k \neq j} \{r_k = 0\}\right) = q_n^j \prod_{k > A_{n,x_2}} (1 - q_n^k).$$

A simple calculus shows

$$\sum_{A_{n,x_2} < j \le A_{n,x_1}} q_n^j = q_n^{\lceil A_{n,x_2} \rceil} \frac{1 - q_n^{\lceil A_{n,x_1} \rceil - \lceil A_{n,x_2} \rceil}}{1 - q_n}$$

$$\to e^{-x_2} - e^{-x_1}, \quad n \to \infty.$$

Also, according to the proof of Theorem 4.9,

$$\lim_{n \to \infty} Q_{q_n} \left(\bigcap_{k > A_{n,x}} \{r_k = 0\} \right) = e^{-e^{-x}}, \quad x \in \mathbb{R}.$$

Therefore it follows from (4.33)

$$\lim_{n \to \infty} Q_{q_n} \left(\frac{c}{\sqrt{n}} \lambda_i - \log \frac{\sqrt{n}}{c} \le x_i, \quad i = 1, 2 \right)$$
$$= e^{-e^{-x_2}} + (e^{-x_2} - e^{-x_1}) e^{-e^{-x_1}}.$$

This is the integral of $p_0(x_1) p(x_1, x_2)$ over the region $\{(x_1, x_2) : x_1 > x_2\}$. The proof is complete. □

Theorem 4.11. *Under* $(\mathcal{P}, Q_{q_n})$

$$X_n \Rightarrow G$$

in terms of finite dimensional distributions. Here G is a centered Gaussian process with the covariance structure

$$Cov(G(s), G(t)) = \int_t^\infty \frac{e^{-cu}}{(1 - e^{-cu})^2} du, \quad s < t.$$

Proof. First, we can prove in a way completely similar to that of Theorem 4.7 that for each $t > 0$

$$X_n(t) \xrightarrow{d} G(t).$$

Next we turn to the 2-dimensional case. Assume $0 < s < t < \infty$. Then for any x_1 and x_2,

$$x_1 X_n(s) + x_2 X_n(t) = \frac{x_1}{n^{1/4}} \left(\sum_{\sqrt{n}s \le k < \sqrt{n}t} r_k - \sqrt{n}(\Psi(s) - \Psi(t)) \right)$$
$$+ \frac{x_1 + x_2}{n^{1/4}} \left(\sum_{k \ge \sqrt{n}t} r_k - \sqrt{n}\Psi(t) \right). \tag{4.34}$$

Since two summands in the right hand side of (4.34) are independent and each converges weakly to a normal random variable, then $x_1 X_n(s) + x_2 X_n(t)$

must converges weakly to a normal random variable with variance $\sigma_{s,t}^2$ given by

$$x_1^2 \int_s^t \frac{e^{-cu}}{(1-e^{-cu})^2} du + (x_1+x_2)^2 \int_t^\infty \frac{e^{-cu}}{(1-e^{-cu})^2} du$$

$$= x_1^2 \int_s^\infty \frac{e^{-cu}}{(1-e^{-cu})^2} du + x_2^2 \int_t^\infty \frac{e^{-cu}}{(1-e^{-cu})^2} du$$

$$+2x_1 x_2 \int_t^\infty \frac{e^{-cu}}{(1-e^{-cu})^2} du.$$

Therefore

$$\big(X_n(s),\, X_n(t)\big) \xrightarrow{d} \big(G(s),\, G(t)\big),$$

where $(G(s), G(t))$ is jointly normally distributed with covariance

$$Cov(G(s),\, G(t)) = \int_t^\infty \frac{e^{-cu}}{(1-e^{-cu})^2} du.$$

The m-dimensional case can be analogously proved. $\qquad\square$

To conclude this section, we investigate the asymptotic behaviours when t_n tends to ∞.

Theorem 4.12. *Assume that t_n is a sequence of positive numbers such that*

$$t_n \to \infty, \quad t_n - \frac{1}{2c}\log n \to -\infty. \tag{4.35}$$

Then under $(\mathcal{P}, Q_{q_n})$,

$$\frac{e^{ct_n/2}}{n^{1/4}}\big(\varphi_\lambda(\sqrt{n}t_n) - \sqrt{n}\Psi(t_n)\big) \xrightarrow{d} N(0,1). \tag{4.36}$$

Proof. First, compute mean and variance of $\varphi_\lambda(\sqrt{n}t_n)$. According to the definition of φ_λ,

$$E_{q_n}\varphi_\lambda(\sqrt{n}t_n) = \sum_{k\geq\sqrt{n}t_n} E_{q_n} r_k = \sum_{k\geq\sqrt{n}t_n} \frac{q_n^k}{1-q_n^k}$$

$$= \sqrt{n}\int_{t_n}^\infty \frac{e^{-cu}}{1-e^{-cu}} du\big(1+O(n^{-1/2})\big)$$

$$= \sqrt{n}\Psi(t_n)\big(1+O(n^{-1/2})\big)$$

and

$$Var_{q_n}\varphi_\lambda(\sqrt{n}t_n) = \sum_{k\geq\sqrt{n}t_n} Var_{q_n}(r_k) = \sum_{k\geq\sqrt{n}t_n} \frac{q_n^k}{(1-q_n^k)^2}$$

$$= \sqrt{n}\int_{t_n}^\infty \frac{e^{-cu}}{(1-e^{-cu})^2} du\big(1+O(n^{-1/2})\big)$$

$$= \sqrt{n}e^{-ct_n}\big(1+O(n^{-1/2})\big).$$

The condition (4.35) guarantees $\sqrt{n}e^{-ct_n} \to \infty$. Next, we verify the Lindeberg condition. For any $\varepsilon > 0$,

$$\sum_{k \geq \sqrt{n}t_n} \sum_{j \geq \varepsilon\sigma_n} j^2 q_n^{jk}(1 - q_n^k) \leq \sum_{k \geq \sqrt{n}t_n} \sum_{j \geq \varepsilon\sigma_n} j^2 q_n^{jk}$$

$$= \sum_{j \geq \varepsilon\sigma_n} j^2 \sum_{k \geq \sqrt{n}t_n} q_n^{jk}$$

$$= \sum_{j \geq \varepsilon\sigma_n} j^2 \frac{q_n^{j\lceil \sqrt{n}t_n \rceil}}{1 - q_n^j}.$$

It is now easy to see

$$\frac{1}{\sigma_n^2} \sum_{j \geq \varepsilon\sigma_n} j^2 \frac{q_n^{j\lceil \sqrt{n}t_n \rceil}}{1 - q_n^j} \to 0$$

under the condition (4.35). So the is satisfied, and we conclude (4.36). □

4.3 Small ensembles

This section is devoted to the proofs of main results given in the Introduction. A basic strategy is to use conditioning argument on the event $\{|\lambda| = n\}$. The following lemma due to Vershik (1996) characterizes the relations between grand ensembles and small ensembles.

Lemma 4.3. *For any $0 < q < 1$ and $n \geq 0$, we have*
(i) $P_{u,n}$ is the conditional probability measure induced on $\mathcal{P}_n$ by Q_q, i.e.,

$$Q_q|_{\mathcal{P}_n} = P_{u,n};$$

(ii) Q_q is a convex combination of measures $P_{u,n}$, i.e.,

$$Q_q = \frac{1}{\mathcal{F}(q)} \sum_{n=0}^{\infty} p(n)q^n P_{u,n}.$$

Let $W_n(\lambda)$ be a function of λ taking values in $\mathbb{R}^{b_n}$ where $b_n < \infty$ or $b_n = \infty$. W_n can be regarded as a random variable in $(\mathcal{P}, Q_{q_n})$. When restricted to $\mathcal{P}_n$, W_n is also a random variable in $(\mathcal{P}_n, P_{u,n})$. Denote by $Q_{q_n} \circ W_n^{-1}$ and $P_{u,n} \circ W_n^{-1}$ the induced measures by W_n respectively. The total variation distance is defined by

$$d_{TV}(Q_{q_n} \circ W_n^{-1}, P_{u,n} \circ W_n^{-1}) = \sup_{B \subset \mathbb{R}^{b_n}} |Q_{q_n} \circ W_n^{-1}(B) - P_{u,n} \circ W_n^{-1}(B)|.$$

Lemma 4.4. *If there exists a sequence of subsets $B_n \subseteq \mathbb{R}^{b_n}$ such that*
(i)

$$Q_{q_n} \circ W_n^{-1}(B_n) \to 1, \tag{4.37}$$

(ii)

$$\sup_{w_n \in B_n} \left| \frac{Q_{q_n}(|\lambda| = n | W_n = w_n)}{Q_{q_n}(|\lambda| = n)} - 1 \right| \to 0, \tag{4.38}$$

then it follows

$$d_{TV}\left(Q_{q_n} \circ W_n^{-1}, P_{u,n} \circ W_n^{-1}\right) \to 0, \quad n \to \infty. \tag{4.39}$$

Proof.   Observe for any $B \subseteq \mathbb{R}^{b_n}$

$$Q_{q_n} \circ W_n^{-1}(B) = Q_{q_n} \circ W_n^{-1}(B \cap B_n) + Q_{q_n} \circ W_n^{-1}(B \cap B_n^c)$$

and

$$P_{u,n} \circ W_n^{-1}(B) = P_{u,n} \circ W_n^{-1}(B \cap B_n) + P_{u,n} \circ W_n^{-1}(B \cap B_n^c).$$

Since $Q_{q_n} \circ W_n^{-1}(B_n^c) \to 0$ by (4.37), then we need only estimate

$$\left| Q_{q_n} \circ W_n^{-1}(B \cap B_n) - Q_{q_n} \circ W_n^{-1}(B \cap B_n | |\lambda| = n) \right|$$

$$\leq \sum_{w_n \in B_n} \left| Q_{q_n}(W_n = w_n) - Q_{q_n}(W_n = w_n | |\lambda| = n) \right|$$

$$= \sum_{w_n \in B_n} Q_{q_n}(W_n = w_n) \left| \frac{Q_{q_n}(|\lambda| = n | W_n = w_n)}{Q_{q_n}(|\lambda| = n)} - 1 \right|. \tag{4.40}$$

It follows from (4.38) that the right hand side of (4.40) tends to 0. Thus we conclude the desired assertion (4.39). $\square$

Lemma 4.5. *Assume that K_n is a sequence of positive integers such that*

$$\sum_{k \in K_n} \frac{k^2 q_n^k}{(1 - q_n^k)^2} = o(n^{3/2}). \tag{4.41}$$

Then for $W_n : \lambda \to (r_k(\lambda), k \in K_n)$

$$d_{TV}\left(Q_{q_n} \circ W_n^{-1}, P_n \circ W_n^{-1}\right) \longrightarrow 0.$$

Proof.   We will construct B_n such that (i) and (ii) of Lemma 4.4 holds. First, observe that there is an a_n such that

$$\sum_{k \in K_n} \frac{k^2 q_n^k}{(1 - q_n^k)^2} = o(a_n^2), \quad a_n = o(n^{3/4}). \tag{4.42}$$

Define

$$B_n = \left\{ (x_k, k \in K_n) : \left| \sum_{k \in K_n} k x_k - \sum_{k \in K_n} k E_{q_n} r_k \right| \le a_n \right\}.$$

Then by (4.42)

$$Q_{q_n} \circ W_n^{-1}(B_n^c) = Q_{q_n} \left(\left| \sum_{k \in K_n} k r_k - \sum_{k \in K_n} k E_{q_n} r_k \right| > a_n \right)$$

$$\le \frac{Var_{q_n} \left(\sum_{k \in K_n} k r_k \right)}{a_n^2}$$

$$= \frac{1}{a_n^2} \sum_{k \in K_n} \frac{k^2 q_n^k}{(1 - q_n^k)^2} \to 0.$$

It remains to show that

$$\frac{Q_{q_n} \left(|\lambda| = n \big| W_n = w_n \right)}{Q_{q_n} (|\lambda| = n)} \to 1 \tag{4.43}$$

uniformly in $w_n \in B_n$.

Fix $w_n = (x_k, k \in K_n)$. Then by independence of the r_k's

$$Q_{q_n} \left(|\lambda| = n \big| W_n = w_n \right)$$

$$= Q_{q_n} \left(\sum_{k=1}^{\infty} k r_k = n \big| r_k = x_k, k \in K_n \right)$$

$$= Q_{q_n} \left(\sum_{k \notin K_n} k r_k = n - \sum_{k \in K_n} k x_k \big| r_k = x_k, k \in K_n \right)$$

$$= Q_{q_n} \left(\sum_{k \notin K_n} k r_k = n - \sum_{k \in K_n} k x_k \right). \tag{4.44}$$

It follows by (4.21) and (4.42)

$$Var_{q_n} \left(\sum_{k \notin K_n} k r_k \right) = \sum_{k=1}^{\infty} k^2 Var_{q_n}(r_k) - \sum_{k \in K_n} k^2 Var_{q_n}(r_k)$$

$$= \sigma_n^2 (1 + o(1)) \to \infty.$$

Hence as in Theorem 4.7 one can prove that under $(\mathcal{P}, Q_{q_n})$

$$\frac{\sum_{k \notin K_n} k(r_k - E_{q_n} r_k)}{\sqrt{Var_{q_n}(\sum_{k \notin K_n} k r_k)}} \longrightarrow N(0, 1)$$

and so

$$\frac{1}{\sigma_n} \sum_{k \notin K_n} k(r_k - E_{q_n} r_k) \longrightarrow N(0, 1).$$

Note that

$$\frac{1}{\sigma_n}\left(n - \sum_{k=1}^{\infty} E_{q_n} r_k - \sum_{k \in K_n} k(x_k - E_{q_n} r_k)\right) \to 0$$

uniformly in $(x_k, k \in K_n) \in B_n$. Then using the inverse formula as in Theorem 4.7, we have

$$Q_{q_n}\left(\sum_{k \notin K_n} k r_k - n - \sum_{k \in K_n} k x_k\right)$$

$$= Q_{q_n}\left(\sum_{k \notin K_n} k(r_k - E_{q_n} r_k) = n - \sum_{k=1}^{\infty} E_{q_n} r_k - \sum_{k \in K_n} k(x_k - E_{q_n} r_k)\right)$$

$$= \frac{1}{(96)^{1/4} n^{3/4}}\big(1 + o(1)\big). \tag{4.45}$$

Combining (4.44), (4.45) and (4.22) yields (4.43), as desired.   □

Now we are ready to prove Theorems 4.3, 4.4 and 4.6.

Proof of Theorems 4.3. As in Theorem 4.9, let $A_{n,x} = \sqrt{n}(\log \sqrt{n}/c + x)/c$. Define $K_n = \{k : k \geq A_{n,x}\}$, then it is easy to see

$$\sum_{k \in K_n} \frac{k^2 q_n^k}{(1 - q_n^k)^2} = o(n^{3/2})$$

since $A_{n,x}/\sqrt{n} \to \infty$. Hence applying Lemma 4.5 to $W_n : \lambda \mapsto (r_k(\lambda), k \in K_n)$ yields

$$Q_{q_n}(\lambda \in \mathcal{P} : r_k = 0, k \in K_n) - P_{u,n}(\lambda \in \mathcal{P}_n : r_k = 0, k \in K_n) \to 0.$$

According to Theorem 4.9, we in turn have for any $x \in \mathbb{R}$

$$P_{u,n}\left(\lambda \in \mathcal{P}_n : \frac{c}{\sqrt{n}}\lambda_1 - \log \frac{\sqrt{n}}{c} \leq x\right) = P_{u,n}(\lambda \in \mathcal{P}_n : r_k = 0, k \in K_n)$$

$$\longrightarrow e^{-e^{-x}}.$$

The proof is complete.   □

Proof of Theorem 4.4. Similar to the proof of Theorem 4.3.   □

Proof of Theorem 4.6 Define $K_n = \{k : k \geq \sqrt{n} t_n\}$, then it is easy to see

$$\sum_{k \in K_n} \frac{k^2 q_n^k}{(1 - q_n^k)^2} = o(n^{3/2})$$

since $t_n \to \infty$. Hence applying Lemma 4.5 to $W_n : \lambda \mapsto (r_k(\lambda), k \in K_n)$ yields

$$Q_{q_n}(\lambda \in \mathcal{P} : W_n \in B) - P_{u,n}(\lambda \in \mathcal{P}_n : W_n \in B) \to 0.$$

where

$$B = \left\{ (x_k, k \in K_n) : \frac{e^{ct_n/2}}{n^{1/4}} \Big(\sum_{k \in K_n} x_k - \sqrt{n}\Psi(t_n) \Big) \leq x \right\}.$$

We now obtain the desired (4.18) according to Theorem 4.12. $\qquad\square$

However, for any fixed $t \geq 0$, the condition (4.41) is not satisfied by $K_n = \{k : k \geq \sqrt{n}t\}$. Thus one cannot directly derive Theorem 4.2 nor Theorem 4.5 from Lemma 4.5.

The rest of this section is devoted to proving Theorems 4.2 and 4.5 following Pittel (1997), and the focus is upon the latter since the other can be proved in a similar and simpler way.

For simplicity of notations, we only consider two dimensional case below. Assume $0 < t_1 < t_2$, we shall prove that for any $x_1, x_2 \in \mathbb{R}$

$$x_1 X_n(t_1) + x_2 X_n(t_2) \xrightarrow{d} N\big(0, \sigma^2_{x_1,x_2}\big), \qquad (4.46)$$

where

$$\sigma^2_{x_1,x_2} = (x_1, x_2)\Sigma_{t_1,t_2} \begin{pmatrix} x_1 \\ x_2 \end{pmatrix}.$$

Here Σ_{t_1,t_2} is a covariance matrix of X given by (4.16). Indeed, it suffices to prove (4.46) holds for

$$\xi_n(\mathbf{x}, \mathbf{t}) := \frac{x_1}{n^{1/4}} \sum_{k \geq \sqrt{n}t_1} (r_k - E_{q_n}r_k) + \frac{x_2}{n^{1/4}} \sum_{k \geq \sqrt{n}t_2} (r_k - E_{q_n}r_k).$$

This will in turn be done by proving

$$E_{u,n}e^{\xi_n(\mathbf{x},\mathbf{t})} \to \exp\Big(\frac{\sigma^2_{x_1,x_2}}{2}\Big), \qquad n \to \infty. \qquad (4.47)$$

In doing this, a main ingredient is to show the following proposition. We need additional notations. Define for any $1 \leq k_1 < k_2 < \infty$

$$m(k_1, k_2) = \sum_{k=k_1}^{k_2-1} \frac{q_n^k}{1 - q_n^k}, \quad \sigma^2(k_1, k_2) = \sum_{k=k_1}^{k_2-1} \frac{q_n^k}{(1 - q_n^k)^2},$$

$$m(k_2) = \sum_{k=k_2}^{\infty} \frac{q_n^k}{1 - q_n^k}, \quad \sigma^2(k_2) = \sum_{k=k_2}^{\infty} \frac{q_n^k}{(1 - q_n^k)^2}$$

and

$$s(k_1, k_2) = \sum_{k=k_1}^{k_2-1} \frac{k q_n^k}{(1 - q_n^k)^2}, \quad s(k_2) = \sum_{k=k_2}^{\infty} \frac{k q_n^k}{(1 - q_n^k)^2}.$$

Proposition 4.1. *For $u_1, u_2 \in \mathbb{R}$*

$$E_{u,n} \exp\left(\frac{u_1}{n^{1/4}} \sum_{k_1 \leq k < k_2} (r_k - E_{q_n} r_k) + \frac{u_2}{n^{1/4}} \sum_{k \geq k_2} (r_k - E_{q_n} r_k)\right)$$

$$= \exp\left(\frac{u_1^2}{2n^{1/2}} \sigma^2(k_1, k_2) + \frac{u_2^2}{2n^{1/2}} \sigma^2(k_2) - \frac{c}{4n^2} (u_1 s(k_1, k_2) + u_2 s(k_2))^2\right)$$

$$\cdot (1 + o(1)). \tag{4.48}$$

The proof of Proposition 4.1 will consist of several lemmas. Note that for any $0 < q < 1$ and z_1, z_2

$$E_q \prod_{k_1 \leq k < k_2} z_1^{r_k} \prod_{k \geq k_2} z_2^{r_k} = \prod_{k_1 \leq k < k_2} E_q z_1^{r_k} \prod_{k \geq k_2} E_q z_2^{r_k}$$

$$= \prod_{k_1 \leq k < k_2} \frac{1 - q^k}{1 - z_1 q^k} \prod_{k \geq k_2} \frac{1 - q^k}{1 - z_2 q^k}.$$

On the other hand, it follows by Lemma 4.3

$$E_q \prod_{k_1 \leq k < k_2} z_1^{r_k} \prod_{k \geq k_2} z_2^{r_k} = \frac{1}{\mathcal{F}(q)} \sum_{n=0}^{\infty} p(n) q^n E_{u,n} \prod_{k_1 \leq k < k_2} z_1^{r_k} \prod_{k \geq k_2} z_2^{r_k}.$$

Thus we have for each $0 < q < 1$

$$\sum_{n=0}^{\infty} p(n) q^n E_{u,n} \prod_{k_1 \leq k < k_2} z_1^{r_k} \prod_{k \geq k_2} z_2^{r_k}$$

$$= \mathcal{F}(q) \prod_{k_1 \leq k < k_2} \frac{1 - z_1 q^k}{1 - q^k} \prod_{k \geq k_2} \frac{1 - q^k}{1 - z_2 q^k}$$

$$=: F(\mathbf{z}; q), \tag{4.49}$$

where $\mathbf{z} = (z_1, z_2)$.

Note that the above equation (4.49) is still valid for all complex number q with $|q| < 1$. Hence using the Cauchy integral formula yields

$$E_{u,n} \prod_{k_1 \leq k < k_2} z_1^{r_k} \prod_{k \geq k_2} z_2^{r_k} = \frac{1}{2\pi p(n)} \int_{-\pi}^{\pi} r^{-n} e^{-in\theta} F(\mathbf{z}; re^{i\theta}) d\theta, \tag{4.50}$$

where r is a free parameter.

Lemma 4.6.

$$\left|F(\mathbf{z}, re^{i\theta})\right| \leq c_5 F(\mathbf{z}, r) \exp\left(\frac{2r(1 + r)(\cos\theta - 1)}{(1 - r)^3 + 2r(1 - r)(1 - \cos\theta)}\right).$$

Proof. Observe an elementary inequality

$$\left|\frac{1}{1-q}\right| \le \frac{1}{1-|q|}\exp(Req - |q|), \quad |q| < 1.$$

We have for $0 < r < 1$

$$\left|F(\mathbf{z}; re^{i\theta})\right| \le F(\mathbf{z}; r)\exp\left(\sum_{k=1}^{\infty} r^k(\cos k\theta - 1)\right)$$

$$= F(\mathbf{z}; r)\exp\left(Re\frac{1}{1 - re^{i\theta}} - \frac{1}{1-r}\right).$$

Also, it is easy to see

$$Re\frac{1}{1 - re^{i\theta}} - \frac{1}{1-r} = \frac{2r(1+r)(\cos\theta - 1)}{(1-r)^3 + 2r(1-r)(1 - \cos\theta)}.$$

The proof is complete. $\square$

We shall asymptotically estimate $r^{-n}F(\mathbf{z}; r)$ below. Define for t and z

$$s^{(1)}(t, z) = \sum_{k=k_1}^{k_2-1} \frac{ke^{-tk}}{(1 - ze^{-tk})(1 - e^{-tk})}$$

and

$$s^{(2)}(t, z) = \sum_{k=k_2}^{\infty} \frac{ke^{-tk}}{(1 - ze^{-tk})(1 - e^{-tk})}.$$

Lemma 4.7. *Let* $r = e^{-\tau}$ *where*

$$\tau = \tau^*\left(1 + \frac{1}{2n}\sum_{b=1}^{2}(z_b - 1)s^{(b)}(\tau^*, z_b)\right), \quad \tau^* = \frac{c}{\sqrt{n}}.$$

Then we have for $z_1 = e^{u_1/\sqrt{n}}$ *and* $z_2 = e^{u_2/\sqrt{n}}$

$$r^{-n}F(\mathbf{z}; r) = \frac{e^{2c\sqrt{n}}}{(24n)^{1/4}}\left(1 + O(n^{-1/4})\right)$$

$$\cdot \exp\left(\frac{u_1}{n^{1/4}}m(k_1, k_2) + \frac{u_1^2}{2n^{1/2}}\sigma^2(k_1, k_2)\right)$$

$$\cdot \exp\left(\frac{u_2}{n^{1/4}}m(k_2) + \frac{u_2^2}{2n^{1/2}}\sigma^2(k_2)\right)$$

$$\cdot \exp\left(-\frac{c}{4n^2}\left(u_1 s(k_1, k_2) + u_2 s(k_2)\right)^2\right).$$

Proof. Let

$$H(\mathbf{z};t) = nt + \frac{c^2}{t} + \sum_{k=k_1}^{k_2-1} \log \frac{1 - e^{-tk}}{1 - z_1 e^{-tk}} + \sum_{k=k_2}^{\infty} \log \frac{1 - e^{-tk}}{1 - z_2 e^{-tk}}.$$

A simple calculus shows

$$H_t(\mathbf{z};t) - n - \frac{c^2}{t^2} - \sum_{b=1}^{2}(z_b - 1)s^{(b)}(t, z_b)$$

and

$$H_{tt}(\mathbf{z};t) = \frac{2c^2}{t^3} - \sum_{b=1}^{2}(z_b - 1)s_t^{(b)}(t, z_b).$$

Making the Taylor expansion at τ, we have

$$H(\mathbf{z};\tau^*) = H(\mathbf{z};\tau) + H_t(\mathbf{z};\tau)(\tau^* - \tau) + \frac{1}{2}H_{tt}(\mathbf{z};\tilde{t})(\tau^* - \tau)^2$$

for a $\tilde{t}$ between τ^* and τ. Hence it follows

$$H(\mathbf{z};\tau) = H(\mathbf{z};\tau^*) - H_t(\mathbf{z};\tau)(\tau^* - \tau) - \frac{1}{2}H_{tt}(\mathbf{z};\tilde{t})(\tau^* - \tau)^2. \quad (4.51)$$

We shall estimate each summand in the right hand side of (4.51) below. Begin with $H(\mathbf{z};\tau^*)$. As in Theorem 4.7, we have by the Taylor expansions

$$\sum_{k=k_1}^{k_2-1} \log \frac{1 - q_n^k}{1 - z_1 q_n^k} = (z_1 - 1)\sum_{k=k_1}^{k_2-1} \frac{q_n^k}{1 - q_n^k} + \frac{(z_1 - 1)^2}{2}\sum_{k=k_1}^{k_2-1} \frac{q_n^{2k}}{(1 - q_n^k)^2}$$

$$+ O\left(|z_1 - 1|^3 \sum_{k=k_1}^{k_2-1} \frac{q_n^{3k}}{(1 - q_n^k)^3}\right)$$

$$= \frac{u_1}{n^{1/4}}\sum_{k=k_1}^{k_2-1} \frac{q_n^k}{1 - q_n^k} + \frac{u_1}{2n^{1/2}}\sum_{k=k_1}^{k_2-1} \frac{q_n^k}{(1 - q_n^k)^2} + O(n^{-1/4})$$

$$= \frac{u_1}{n^{1/4}}m(k_1, k_2) + \frac{u_1}{2n^{1/2}}\sigma^2(k_1, k_2) + O(n^{-1/4}).$$

Similarly,

$$\sum_{k=k_2}^{\infty} \log \frac{1 - q_n^k}{1 - z_2 q_n^k} = \frac{u_2}{n^{1/4}}m(k_2) + \frac{u_2}{2n^{1/2}}\sigma^2(k_2) + O(n^{-1/4}).$$

This immediately gives

$$H(\mathbf{z};\tau^*) = 2c\sqrt{n} + \frac{u_1}{n^{1/4}}m(k_1, k_2) + \frac{u_1}{2n^{1/2}}\sigma^2(k_1, k_2)$$

$$+ \frac{u_2}{n^{1/4}}m(k_2) + \frac{u_2}{2n^{1/2}}\sigma^2(k_2) + O(n^{-1/4}). \quad (4.52)$$

Turn to the second term $H_t(\mathbf{z}; \tau)$. Note for $b = 1, 2$

$$z_b - 1 = \frac{u_b}{n^{1/4}} + O(n^{-1/2})$$

and

$$s^{(b)}(\tau^*, z_b) = O(n), \quad s^{(b)}(\tau, z_b) = O(n).$$

Then we have

$$H_t(\mathbf{z}, \tau) = n - n\left(1 + \frac{1}{2n}\sum_{b=1}^{2}(z_b - 1)s^{(b)}(\tau^*, z_b)\right)^{-2}$$

$$- \sum_{b=1}^{2}(z_b - 1)s^{(b)}(\tau, z_b)$$

$$= \sum_{b=1}^{2}(z_b - 1)\left(s^{(b)}(\tau^*, z_b) - s^{(b)}(\tau, z_b)\right) + O(n^{1/2})$$

$$= O(n^{1/2}). \tag{4.53}$$

To evaluate the third term, note for any $\tilde{t}$ between τ and τ^*

$$H_{tt}(\mathbf{z}; \tau) = \frac{2}{c}n^{3/2} + O(n^{5/4})$$

and

$$\tau^* - \tau = \frac{\tau^*}{2n}\sum_{b=1}^{2}(z_b - 1)s^{(b)}(\tau^*, z_b)$$

$$= \frac{c}{2n^{7/4}}\sum_{b=1}^{2}u_b s^{(b)} + O(n^{-1}).$$

We have

$$H_{tt}(\mathbf{z}; \tilde{t})(\tau^* - \tau)^2 = \frac{c}{2n^2}\left(\sum_{b=1}^{2}u_b s^{(b)}\right)^2 + O(n^{-1/2}). \tag{4.54}$$

Inserting (4.52)-(4.54) into (4.51) yields

$$H(\mathbf{z}; \tau) = 2c\sqrt{n} + \frac{u_1}{n^{1/4}}m(k_1, k_2) + \frac{u_1}{2n^{1/2}}\sigma^2(k_1, k_2)$$

$$+ \frac{u_2}{n^{1/4}}m(k_2) + \frac{u_2}{2n^{1/2}}\sigma^2(k_2)$$

$$- \frac{c}{4n^2}\left(u_1 s^{(1)} + u_2 s^{(2)}\right)^2 + O(n^{-1/4}).$$

Finally, with the help of $\tau - \tau^* = O(n^{-3/4})$, we have

$$\frac{1}{2}\log\frac{\tau}{2\pi} = \log\frac{1}{(24n)^{1/4}} + O(n^{-1/4}),$$

and so

$$\mathcal{F}(r) = \mathcal{F}(e^{-\tau}) = \frac{e^{c\sqrt{n}}}{(24n)^{1/4}}(1 + O(n^{-1/4})).$$

To conclude the proof, we need only to note

$$r^{-n}F(\mathbf{z};r) = \mathcal{F}(r)\exp\left(-\frac{c^2}{\tau} + H(\mathbf{z};\tau)\right).$$ $\square$

To estimate the integral over $(-\pi, \pi)$, we split the interval $(-\pi, \pi)$ into two subsets: $|\theta| \leq \theta_n$ and $|\theta| \geq \theta_n$, where $\theta_n = n^{-3/4}\log n$. The following lemma shows that the overall contribution to the value of integral in (4.50) made by *large* θ's is negligible.

Lemma 4.8. *Assume $r = e^{-\tau}$ is as in Lemma 4.7. Then*

$$r^{-n}\int_{|\theta|\geq\theta_n}\left|F(\mathbf{z};re^{i\theta})\right|d\theta \leq r^{-n}F(\mathbf{z};r)\exp\left(-c_7\log^2 n\right). \quad (4.55)$$

Proof. If $r = e^{-\tau}$, then for all $n \geq 1$

$$\frac{2r(1+r)(\cos\theta-1)}{(1-r)^3 + 2r(1-r)(1-\cos\theta)} \geq -\frac{c_6\theta^2}{n^{-3/2} + \theta^2 n^{-1/2}}.$$

By Lemma 4.6,

$$\int_{|\theta|\geq\theta_n}r^{-n}\left|F(\mathbf{z};re^{i\theta})\right|d\theta \leq r^{-n}F(\mathbf{z};r)\int_{|\theta|\geq\theta_n}\exp\left(-\frac{c_6\theta^2}{n^{-3/2} + \theta^2 n^{-1/2}}\right)d\theta.$$

To estimate the integral, we consider two cases separately: $\theta_n \leq |\theta| \leq n^{-1/2}$ and $|\theta| > n^{-1/2}$.

$$\int_{\theta_n\leq|\theta|\leq n^{-1/2}}\exp\left(-\frac{c_6\theta^2}{n^{-3/2} + \theta^2 n^{-1/2}}\right)d\theta \leq \int_{\theta_n\leq|\theta|\leq n^{-1/2}}e^{-c_6 n^{1/2}/2}d\theta$$
$$= \exp(-c_6\log^2 n/3)$$

and

$$\int_{|\theta|\geq n^{-1/2}}\exp\left(-\frac{c_6\theta^2}{n^{-3/2} + \theta^2 n^{-1/2}}\right)d\theta \leq \int_{|\theta|\geq n^{-1/2}}e^{-c_6 n^{3/2}\theta^2/2}d\theta$$
$$= \exp(-c_6\log^2 n/3).$$

In combination, (4.55) holds for a new constant $c_7 > 0$. $\square$

Turn now to the major contribution to the integral value of *small* θ's.

Lemma 4.9. *Assume $r = e^{-\tau}$ is as in Lemma 4.7. Then*

$$\int_{|\theta|\leq\theta_n}r^{-n}e^{-in\theta}F(\mathbf{z};re^{i\theta})d\theta = r^{-n}F(\mathbf{z};r)\frac{\pi}{6^{1/4}n^{3/4}}(1 + o(1)). \quad (4.56)$$

Proof. First, observe

$$r^{-n}e^{-in\theta}F(\mathbf{z};re^{i\theta}) = \mathcal{F}(e^{\tau-i\theta})\exp\left(-\frac{c^2}{\tau-i\theta}+H(\mathbf{z};\tau-i\theta)\right). \quad (4.57)$$

Also, it follows from (4.5)

$$\mathcal{F}(e^{\tau-i\theta}) = \exp\left(\frac{c^2}{\tau-i\theta}+\frac{1}{2}\log\frac{\tau-i\theta}{2\pi}+O(|\tau-i\theta|)\right). \quad (4.58)$$

It is easy to see

$$\sup\{|H_{tt}(\mathbf{z};t)| : |t-\tau| \le \theta_n\} = O(n^2).$$

Then the Taylor expansion at τ gives

$$H(\mathbf{z};\tau-i\theta) = H(\mathbf{z};\tau) - iH_t(\mathbf{z};\tau)\theta - \frac{1}{2}H_{tt}(\mathbf{z};\tau)\theta^2 + O(n^2\theta_n^3). \quad (4.59)$$

Hence combining (4.57)-(4.59) together implies

$$r^{-n}e^{-in\theta}F(\mathbf{z};re^{i\theta})$$

$$= r^{-n}F(\mathbf{z};r)\exp\left(-i\theta H_t(\mathbf{z},t) - \frac{\theta^2}{2}H_{tt}(\mathbf{z},t)\right)(1+o(1)),$$

which in turn gives

$$\int_{|\theta|\le\theta_n} r^{-n}e^{-in\theta}F(\mathbf{z};re^{i\theta})\,d\theta$$

$$= r^{-n}F(\mathbf{z},r)\int_{|\theta|\le\theta_n}\exp\left(-i\theta H_t(\mathbf{z},t) - \frac{\theta^2}{2}H_{tt}(\mathbf{z},t)\right)d\theta(1+o(1)).$$

Note for a $c_8 > 0$

$$\frac{H_t^2(\mathbf{z};\tau)}{H_{tt}(\mathbf{z};t)} = O(n^{-1/2}), \quad \theta_n^2 H_{tt}(\mathbf{z},t) \ge c_8\log^2 n.$$

Hence it follows

$$\int_{|\theta|\le\theta_n}\exp\left(-i\theta H_t(\mathbf{z};\tau) - \frac{\theta^2}{2}H_{tt}(\mathbf{z};\tau)\right)d\theta$$

$$= \left(\int_{-\infty}^{\infty} - \int_{|\theta|>\theta_n}\right)\exp\left(-i\theta H_t(\mathbf{z};\tau) - \frac{\theta^2}{2}H_{tt}(\mathbf{z};\tau)\right)d\theta$$

$$= \sqrt{\frac{2\pi}{H_{tt}(\mathbf{z};\tau)}}\exp\left(-\frac{H_t^2(\mathbf{z};\tau)}{2H_{tt}(\mathbf{z};\tau)}\right) + O\left(H_{tt}^{-1/2}e^{-\theta_n^2 H_{tt}(\mathbf{z};\tau)/2}\right)$$

$$= \frac{\pi}{6^{1/4}n^{3/4}}(1+o(1)),$$

where in the second equation we used a standard normal integral formula. Likewise, it follows

$$\int_{|\theta|\le\theta_n}\left|\exp\left(-i\theta H_t(\mathbf{z};\tau) - \frac{\theta^2}{2}H_{tt}(\mathbf{z};\tau)\right)\right|d\theta = \frac{\pi}{6^{1/4}n^{3/4}}(1+o(1)).$$

Hence

$$\int_{|\theta|\le\theta_n} r^{-n}e^{-in\theta}F(\mathbf{z},re^{i\theta})\,d\theta = r^{-n}F(\mathbf{z},r)\frac{\pi}{6^{1/4}n^{3/4}}(1+o(1)). \qquad \square$$

Proof of Proposition 4.1. Let $z_1 = e^{u_1/n^{1/4}}$ and $z_2 = e^{u_2/n^{1/4}}$, and choose $r = e^{-\tau}$ in (4.50). Combining (4.55) and (4.56) yields

$$\int_{-\pi}^{\pi} r^{-n} e^{-in\theta} F(\mathbf{z}; re^{i\theta}) d\theta = \left(\int_{|\theta| \le \theta_n} + \int_{|\theta| \ge \theta_n} \right) r^{-n} e^{-in\theta} F(\mathbf{z}; re^{i\theta}) d\theta$$

$$= r^{-n} F(\mathbf{z}, r) \frac{\pi}{6^{1/4} n^{3/4}} \big(1 + o(1)\big).$$

Taking Theorem 4.1 and Lemma 4.7 into account, we conclude (4.48) as desired.　　　　　　　　　　　　　　　　　　　　　　　　　　□

Proof of Theorem 4.5. Take $u_1 = x_1$ and $u_2 = x_1 + x_2$ and $k_1 = \sqrt{n} t_1$ and $k_2 = \sqrt{n} t_2$. Note

$$\sigma^2(k_1, k_2) = \sqrt{n} \int_{t_1}^{t_2} \frac{e^{-cu}}{(1 - e^{-cu})^2} du (1 + o(1)),$$

$$\sigma^2(k_2) = \sqrt{n} \int_{t_2}^{\infty} \frac{e^{-cu}}{(1 - e^{-cu})^2} du (1 + o(1)),$$

$$s(k_1, k_2) = n \int_{t_1}^{t_2} \frac{ue^{-cu}}{(1 - e^{-cu})^2} du (1 + o(1)),$$

$$s(k_2) = n \int_{t_2}^{\infty} \frac{ue^{-cu}}{(1 - e^{-cu})^2} du (1 + o(1)).$$

Substituting these into (4.48) of Proposition 4.1, we easily get (4.47), as required. We conclude the proof.　　　　　　　　　　　　　　□

4.4　A functional central limit theorem

In this section we shall first prove a theorem that may allow us to get the distributional results in the case when the functional of λ depends primarily on the moderate-sized parts. Then we shall use the functional central limit theorem to prove the asymptotic normality for character ratios and the log-normality for d_λ.

Introduce an integer-valued function

$$k_n(t) = \left\lceil \frac{\sqrt{n}}{c} \log \frac{1}{1 - t} \right\rceil, \quad t \in (0, 1).$$

Let

$$t_0(n) = \frac{c}{2\sqrt{n}}, \quad t_1(n) = n^{-\delta_0}, \quad t_2(n) = 1 - e^{-c/\sqrt{n}}$$

where $\delta_0 \in (0, 1/8)$. Define

$$
Y_n(t) = \begin{cases} \frac{t}{n^{1/4}} \sum_{k \geq k_n(t)} (r_k - E_{q_n} r_k), & t \in [t_0(n), 1), \\ 0, & 0 \leq t < t_0(n) \text{ or } t = 1. \end{cases}
$$

Let $Y(t), 0 \leq t \leq 1$ be a separable centered Gaussian process with the covariance function given by

$$
EY(s)Y(t) = \frac{1}{c}\Big(s(1-t) - \frac{1}{2}l(s)l(t)\Big), \quad 0 < s \leq t < 1 \tag{4.60}
$$

where

$$
l(t) = \frac{1}{c}\Big(t \log t - (1-t) \log(1-t)\Big).
$$

The so-called functional central limit theorem reads as follows.

Theorem 4.13. (i) *With probability 1, $Y(t)$ is uniformly continuous on $[0, 1]$.*
(ii) *Y_n converges to Y in terms of finite dimensional distributions.*
(iii) *Let $g(t, x)$ be continuous for $(t, x) \in D := (0, 1) \times \mathbb{R}$ and such that*

$$
|g(t, x)| \leq c_{10} \frac{|x|^\gamma}{t^\alpha (1-t)^\beta} \tag{4.61}
$$

for some $\gamma > 0$, $\alpha < 1 + \gamma/2$, $\beta < 1 + \gamma/6$, uniformly for $(t, x) \in D$. Then under $(\mathcal{P}_n, P_{u,n})$

$$
\int_0^1 g(t, Y_n(t)) dt \xrightarrow{d} \int_0^1 g(t, Y(t)) dt.
$$

We shall prove the theorem following the line of Pittel (2002) by applying the Gikhman-Skorohod theorem, namely Theorem 1.16. A main step is to verify the stochastic equicontinuity for $Y_n(t), 0 \leq t \leq 1$. As the reader may notice, a significant difference between Y_n and X_n is that there is an extra factor t in Y_n besides parametrization. This factor is added to guarantee that Y_n satisfies the stochastic equicontinuity property and so the limit process Y has continuous sample paths.

Lemma 4.10. *For $u \in \mathbb{R}$*

$$
E_{q_n} \exp\Big(\frac{u}{n^{1/4}} \sum_{k_1 \leq k < k_2} (r_k - E_{q_n} r_k)\Big)
$$

$$
= \exp\Big(\frac{u^2}{2n^{1/2}} \sum_{k_1 \leq k < k_2} \frac{q_n^k}{(1-q_n^k)^2} + O(n^{3/4-2\delta})\Big), \tag{4.62}
$$

where the error term holds uniformly over $n^\delta \leq k_1 < k_2 \leq \infty$ with $\delta < 1/2$.

Proof.   The proof is completely similar to that of Theorem 4.7. It follows by independence

$$E_{q_n} \exp\left(\frac{u}{n^{1/4}} \sum_{k_1 \le k < k_2} r_k\right) = \prod_{k_1 \le k < k_2} E_{q_n} \exp\left(\frac{u}{n^{1/4}} r_k\right)$$

$$= \prod_{k_1 \le k < k_2} \frac{1 - q_n^k}{1 - e^{u/n^{1/4}} q_n^k}$$

$$= \exp\left(-\sum_{k_1 \le k < k_2} \log \frac{1 - e^{u/n^{1/4}} q_n^k}{1 - q_n^k}\right).$$

Using the Taylor expansion, we obtain

$$\sum_{k_1 \le k < k_2} \log \frac{1 - e^{u/n^{1/4}} q_n^k}{1 - q_n^k} = \left(e^{u/n^{1/4}} - 1\right) \sum_{k_1 \le k < k_2} \frac{q_n^k}{1 - q_n^k}$$

$$+ \frac{1}{2}\left(e^{u/n^{1/4}} - 1\right)^2 \sum_{k_1 \le k < k_2} \frac{q_n^{2k}}{(1 - q_n^k)^2}$$

$$+ O\left(|e^{u/n^{1/4}} - 1|^3\right) \sum_{k_1 \le k < k_2} \frac{q_n^{3k}}{(1 - q_n^k)^3}.$$

Note

$$|e^{u/n^{1/4}} - 1|^3 \sum_{k_1 \le k < k_2} \frac{q_n^{3k}}{(1 - q_n^k)^3} = O\left(\frac{|u|^3}{n^{1/4}} \int_{k_1/\sqrt{n}}^{\infty} \frac{e^{-3cx}}{(1 - e^{-cx})^2} dx\right)$$

$$= O\left(n^{3/4 - 2\delta}\right),$$

and the contribution proportional to u^3 that comes from the first two terms is of lesser order of magnitude. We conclude the proof.   □

Lemma 4.11. *For* $u \in \mathbb{R}$

$$E_{u,n} \exp\left(\frac{u}{n^{1/4}} \sum_{k_1 \le k < k_2} (r_k - E_{q_n} r_k)\right)$$

$$\le c_{11} \exp\left(\frac{u^2}{2n^{1/2}} \sum_{k_1 \le k < k_2} \frac{q_n^k}{(1 - q_n^k)^2} + O\left(n^{3/4 - 2\delta}\right)\right), \qquad (4.63)$$

where the error term holds uniformly over $n^\delta \le k_1 < k_2 \le \infty$ *with* $\delta > 3/8$.

Proof.   This can be proved by a slight modification of Theorem 4.1. For any $0 < q < 1$,

$$E_q x^{\sum_{k_1 \le k < k_2} r_k} = \frac{1}{\mathcal{F}(q)} \sum_{n=0}^{\infty} q^n p(n) E_{u,n} x^{\sum_{k_1 \le k < k_2} r_k}.$$

On the other hand,

$$E_q x^{\sum_{k_1 \leq k < k_2} r_k} = \prod_{k_1 \leq k < k_2} \frac{1 - q^k}{1 - xq^k}.$$

Thus we have for any $0 < q < 1$

$$\sum_{n=0}^{\infty} q^n p(n) E_{u,n} x^{\sum_{k_1 \leq k < k_2} r_k} = \mathcal{F}(q) \prod_{k_1 \leq k < k_2} \frac{1 - q^k}{1 - xq^k}.$$

Indeed, the above equation holds for any complex number $|q| < 1$. Using the Cauchy integral formula, we obtain

$$E_{u,n} x^{\sum_{k_1 \leq k < k_2} r_k} = \frac{1}{2\pi p(n)} \int_{-\pi}^{\pi} r^{-n} e^{-in\theta} F_n(x; re^{i\theta}) d\theta$$

where

$$F_n(x; q) := \mathcal{F}(q) \prod_{k_1 \leq k < k_2} \frac{1 - q^k}{1 - xq^k}.$$

Choose $r = q_n = e^{-c/\sqrt{n}}$, with the help of (4.5) we further obtain

$$E_{u,n} x^{\sum_{k_1 \leq k < k_2} r_k} \leq c_8 \prod_{k_1 \leq k < k_2} \frac{1 - q_n^k}{1 - xq_n^k}.$$

Letting $x = e^{u/\sqrt{n}}$, we have by (4.62)

$$E_{u,n} \exp\Big(\frac{u}{n^{1/4}} \sum_{k_1 \leq k < k_2} r_k\Big) \leq c_8 \exp\Big(\frac{u}{n^{1/4}} \sum_{k_1 \leq k < k_2} E_{q_n} r_k\Big)$$

$$\cdot \exp\Big(\frac{u^2}{2n^{1/2}} \sum_{k_1 \leq k < k_2} \frac{q_n^k}{(1 - q_n^k)^2} + O\big(n^{3/4 - 2\delta}\big)\Big),$$

which immediately implies (4.63). The proof is complete. $\qquad\square$

Lemma 4.12. *(i) For $\varepsilon > 0$,*

$$\lim_{\delta \to 0} \lim_{n \to \infty} \sup_{|t-s| \leq \delta} P_{u,n}\big(|Y_n(t) - Y_n(s)| > \varepsilon\big) = 0. \qquad (4.64)$$

(ii) For any $m \geq 1$ and $0 < \rho < 1/12 - 2\delta_0/3$,

$$E_{u,n} |Y_n(t)|^m \leq c_9\big(t^{m/2}(1 - t)^{m/2} + n^{-m\rho}\big), \qquad t \geq t_0(n).$$

Proof. Observe that if $t > t_2(n)$ then $k_n(t) > n$ so $Y_n(t) = 0$; while if $t < t_0(n)$ then $Y_n(t) = O_p(n^{1/4-\delta_0} \log n)$. So it suffices to prove (4.64) uniformly for $t, s \in [t_1(n), t_2(n)]$. Assume $s < t$. We use Lemma 4.11 to obtain

$$E_{u,n} \exp \left(\frac{us}{n^{1/4}} \sum_{k_n(s) \leq k < k_n(t)} (r_k - E_{q_n} r_k) \right)$$

$$\leq c_9 \exp \left(\frac{u^2 s^2}{2n^{1/2}} \sum_{k_n(s) \leq k < k_n(t)} \frac{q_n^k}{(1 - q_n^k)^2} \right)$$

$$\leq c_{10} \exp \left(\frac{u^2}{2c} (t - s) \right) \tag{4.65}$$

and

$$E_{u,n} \exp \left(\frac{u(t-s)}{n^{1/4}} \sum_{k \geq k_n(t)} (r_k - E_{q_n} r_k) \right)$$

$$\leq c_{11} \exp \left(\frac{u^2 (t-s)^2}{2n^{1/2}} \sum_{k \geq k_n(t)} \frac{q_n^k}{(1 - q_n^k)^2} \right)$$

$$\leq c_{12} \exp \left(\frac{u^2}{2c} (t - s) \right). \tag{4.66}$$

Note

$$Y_n(t) - Y_n(s) = \frac{t - s}{n^{1/4}} \sum_{k \geq k_n(t)} (r_k - E_{q_n} r_k)$$

$$- \frac{s}{n^{1/4}} \sum_{k_n(s) \leq k < k_n(t)} (r_k - E_{q_n} r_k).$$

It follows by the Cauchy-Schwarz inequality, (4.65) and (4.66)

$$E_{u,n} \exp \left(u(Y_n(t) - Y_n(s)) \right)$$

$$\leq \left(E_{u,n} \exp \left(-\frac{2us}{n^{1/4}} \sum_{k_n(s) \leq k < k_n(t)} (r_k - E_{q_n} r_k) \right) \right)^{1/2}$$

$$\cdot \left(E_{u,n} \exp \left(\frac{2u(t-s)}{n^{1/4}} \sum_{k \geq k_n(t)} (r_k - E_{q_n} r_k) \right) \right)^{1/2}$$

$$\leq c_{13} \exp \left(\frac{2u^2}{c} (t - s) \right).$$

A standard argument now yields

$$P_{u,n} \left(|Y_n(t) - Y_n(s)| > \varepsilon \right) \leq \begin{cases} e^{-c\varepsilon^2/8(t-s)}, & \varepsilon \leq n^\rho (t - s), \\ e^{-cn^\rho \varepsilon/8}, & \varepsilon \geq n^\rho (t - s). \end{cases}$$

Therefore we have

$$P_{u,n}\big(|Y_n(t) - Y_n(s)| > \varepsilon\big) \le e^{-c\varepsilon^2/8\delta} + e^{-cn^\rho\varepsilon/8}$$

uniformly for $s, t \in [t_1(n), t_2(n)]$ with $|t-s| \le \delta$. This verifies the stochastic equicontinuity property (4.64).

We can analogously obtain

$$P_{u,n}\big(|Y_n(t)| > x\big) \le \begin{cases} e^{-cx^2/8(t-s)}, & x \le n^\rho t(1-t)/c, \\ e^{-cn^\rho x/8}, & x \ge n^\rho t(1-t)/c. \end{cases}$$

Therefore it follows by integral formula by parts

$$E_{u,n}|Y_n(t)|^m = m \int_0^\infty x^{m-1} P_{u,n}\big(|Y_n(t)| > x\big) dx$$

$$\le c_{15}\big(t^{m/2}(1-t)^{m/2} + n^{-m\rho}\big),$$

as desired. □

Proof of Theorem 4.13. Begin with the continuity of sample paths of $Y(t)$. Note

$$E\big(Y(t) - Y(s)\big)^2 = EY(t)^2 - 2EY(s)Y(t) + EY(s)^2$$

$$= \frac{1}{c}\Big(t - s - (t-s)^2 - \frac{1}{2}(l(t) - l(s))^2\Big)$$

$$\le \frac{1}{c}(t-s).$$

Since $Y(t) - Y(s)$ is Gaussian with zero mean, we have

$$E\big(Y(t) - Y(s)\big)^4 \le \frac{3}{c^2}(t-s)^2.$$

This implies by Kolmogorov's continuity criterion that there exists a separable continuous version of $Y(\cdot)$ on $[0, 1]$.

Turn to the proof of (ii). The asymptotic normality directly follows from Theorem 4.5. Indeed, making a change of time parameter, we see that $\big(tX_n(\frac{\sqrt{n}}{c}\log\frac{1}{1-t}), 0 < t < 1\big)$ converges weakly to $\big(tX(\frac{1}{c}\log\frac{1}{1-t}), 0 < t < 1\big)$ in terms of finite dimensional distributions. If letting $Y(t) = tX(\frac{1}{c}\log\frac{1}{1-t})$, then a simple calculus shows that $Y(t), 0 < t < 1$ has the desired covariance structure (4.60).

Finally, we show (iii). Fix g. Without loss of generality, we assume that $\alpha, \beta > 1$. Introduce $\varepsilon_m = 1/m$, $m \ge 1$, and break up the integration interval into three subsets: $(0, \varepsilon_m)$, $(\varepsilon_m, 1 - \varepsilon_m)$, and $(1 - \varepsilon_m, 1)$. Let

$$Z_n = \int_0^1 g(t, Y_n(t)) dt, \quad Z_{n,m} = \int_{\varepsilon_m}^{1-\varepsilon_m} g(t, Y_n(t)) dt$$

and

$$Z = \int_0^1 g(t, Y(t))dt, \quad Z_m = \int_{\varepsilon_m}^{1-\varepsilon_m} g(t, Y(t))dt.$$

Then using Lemma 4.12 and Theorem 1.16, it is not difficult to check the following three statements:

(a) for any $\varepsilon > 0$,

$$\lim_{m \to \infty} \lim_{n \to \infty} P_n(|Z_n - Z_{n,m}| > \varepsilon) = 0;$$

(b) for each $m \geq 1$,

$$Z_{n,m} \xrightarrow{d} Z_m;$$

(c) for any $\varepsilon > 0$,

$$\lim_{m \to \infty} P(|Z_m - Z| > \varepsilon) = 0, \quad m \to \infty.$$

Here we prefer to leave the detailed computation to the reader, see also Pittel (2002).

Having (a), (b) and (c) above, Theorem 4.2, Chapter 1 of Billingsley (1999a) guarantee $Z_n \xrightarrow{d} Z$, which concludes the proof. $\qquad\square$

To illustrate, we shall give two examples. The first one treats the character ratios in the symmetric group $\mathcal{S}_n$. Fix a transposition $\tau \in \mathcal{S}_n$. Define the character ratio by

$$\gamma_\tau(\lambda) = \frac{\chi_\lambda(\tau)}{d_\lambda}, \quad \lambda \in \mathcal{P}_n$$

where χ_λ be an irreducible representation associated with the partition $\lambda \mapsto n$, d_λ is the dimension of χ_λ, i.e., $d_\lambda = \chi_\lambda(1^n)$.

The ratio function played an important role in the well-known analysis of the card-shuffling problem performed by Diaconis and Shahshahani (1981). In fact, Diaconis and Shahshahani proved that the eigenvalues for this random walk are the character ratios each occurring with multiplicity d_λ^2. Character ratios also play a crucial role in the work on moduli spaces of curves, see Eskin and Okounkov (2001), Okounkov and Pandharipande (2004). The following theorem can be found in the end of the paper of Diaconis and Shahshahani (1981).

Theorem 4.14. *Under* $(\mathcal{P}_n, P_{u,n})$,

$$n^{3/4}\gamma_\tau(\lambda) \xrightarrow{d} N(0, \sigma_\tau^2),$$

where σ_τ^2 *is given by*

$$\sigma_\tau^2 = \frac{4}{c^4} \int_0^1 \int_0^1 \frac{EY(s)Y(t)}{s(1-s)t(1-t)} \log \frac{1-s}{s} \log \frac{1-t}{t} ds dt$$

where $EY(s)Y(t)$ *is given by (4.60).*

Proof.   Recall the following classic identity due to Frobenius (1903)

$$\gamma_\tau(\lambda) = \frac{1}{n(n-1)} \sum_k \left(\lambda_k^2 - (2k-1)\lambda_k \right)$$

$$= \frac{1}{\binom{n}{2}} \sum_k \left(\binom{\lambda_k}{2} - \binom{\lambda_k'}{2} \right).$$

It follows from the second equation that $\gamma_\tau(\lambda) \overset{d}{=} \gamma_\tau(\lambda')$ and $E_{u,n}\gamma_\tau(\lambda) = 0$ since $P_{u,n}(\lambda) = P_{u,n}(\lambda')$.

To prove the central limit theorem, we observe

$$\sum_k \left(\lambda_k'^2 - (2k-1)\lambda_k' \right) = n + U_{n,1} + U_{n,2} + U_{n,3},$$

where

$$U_{n,1} = \sum_k \left(m(k)^2 - 2km(k) \right),$$

$$U_{n,2} = \sum_k \left(\varphi_\lambda(k) - m(k) \right)^2,$$

$$U_{n,3} = 2 \sum_k \left(m(k) - k \right)\left(\varphi_\lambda(k) - m(k) \right).$$

It is easy to check that

$$U_{n,1} = O_p(n \log^2 n), \quad U_{n,2} = O_p(n \log^2 n).$$

Turn to $U_{n,3}$. Switching to integration, we get

$$U_{n,3} = 2 \int_0^\infty \left(m(x) - x \right)\left(\varphi_\lambda(x) - m(x) \right)dx + O_p\left(n \log^2 n \right).$$

Substituting $x = \sqrt{n}|\log(1-t)|/c$ and a simple calculus shows that

$$\int_0^\infty \left(m(x) - x \right)\left(\varphi_\lambda(x) - m(x) \right)dx$$

$$= \frac{n^{5/4}}{c^2} \int_0^1 \frac{Y_n(t)}{t(1-t)} \log \frac{1-t}{t} dt + O_p\left(n^{3/4} \log n \right).$$

Set

$$g(t,x) = \frac{x}{t(1-t)} \log \frac{1-t}{t} dt.$$

Then $g(t,x)$ obviously satisfies the condition (4.61) of Theorem 4.13 with parameters $\mu = 1$ and $\alpha = \beta = 3/2$. So we have

$$\int_0^1 \frac{Y_n(t)}{t(1-t)} \log \frac{1-t}{t} dt \overset{d}{\longrightarrow} \int_0^1 \frac{Y(t)}{t(1-t)} \log \frac{1-t}{t} dt.$$

Note that the limit variable is a centered Gaussian random variable with variance

$$\int_0^1 \int_0^1 \frac{EY(s)Y(t)}{s(1-s)t(1-t)} \log \frac{1-s}{s} \log \frac{1-t}{t} ds dt,$$

where $EY(s)Y(t)$ is given by (4.60).

In combination, we obtain

$$n^{3/4} \gamma_\tau(\lambda) \xrightarrow{d} N(0, \sigma_\tau^2),$$

as desired.   □

The second example we shall treat is d_λ. It turns out that the logarithm of d_λ satisfies the central limit theorem. Introduce

$$\kappa(t) = \int_0^\infty \frac{\log |\log x|}{(1-t-tx)^2} dx. \tag{4.67}$$

Theorem 4.15. *Under* $(\mathcal{P}_n, P_{u,n})$,

$$\frac{1}{n^{3/4}} \left(\log d_\lambda - \frac{1}{2} n \log n + An \right) \xrightarrow{d} N(0, \sigma_d^2).$$

Here A *and* σ_d^2 *are given by*

$$A = 1 - \log c + \frac{1}{c^2} \int_0^\infty \frac{y \log y}{e^y - 1} dy$$

and

$$\sigma_d^2 = \frac{1}{c^2} \int_0^1 \int_0^1 EY(s)Y(t)\kappa(s)\kappa(t) ds dt, \tag{4.68}$$

where $EY(s)Y(t)$ *is given by (4.60). Numerically,* $\sigma_d^2 = 0.3375 \cdots$.

The theorem was first proved by Pittel (2002). The proof will use the following classic identities (see also Chapter 5):

$$d_\lambda = n! \frac{\prod_{1 \le i < j \le l}(\lambda_i - \lambda_j + j - i)}{\prod_{1 \le i \le l}(\lambda_i - i + l)!} \tag{4.69}$$

and

$$d_\lambda = \frac{n!}{H_\lambda} := \frac{n!}{\prod_{\square \in \lambda} h_\square}, \tag{4.70}$$

where $h_\square = \lambda_i - i + \lambda_j' - j + 1$, the hook length of the (i, j) square.

It follows directly from (4.70) and (4.69) that

$$d_\lambda = d_{\lambda'} = n! \frac{\prod_{1 \le i < j \le \lambda_1}(\lambda_i' - \lambda_j' + j - i)}{\prod_{1 \le i \le \lambda_1}(\lambda_i' - i + \lambda_1)!}.$$

Consequently, we obtain

$$\log d_\lambda - \log n!$$
$$= \sum_{1 \le i < j \le \lambda_1} \log \left(\lambda'_i - \lambda'_j + j - i \right) - \sum_{1 \le i \le \lambda_1} \log \left(\lambda'_i - i + \lambda_1 \right)!$$
$$=: M_n - N_n. \tag{4.71}$$

The bulk of the argument consists of computing M_n and N_n. We need some basic estimates about $\lambda'_i - \lambda'_j$ below. Let $0 < \delta < 1/2$ and define

$$K = \left\{ (k_1, k_2) : n^\delta \le k_1 \le k_2 - n^\delta \right\}.$$

Lemma 4.13. *Let*

$$\ell(x) = \frac{\sqrt{n}}{c} \log \frac{1}{1 - e^{cx/\sqrt{n}}}, \quad x > 0$$

and denote $\ell(x, y) = \ell(x) - \ell(y)$ *for any* $0 < x \le y$. *Then we have*
(i)

$$m(k_1, k_2) = \left(1 + O(n^{-\delta}) \right) \ell(k_1, k_2)$$

uniformly for $(k_1, k_2) \in K$, *and for* $x_i = c k_i / \sqrt{n}$

$$\sigma^2(k_1, k_2) = \left(1 + O(n^{-\delta}) \right) \frac{\sqrt{n}}{c} \left(\frac{1}{1 - e^{-x_1}} - \frac{1}{1 - e^{-x_2}} \right);$$

(ii)

$$\frac{\sigma(k_1, k_2)}{m(k_1, k_2)} = O\left(n^{-(\delta-a)/2} \right)$$

uniformly for $(k_1, k_2) \in K$ *and* $k_1 \le a \sqrt{n} \log n / c$ *where* $a < \delta$;
(iii)

$$\lim_{n \to \infty} \frac{\sigma^2(k_1, k_2)}{m(k_1, k_2)} = 1$$

uniformly for $(k_1, k_2) \in K$ *and* $k_1 \ge a \sqrt{n} \log n / c$ *where* $a < \delta$.

Proof. See Lemmas 1 and 2 of Pittel (2002). □

Lemma 4.14. *Given* $\varepsilon > 0$ *and* $0 < a < \delta < 1/2$, *there is an* $n_0(a, \delta, \varepsilon) \ge 1$ *such that*
(i)

$$Q_{q_n} \left(|\lambda'_{k_1} - \lambda'_{k_2} - m(k_1, k_2)| > \sigma(k_1, k_2) \sqrt{\varepsilon \log n} \right) \le n^{-\varepsilon/3}$$

uniformly for $(k_1, k_2) \in K$ *and* $k_1 \le a \sqrt{n} \log n / c$;
(ii)

$$Q_{q_n} \left(|\lambda'_{k_1} - \lambda'_{k_2} - m(k_1, k_2)| > (\sigma(k_1, k_2) + \log n) \sqrt{\varepsilon \log n} \right) \le n^{-\varepsilon/3}$$

uniformly for $(k_1, k_2) \in K$ *and* $k_1 \ge a \sqrt{n} \log n / c$.

Proof.   Let η be small enough to ensure that $e^\eta q_n^k < 1$ for all $k \in [k_1, k_2)$. For instance, select $|\eta| = \sigma(k_1, k_2)\sqrt{\varepsilon \log n}$. We have

$$E_{q_n} \exp\left(\eta(\lambda'_{k_1} - \lambda'_{k_2})\right) = \exp\left(\eta m(k_1, k_2) + \frac{\eta^2}{2}\sigma^2(k_1, k_2)\right)$$

$$\cdot \exp\left(O\left(|\eta|^3 \sum_{k_1 \leq k < k_2} \frac{q_n^{3k}}{(1 - q_n)^{3k}}\right)\right). \quad (4.72)$$

Moreover, a delicate analysis shows the remainder term is indeed of order $n^{-\delta/2}\log^{3/2} n = o(1)$. Using (4.72) and Markov's inequality, we easily get

$$Q_{q_n}\left(|\lambda'_{k_1} - \lambda'_{k_2} - m(k_1, k_2)| \geq \sigma(k_1, k_2)\sqrt{\varepsilon \log n}\right)$$

$$\leq 2\exp\left(-\frac{\varepsilon}{2}\log n + O(n^{-\delta/2}\log^{3/2} n)\right)$$

$$\leq n^{-\varepsilon/3}.$$

This concludes the proof of (i). Turn to (ii). Set

$$|\eta| = \frac{\sqrt{\varepsilon \log n}}{\sigma(k_1, k_2) + \log n}.$$

Then $|\eta| \to 0$. Using only the first order expansion, we obtain

$$E_{q_n} \exp\left(\eta(\lambda'_{k_1} - \lambda'_{k_2})\right) = \exp\left((e^\eta - 1)m(k_1, k_2)\right)$$

$$\cdot \exp\left(O\left(\eta^2 \sum_{k_1 \leq k < k_2} \frac{q_n^{2k}}{(1 - q_n)^{2k}}\right)\right). \quad (4.73)$$

Note

$$\eta^2 \sum_{k_1 \leq k < k_2} \frac{q_n^{2k}}{(1 - q_n)^{2k}} = O(n^{-a}\log n).$$

Using (4.73) and the Markov inequality, and noting $\sigma^2(k_1, k_2)/m(k_1, k_2) = 1 + o(1)$, we have

$$Q_{q_n}\left(|\lambda'_{k_1} - \lambda'_{k_2} - m(k_1, k_2)| \geq (\sigma(k_1, k_2) + \log n)\sqrt{\varepsilon \log n}\right) \leq n^{-\varepsilon/3}.$$

The proof is complete.   □

Lemma 4.14 can be used to obtain the following concentration-type bound for $\lambda'_i - \lambda'_j$ under $(\mathcal{P}_n, P_{u,n})$.

Proposition 4.2. *With $P_{u,n}$-probability $1 - O(n^{-1/4})$ at least,*

$$|\lambda'_{k_1} - \lambda'_{k_2} - m(k_1, k_2)| \leq 3(\sigma(k_1, k_2) + \log n)(\log n)^{1/2} \quad (4.74)$$

uniformly for $(k_1, k_2) \in K$. If, in addition, $k_1 \leq an^{1/2}\log n/c$ where $a < \delta$, then the summand $\log n$ can be dropped.

Proof. Take $\varepsilon = 9$ in Lemma 4.14 to conclude

$$Q_{q_n}\left(\left|\lambda'_{k_1} - \lambda'_{k_2} - m(k_1, k_2)\right| \geq 3(\sigma(k_1, k_2) + \log n)(\log n)^{1/2}\right) \leq n^{-3}.$$

On the other hand, note by Lemma 4.2

$$P_{u,n}(B) = Q_{q_n}\left(B\,\big|\,|\lambda| = n\right)$$
$$\leq \frac{Q_{q_n}(B)}{Q_{q_n}(|\lambda| = n)}.$$

Then according to (4.22)

$$P_{u,n}\left(\left|\lambda'_{k_1} - \lambda'_{k_2} - m(k_1, k_2)\right| \geq 3(\sigma(k_1, k_2) + \log n)(\log n)^{1/2}\right) \leq n^{-9/4}$$

for each $(k_1, k_2) \in K$ and $1 \leq k_1, k_2 \leq n$. This immediately implies (4.74), as asserted. $\square$

Having these basic estimates, we are now ready to compute M_n and N_n. Let us start with N_n. Define

$$\mu_k = \lceil \ell(k) \rceil,$$

$$\mathcal{N}_n = \sum_{1 \leq k \leq \lambda_1} \log \left(\mu_k - k + \lambda_1\right)!,$$

$$R_n = \sum_{l_n \leq k < k_n} \left(\lambda'_k - m(k)\right) \log \left(m(k) - k + \lambda_1\right),$$

where $l_n = [n^\delta]$ and $k_n = \left[\sqrt{n}\log n / c\right]$.

Lemma 4.15. *Under $(\mathcal{P}_n, P_{u,n})$, we have for $1/8 < \delta < 1/4$*

$$N_n = \mathcal{N}_n + R_n + o_p(n^{3/4}). \tag{4.75}$$

Proof. Let $\phi(x) = x \log x - x$. Then by the Stirling formula for factorial,

$$N_n = \sum_{k=1}^{\lambda_1} \phi(\lambda'_k - k + \lambda_1) + \Delta(\lambda_1), \tag{4.76}$$

where $\Delta(\lambda_1) = O_p\left(n^{1/2}\log^2 n\right)$. Define

$$\bar{N}_n = \sum_{k=1}^{\lambda_1} \phi(m(k) - k + \lambda_1).$$

We shall compare N_n with $\bar{N}_n$ below. For this, we break up the sum in (4.76) into $N_n^{(1)}$, $N_n^{(2)}$ and $N_n^{(3)}$ for $k \in [1, l_n)$, $k \in [l_n, k_n]$ and $k \in (k_n, \infty)$ respectively. We similarly define $\bar{N}_n^{(1)}$, $\bar{N}_n^{(2)}$ and $\bar{N}_n^{(3)}$.

Observe that uniformly for $1 \le k \le \lambda_1$

$$\log\left(\lambda_k' - k + \lambda_1\right) = O_p(\log n), \quad \log\left(m(k) - k + \lambda_1\right) = O_p(\log n),$$

and so

$$\phi(\lambda_k' - k + \lambda_1) = O_p(n^{1/2}\log^2 n), \quad \phi(m(k) - k + \lambda_1) = O_p(n^{1/2}\log^2 n).$$

Then we have

$$N_n^{(1)} = O_p(n^{1/2+\delta}\log^2 n), \quad \bar{N}_n^{(1)} = O_p(n^{1/2+\delta}\log^2 n),$$

which implies

$$N_n^{(1)} - \bar{N}_n^{(1)} = O_p(n^{1/2+\delta}\log^2 n) = o_p(n^{3/4}). \tag{4.77}$$

Turn to $N_n^{(2)} - \bar{N}_n^{(2)}$. With the help of (4.74), we expand $\phi(x)$ at $x = m(k) - k + \lambda_1$:

$$\phi(\lambda_k' - k + \lambda_1) = \phi(m(k) - k + \lambda_1) + (\lambda_k' - m(k))\log(m(k) - k + \lambda_1)$$
$$+ O_p\left(\frac{\sigma^2(k)\log n}{m(k) - k + \lambda_1}\right).$$

It follows from Lemma 4.13 that the remainder term is controlled by $O_p(\log n)$. Hence

$$N_n^{(2)} - \bar{N}_n^{(2)} = \sum_{k=l_n}^{k_n} (\lambda_k' - m(k))\log(m(k) - k + \lambda_1)$$
$$+ O_p(n^{1/2}\log^2 n). \tag{4.78}$$

As for the third term, we analogously use the Taylor expansion to obtain

$$N_n^{(3)} - \bar{N}_n^{(3)} = \sum_{k>k_n} (\lambda_k' - m(k))\log x_k^*,$$

where x_k^* is between $m(k) - k + \lambda_1$ and $\lambda_k' - k + \lambda_1$.

It follows from (4.74)

$$N_n^{(3)} - \bar{N}_n^{(3)} = o_p(n^{3/4}). \tag{4.79}$$

Putting (4.77)-(4.79) together yields

$$N_n - \bar{N}_n = R_n + o_p(n^{3/4}).$$

To conclude the proof, we observe

$$\bar{N}_n = \mathcal{N}_n + O_p(n^{1/2}\log n).$$

Now the assertion (4.75) is valid. $\qquad\square$

We shall next turn to compute the M_n. Let

$$g(y) = \log(e^y - 1),$$

$$v(x) = -\int_0^\infty e^{-y} \log|g(y) - g(x)|dy, \quad x > 0$$

$$\mathcal{M}_n = \sum_{1 \le i < j \le \lambda_1} \log\big(\mu(i,j) + j - i\big),$$

$$S_n = \sum_{2l_n \le k \le \lambda_1 - l_n} \big(\lambda'_k - m(k)\big)\big(v(y_k) + \log(g(y_{\lambda_1}) - g(y_k))\big),$$

where $y_k = ck/\sqrt{n}$ and $\mu(i,j) = \sum_{l=i}^{j-1} \mu_l$.

Lemma 4.16. *Under $(\mathcal{P}_n, P_{u,n})$, we have*

$$M_n = \mathcal{M}_n + S_n + o_p(n^{3/4}). \qquad (4.80)$$

Proof. Denote $K = \{(k_1, k_2) : l_n \le k_1 \le k_2 - l_n\} \subseteq [1, \lambda_1] \times [1, \lambda_1]$. Obviously, it follows

$$M_n = \sum_{(k_1,k_2)\in K} \log\big(\lambda'_{k_1} - \lambda'_{k_2} + k_2 - k_1\big) + O_p(n^{1/2+\delta}\log^2 n). \quad (4.81)$$

By (4.74), with high probability

$$\frac{|\lambda'_{k_1} - \lambda'_{k_2} - m(k_1, k_2)|}{m(k_1, k_2) + k_2 - k_1} \le 3\frac{\sigma(k_1, k_2)}{m(k_1, k_2) + k_2 - k_1}$$
$$= o(1)$$

for all $(k_1, k_2) \in K$. So uniformly for $(k_1, k_2) \in K$

$$\log\big(\lambda'_{k_1} - \lambda'_{k_2} + k_2 - k_1\big) = \log\big(m(k_1, k_2) + k_2 - k_1\big)$$
$$+ \frac{\lambda'_{k_1} - \lambda'_{k_2} + m(k_1, k_2)}{m(k_1, k_2) + k_2 - k_1}$$
$$+ O_p\Big(\frac{\sigma^2(k_1, k_2)\log^2 n + \log^4 n}{(m(k_1, k_2) + k_2 - k_1)^2}\Big). \quad (4.82)$$

Take a closer look at each term in the right hand side of (4.82). First, observe

$$\sum_{(k_1,k_2)\in K} \frac{\sigma^2(k_1, k_2)\log^2 n + \log^4 n}{(m(k_1, k_2) + k_2 - k_1)^2} = o_p(n^{3/4}). \qquad (4.83)$$

Second, a simple algebra shows

$$\sum_{(k_1,k_2)\in K} \frac{\lambda'_{k_1} - \lambda'_{k_2} + m(k_1,k_2)}{m(k_1,k_2) + k_2 - k_1} = S_n^{(1)} + S_n^{(2)}, \qquad (4.84)$$

where

$$S_n^{(1)} = \sum_{k=l_n}^{2l_n} \left(\lambda'_k - m(k)\right) \sum_{j=k+l_n}^{\lambda_1} \frac{1}{m(j,k) + j - k}$$

and

$$S_n^{(2)} = \sum_{k=2l_n}^{\lambda_1} \left(\lambda'_k - m(k)\right) \sum_{l_n \le j \le \lambda_1, |j-k| \ge l_n} \frac{1}{m(j,k) + j - k}.$$

It follows from Proposition 4.2 that with high probability $S_n^{(1)}$ is of smaller order than $n^{3/4}$. To study $S_n^{(2)}$, we need a delicate approximation (see pp. 200-202 of Pittel (2002) for lengthy and laborious computation):

$$\sum_{l_n \le j \le \lambda_1, |j-k| \ge l_n} \frac{1}{m(j,k) + j - k} = v(y_k) + \log\left(g(y_{\lambda_1}) - g(y_k)\right)$$

$$+ O_p\left(n^{-1/2}\log n\right).$$

Then (4.84) becomes

$$\sum_{k=2l_n}^{\lambda_1} \left(\lambda'_k - m(k)\right)\left(v(y_k) + \log(g(y_{\lambda_1}) - g(y_k))\right) + o_p\left(n^{3/4}\right). \quad (4.85)$$

Inserting (4.82) into (4.81) and noting (4.83) and (4.85),

$$M_n = \sum_{(k_1,k_2)\in K} \log\left(m(k_1,k_2) + k_2 - k_1\right)$$

$$+ \sum_{k=2l_n}^{\lambda_1} \left(\lambda'_k - m(k)\right)\left(v(y_k) + \log(g(y_{\lambda_1}) - g(y_k))\right) + o_p\left(n^{3/4}\right).$$

To conclude the proof, we observe

$$\sum_{(k_1,k_2)\in K^c} \log\left(m(k_1,k_2) + k_2 - k_1\right) = O_p(n^{1/2+\delta}\log^2 n)$$

and

$$\sum_{1 \le k_1 < k_2 \le \lambda_1} \left|\log\left(m(k_1,k_2) + k_2 - k_1\right) - \log\left(\mu(k_1,k_2) + k_2 - k_1\right)\right|$$

$$\le 2 \sum_{1 \le k_1 < k_2 \le \lambda_1} \frac{1}{k_2 - k_1} = O_p(n^{1/2}\log^2 n).$$

□

Lemma 4.17. *Under* $(\mathcal{P}_n, P_{u,n})$, *we have*

$$\log d_\lambda - \log n! = \log f(\mu) + T_n + o_p(n^{3/4}),\qquad(4.86)$$

where

$$f(\mu) = \frac{\prod_{1\leq i<j\leq\lambda_1}(\mu_i - \mu_j + j - i)}{\prod_{1\leq i\leq\lambda_1}(\mu_i - i + \lambda_1)!}$$

and

$$T_n = \sum_{k=2l_n}^{\lambda_1} v(y_k)(\lambda_k' - m(k)).\qquad(4.87)$$

Proof. By (4.16), (4.75) and (4.80), we have

$$\begin{aligned}
\log d_\lambda - \log n! &= M_n - N_n \\
&= \mathcal{M}_n - \mathcal{N}_n + S_n - R_n + o_p(n^{3/4}) \\
&= \log f(\mu) + S_n - R_n + o_p(n^{3/4}).
\end{aligned}$$

On the other hand, it trivially follows

$$\begin{aligned}
S_n - R_n - T_n &= \sum_{k>\lambda_1-l_n} v(y_k)(\lambda_k' - m(k)) \\
&+ \sum_{2l_n\leq k\leq\lambda_1-l_n} \log\big(g(y_{\lambda_1}) - g(y_k)\big)(\lambda_k' - m(k)) \\
&- \sum_{l_n\leq k\leq k_n} \log\big(m(k) - k + \lambda_1\big)(\lambda_k' - m(k)).\quad(4.88)
\end{aligned}$$

Therefore we need only prove that the right hand side of (4.88) are negligible. First, according to Lemma 5 of Pittel (2002), there is a constant $c_6 > 0$

$$|v(x)| = c_6 \log\left(x + \frac{1}{x}\right), \quad |v'(x)| \leq \frac{c_6}{x}.\qquad(4.89)$$

We easily get

$$\sum_{k>\lambda_1-l_n} v(y_k)\lambda_k' = O(n^\delta \lambda_1 \log n) = O_p(n^{1/2+\delta}\log^2 n).$$

Also, it is even simpler to check

$$\sum_{k>\lambda_1-l_n} v(y_k)m(k) = O_p(n^{1/2}\log^2 n).$$

Thus we have

$$\sum_{k>\lambda_1-l_n} v(y_k)(\lambda_k' - m(k)) = o_p(n^{3/4}).$$

Second, observe the following simple facts:

$$\sum_{k=1}^{\lambda_1} \lambda_k' = n$$

and

$$\sum_{k=1}^{\lambda_1} m(k) = \frac{n}{c} \int_0^\infty \log \frac{1}{1 - e^{-y}} dy + O_p(n^{1/2} \log n)$$

$$= n + O_p(n^{1/2} \log n).$$

Then we easily have

$$\sum_{k=1}^{\lambda_1} (\lambda_k' - m(k)) = O_p(n^{1/2} \log n). \tag{4.90}$$

On the other hand, it follows from (4.74)

$$\sum_{k \notin [2l_n, k_n]} |\lambda_k' - m(k)| = O_p(n^{3/4-\varepsilon}),$$

which together with (4.90) in turn implies

$$\sum_{2l_n \le k \le k_n} (\lambda_k' - m(k)) = O_p(n^{3/4}). \tag{4.91}$$

Besides, we have by (4.74)

$$\sum_{2l_n \le k \le k_n} |\lambda_k' - m(k)| = O_p(n^{3/4} \log n). \tag{4.92}$$

By the definition of $g(x)$ and $m(k)$,

$$\frac{\sqrt{n}}{c}(g(y_{\lambda_1}) - g(y_k)) = m(k) - k + \lambda_1 + \frac{\sqrt{n}}{c}(1 - e^{-c\lambda_1/\sqrt{n}})$$

$$= m(k) - k + \lambda_1 + O_p(1).$$

Therefore for $k \le k_n$

$$\log(g(y_{\lambda_1}) - g(y_k)) - \log(m(k) - k + \lambda_1)$$

$$= \log \frac{c}{\sqrt{n}} + O_p(n^{-1/2} \log^{-1} n).$$

Thus by (4.91) and (4.92)

$$\sum_{2l_n \le k \le k_n} (\log(g(y_{\lambda_1}) - g(y_k)) - \log(m(k) - k + \lambda_1))(\lambda_k' - m(k))$$

$$= \log \frac{c}{\sqrt{n}} \sum_{2l_n \le k \le k_n} (\lambda_k' - m(k))$$

$$+ O_p(n^{-1/2} \log^{-1} n) \sum_{2l_n \le k \le k_n} |\lambda_k' - m(k)|$$

$$= o_p(n^{3/4}).$$

The proof is complete. $\qquad\qquad\qquad\qquad\qquad\qquad\qquad\qquad\square$

To proceed, we need to treat the constant term $\log f_\mu$.

Lemma 4.18. *Under* $(\mathcal{P}_n, P_{u,n})$, *we have*

$$\log f(\mu) = -\frac{1}{2} n \log n - n \int_0^\infty \frac{t \log t}{e^{ct} - 1} + o_p(n^{3/4}).$$

Proof. Form a Young diagram $\mu = (\mu(1), \mu(2), \cdots, \mu(\lambda_1))$ and denote its dual by $\nu = (\nu(1), \nu(2), \cdots, \nu(\mu(1)))$, where

$$\nu(i) = \max\{1 \le k \le \lambda_1 : \mu(k) \ge i\}, \quad 1 \le k \le \mu(1).$$

We remark that ν can be viewed as an approximation to the random diagram λ since μ is an approximation of λ'. Now apply the hook formula (4.70) to the diagram μ to obtain

$$\log f(\mu) = -\sum_{i \le \mu(1), k \le \nu(i)} \log\big(\nu(i) - k + \mu(k) - i + 1\big). \qquad (4.93)$$

As a first step, we need to replace asymptotically $\mu(\cdot)$ and $\nu(\cdot)$ by $\ell(\cdot)$ in (4.93). Indeed, by the definition of $\ell(\cdot)$ and $\mu(\cdot)$, it follows

$$\ell(k) - 1 \le \nu(k) \le \ell(k-1), \quad 1 < k \le \mu(1).$$

Define

$$D = \big\{(i,k) : i \le \min\{\mu(k), \ell(k)\}, \ k \le \min\{\nu(i), \ell(i)\}\big\},$$

then

$$\log f(\mu) = -\sum_{(i,k) \in D} \log\big(\nu(i) - k + \mu(k) - i + 1\big) + O_p(n^{1/2} \log^2 n).$$

Moreover,

$$-\sum_{(i,k) \in D} \log\big(\nu(i) - k + \mu(k) - i + 1\big)$$

$$= -\sum_{(i,k) \in D} \log\big(\ell(i) - k + \ell(k) - i + 1\big)$$

$$+ O\Big(\sum_{(i,k) \in D} \frac{|\ell(i) - \nu(i)|}{\min\{\nu(i), \ell(i)\} - k + 1} + \frac{|\ell(k) - \mu(k)|}{\min\{\mu(k), \ell(k)\} - i + 1}\Big)$$

$$= -\sum_{(i,k) \in D} \log\big(\ell(i) - k + \ell(k) - i + 1\big) + O_p(n^{1/2} \log^2 n).$$

The same argument results in another $O_p(n^{1/2} \log^2 n)$ error term if we replace further D by $D^* = \{(i,k) : i \le \ell(k), k \le \ell(i)\}$. Thus

$$\log f(\mu) = -\sum_{(i,k) \in D^*} \log\big(\ell(i) - k + \ell(k) - i + 1\big) + O_p(n^{1/2} \log^2 n).$$

The next step is to switch the sum into an integral. Let

$$H_n = \{(x,y) : 0 < x, y \le \ell(1), \, x \le \ell(y), \, y \le \ell(x)\},$$

then

$$\log f(\mu) = -\int\int_{(x,y)\in H_n} \log\big(\ell(x) - y + \ell(y) - x + 1\big)dxdy$$
$$+ O\big(n^{1/2}\log^2 n\big).$$

Furthermore, if letting

$$H_\infty = \{(x,y) : x, y > 0, \, x \le \ell(y), \, y \le \ell(x)\},$$

then we have

$$\log f(\mu) = -\int\int_{(x,y)\in H_\infty} \log\big(\ell(x) - y + \ell(y) - x + 1\big)dxdy$$
$$+ O\big(n^{1/2}\log^2 n\big). \tag{4.94}$$

To see this, make a change of variables

$$u = \frac{\ell(x) - y}{n^{1/2}}, \qquad v = \frac{\ell(y) - x}{n^{1/2}}.$$

Then in terms of u, v, the domain H_∞ becomes $\{(u,v) : u \ge 0, v \ge 0\}$, and the inverse transform is

$$x = \frac{\sqrt{n}}{c}\log\frac{e^{c(u+v)} - 1}{e^{c(u+v)} - e^{cv}}, \qquad y = \frac{\sqrt{n}}{c}\log\frac{e^{c(u+v)} - 1}{e^{c(u+v)} - e^{cu}}.$$

So the Jacobian determinant is

$$\det\begin{pmatrix} \frac{\partial x}{\partial u} & \frac{\partial x}{\partial v} \\ \frac{\partial y}{\partial u} & \frac{\partial y}{\partial v} \end{pmatrix} = \frac{n}{e^{c(u+v)} - 1}.$$

The difference between the integrals over H_∞ and H_n is the double integral over $H_\infty \setminus H_n$:

$$\int\int_{(x,y)\in H_\infty\setminus H_n} \log\big(\ell(x) - y + \ell(y) - x + 1\big)dxdy$$
$$\le n\int\int_{0\le u\le n^{-1/2}, v\ge 0} \frac{\log(\sqrt{n}(u+v) + 1)}{e^{c(u+v)} - 1}dudv$$
$$= O\big(n^{1/2}\log^2 n\big).$$

The last step is to explicitly calculate the double integral value over H_∞ in (4.94). Via the substitutions, the integral is equal to

$$\frac{1}{2}n\log n + n\int_0^\infty \frac{t\log t}{e^{ct} - 1}dt + O\big(n^{1/2}\big),$$

as claimed. $\qquad\qquad\qquad\qquad\qquad\qquad\qquad\qquad\qquad\qquad\qquad\quad \square$

To conclude, we shall show that the linearised weighted sum T_n in (4.87) admits an integral representation up to a negligible error term, and so converges in distribution to a normal random variable.

Lemma 4.19. *Under* $(\mathcal{P}_n, P_{u,n})$, *we have*

$$\frac{T_n}{n^{3/4}} \xrightarrow{d} N(0, \sigma_T^2),$$

where $\sigma_T^2 = \sigma_d^2$ *given by (4.68).*

Proof. Start with an integral representation for T_n. Using the second inequality of (4.89),

$$v(y_k) - v(x) = O\left(\frac{y_k - y_{k-1}}{y_k}\right) = O(n^{-\delta})$$

uniformly for $x \in [k-1, k)$ and $k \geq 2l_n$.

Also, it is easy to see

$$m(k) - m(x) = O\left((e^{y_k} - 1)^{-1/2}\right)$$

uniformly for $x \in [k-1, k)$ and $k \geq 2l_n$.

Therefore we have

$$T_n = \int_{2l_n - 1}^{\infty} v(x)\left(\sum_{k \geq x}(r_k - E_{q_n} r_k)\right) dx + \Delta T_n, \tag{4.95}$$

where ΔT_n is of order

$$n^{-\delta} \sum_{k=2l_n}^{\infty} |\lambda_k' - m(k)| + \sum_{k=2l_n}^{\infty} \frac{\log(y_k + y_k^{-1}) + n^\delta}{e^{y_k} - 1}.$$

By virtue of Proposition 4.2, the whole order is actually $O_p(n^{3/4-\delta}\log^{1/2}n)$. Neglecting ΔT_n, we can equate T_n with the integral on the right side in (4.95). Furthermore, we also extend the integration to $[x_n, \infty)$, where

$$x_n = -\frac{\sqrt{n}}{c}\log\left(1 - \frac{c}{2\sqrt{n}}\right).$$

Making the substitution

$$x = \frac{\sqrt{n}}{c}\log\frac{1}{1-t}$$

in the last integral and using the definition of the process $Y_n(t)$ we obtain

$$T_n = \frac{n^{3/4}}{c}\int_0^1 \frac{v(-\log(1-t))}{t(1-t)}Y_n(t)dt + O(n^{3/4-\varepsilon}).$$

Now we are in a position to apply Theorem 4.13 to the function

$$g(t, x) := \frac{v(-\log(1 - t))}{t(1 - t)} x,$$

which clearly meets the condition (4.61) with $\mu = \alpha = \beta = 1$. Therefore

$$\frac{T_n}{n^{3/4}} \xrightarrow{d} \frac{1}{c} \int_0^1 \frac{v(-\log(1 - t))}{t(1 - t)} Y(t) dt.$$

The limit variable in the right hand side is a centered Gaussian random variable with variance

$$\sigma_T^2 := \frac{1}{c^2} \int_0^1 \int_0^1 EY(s)Y(t) \frac{v(-\log(1 - s))}{s(1 - s)} \frac{v(-\log(1 - t))}{t(1 - t)} ds dt.$$

To conclude the proof, we note

$$v(-\log(1 - t)) = -\int_0^\infty e^{-y} \log \left| \log \frac{(1 - t)}{t} + \log(e^y - 1) \right| dy$$
$$= -t(1 - t)\kappa(t),$$

where $\kappa(t)$ is given by (4.67). The proof is complete. $\qquad\square$

Proof of Theorem 4.15. Putting Lemmas 4.17, 4.18 and 4.19 all together, we can conclude the proof. $\qquad\square$

4.5 Random multiplicative partitions

In this section we shall introduce a class of multiplicative measures as extension of uniform measure and describe briefly the corresponding limit shape and second order fluctuation around the shape. The reader is referred to Su (2014) for detailed proofs and more information.

Consider a sequence of functions $g_k(z), k \geq 1$, analytic in the open disk $\mathbb{D}_\varrho = \{z \in \mathbb{C} : |z| < \varrho\}$, $\varrho = 1$ or $\varrho = \infty$, such that $g_k(0) = 1$. Assume further that
(i) the Taylor series

$$g_k(z) = \sum_{j=0}^\infty s_k(j) z^j$$

have all coefficients $s_k(j) \geq 0$ and
(ii) the infinite product

$$\mathcal{G}(z) = \prod_{k=1}^\infty g_k(z^k)$$

converges in $\mathbb{D}_\varrho$.

Now define the measure $P_{m,n}$ on $\mathcal{P}_n$ by

$$P_{m,n}\big(\lambda \in \mathcal{P}_n : r_k(\lambda) = j\big) = \frac{s_k(j)}{Z_{m,n}}$$

and

$$P_{m,n}(\lambda) = \frac{\prod_{k=1}^{\infty} s_k(r_k)}{Z_{m,n}}, \quad \lambda \in \mathcal{P}_n,$$

where

$$Z_{m,n} = \sum_{\lambda \in \mathcal{P}_n} \prod_{k=1}^{\infty} s_k(r_k).$$

Here m in the subscript stands for multiplicative.

We also define a family of probability measures $Q_{m,q}, q \in (0, \varrho)$ on $\mathcal{P}$ in the following way:

$$Q_{m,q}(\lambda) = \frac{\prod_{k=1}^{\infty} s_k(r_k)}{\mathcal{G}(q)} q^{|\lambda|}, \quad \lambda \in \mathcal{P}.$$

It is easy to see

$$Q_{m,q}\big(\lambda \in \mathcal{P} : r_k(\lambda) = j\big) = \frac{s_k(j)q^{kj}}{g_k(q^k)}, \quad j \geq 0, \quad k \geq 1$$

and so different occupation numbers are independent. The measure $Q_{m,q}$ is called multiplicative.

According to Vershik (1996), analog of Lemma 4.3 is valid for $Q_{m,q}$ and $P_{m,n}$. This will enable us make full use of conditioning argument.

Note that the generating function $\mathcal{G}(z)$, along with its decomposition $\mathcal{G}(z) = \prod_{k=1}^{\infty} g_k(z^k)$, completely determines such a family. It actually contain many important examples, see Vershik (1996), Vershik and Yakubovich (2006). A particularly interesting example is the $\mathcal{G}(z)$ is generated by $g_k(z) = 1/(1 - z)^{k^\beta}, \beta > -1$. In such a special case, the convergence radius of g_k and $\mathcal{G}$ is $\varrho = 1$. We also write $Q_{\beta,q}, P_{\beta,n}$ for probabilities and $E_{\beta,q}, E_{\beta,n}$ for expectations respectively. Set

$$\mathcal{G}_\beta(z) = \prod_{k=1}^{\infty} \frac{1}{(1 - z^k)^{k^\beta}}.$$

Vershik (1996), Vershik and Yakubovich (2006) treat $Q_{\beta,q}$ and $P_{\beta,n}$ as generalized Bose-Einstein models of ideal gas; while in combinatorics and number theory they are well known for a long time as weighted partitions.

Remark 4.1. $P_{0,n}$ corresponds to the uniform measure $P_{u,n}$ on $\mathcal{P}_n$, and $Z_{0,n}$ is the Euler function $p(n)$: the number of partitions of n. In the case of $\beta = 1$, the $\mathcal{G}_\beta(z)$ is the generating function for the numbers $p_3(n)$ of 3-dimensional plane partitions of n (see Andrews (1976)):

$$\sum_{n \geq 0} p_3(n) z^n = \prod_{k \geq 1} \frac{1}{(1 - z^k)^k}.$$

However, the $P_{1,n}$ is completely different from the uniform measure on 3-dimensional plane partitions of n.

Vershik (1996), in an attempt to capture various limiting results concerning particular functionals in a unified framework, posed the question of evaluating the limit shape for $\varphi_\lambda(t)$ under $P_{\beta,n}$. In particular, we have

Theorem 4.16. *Assume $\beta \geq 0$, let $h_n = (\frac{\Gamma(\beta+2)\zeta(\beta+2)}{n})^{1/(\beta+2)}$. Consider the scaled function*

$$\varphi_{\beta,n}(t) = h_n^{\beta+1} \varphi_\lambda\Big(\frac{t}{h_n}\Big), \quad t \geq 0.$$

Then it follows

$$\varphi_{\beta,n} \to \Psi_\beta$$

in the sense of uniform convergence on compact sets, where Ψ_β is the function defined by

$$\Psi_\beta(t) = \int_t^\infty \frac{u^\beta e^{-u}}{1 - e^{-u}} du.$$

More precisely, for any $\varepsilon > 0$ and $0 < a < b < \infty$, there exists an n_0 such that for $n > n_0$ we have

$$P_{\beta,n}\Big(\lambda \in \mathcal{P}_n : \sup_{a \leq t \leq b} |\varphi_{\beta,n}(t) - \Psi_\beta(t)| > \varepsilon\Big) < \varepsilon.$$

Remark 4.2. The value of h_n is in essence determined so that $E_{\beta,q}|\lambda| \sim n$, where $q = e^{-h_n}$. For $\beta = 0$, the scaling constants along both axes are $c\sqrt{n}$. Moreover $\Psi_0(t)$ is equal to $\Psi(t)$ of (4.12). While for $\beta > 0$, two distinct scaling constants must be adapted. In fact, the value on the s axis is more compressed than the indices on the t axis. Also, it is worth noting

$$\Psi_\beta(0) = \int_0^\infty \frac{u^\beta e^{-u}}{1 - e^{-u}} du < \infty$$

by virtue of $\beta > 0$.

Having the limit shape, we will continue to further study the second order fluctuation of Young diagrams around it. This will separately discussed according to two cases: at the edge and in the bulk. First, let us look at the asymptotic distribution of the largest part of a partition under $P_{\beta,n}$. The following result, due to Vershik and Yakubovich (2006), is an extension of Erdös and Lehner's theorem

Theorem 4.17.

$$\lim_{n\to\infty} P_{\beta,n}\left(\lambda \in \mathcal{P}_n : \lambda_1 - \frac{A_n}{h_n} \leq \frac{x}{h_n}\right) = e^{-e^{-x}},$$

where

$$A_n = \frac{\beta+1}{\beta+2} \log n + \beta \log\log n + \beta \log\frac{\beta+1}{\beta+2} - \frac{\beta+1}{\beta+2} \log \Gamma(\beta+2)\zeta(\beta+2).$$

Remark 4.3. As known to us, $\lambda \overset{d}{=} \lambda'$, and so λ_1' and λ_1 have the same asymptotic distribution under $(\mathcal{P}_n, P_{u,n})$. But such an elegant property is no longer valid under $(\mathcal{P}_n, P_{\beta,n})$. In fact, we have the following asymptotic normality for λ_1' instead of Gumbel distribution.

Let $\sigma_n^2 = h_n^{-(\beta+1)}$ and define for $k \geq 1$

$$\mu_{n,k} = \sum_{j=k}^{\infty} j^\beta \frac{e^{-h_n j}}{1 - e^{-h_n j}}, \quad \sigma_{n,k}^2 = \sum_{j=k}^{\infty} j^\beta \frac{e^{-h_n j}}{(1 - e^{-h_n j})^2},$$

where h_n is as in Theorem 4.16.

Theorem 4.18. *(i) Under $P_{\beta,n}$ with $\beta > 1$, we have as $n \to \infty$,*

$$\frac{\lambda_k' - \mu_{n,k}}{\sigma_n} \overset{d}{\longrightarrow} N\left(0, \kappa_\beta^2(0)\right),$$

where

$$\kappa_\beta^2(0) = \Gamma(\beta+1)\zeta(\beta+1,0) - \frac{\Gamma(\beta+2)\zeta^2(\beta+2,0)}{\zeta(\beta+2)}$$

and

$$\zeta(r+1,0) := \frac{1}{\Gamma(r+1)} \int_0^\infty \frac{u^r e^{-u}}{(1-e^{-u})^2} du \quad \text{for } r > 1.$$

(ii) Under $P_{1,n}$, we have as $n \to \infty$,

$$\frac{\lambda_k' - \mu_{n,k}}{\sigma_n\sqrt{|\log h_n|}} \overset{d}{\longrightarrow} N(0,1).$$

Theorem 4.18 corresponds to the end of partitions. We consider the fluctuations in the deep bulk of partitions below. Let

$$X_{\beta,n}(t) = \frac{1}{\sigma_n}\left(\varphi_\lambda(\frac{t}{h_n}) - \mu_{n,\lceil\frac{t}{h_n}\rceil}\right), \quad t \geq 0.$$

Theorem 4.19. *Under $P_{\beta,n}$ with $\beta > -1$, we have as $n \to \infty$*
(i) for each $t > 0$,

$$X_{\beta,n}(t) \xrightarrow{d} X_\beta(t),$$

where $X_\beta(t)$ is a normal random variable with zero mean and variance

$$\kappa_\beta^2(t) = \sigma_\beta^2(t) - \frac{1}{\Gamma(\beta+3)\zeta(\beta+2)}\left(\sigma_{\beta+1}^2(t)\right)^2;$$

(ii) for $0 < t_1 < t_2 < \cdots < t_m < \infty$,

$$\left(X_{\beta,n}(t_1), X_{\beta,n}(t_2), \cdots, X_{\beta,n}(t_m)\right) \xrightarrow{d} \left(X_\beta(t_1), X_\beta(t_2), \cdots, X_\beta(t_m)\right),$$

where $\left(X_\beta(t_1), X_\beta(t_2), \cdots, X_\beta(t_m)\right)$ is a Gaussian vector with covariance structure

$$Cov\left(X_\beta(s), X_\beta(t)\right) = \sigma_\beta^2(t) - \frac{\sigma_{\beta+1}^2(s)\sigma_{\beta+1}^2(t)}{\Gamma(\beta+3)\zeta(\beta+2)}, \quad s < t;$$

(iii) Each separable version of X_β is continuous in $(0,\infty)$.

Next we give the limiting distribution of d_λ after properly scaled.

Theorem 4.20. *Under $(\mathcal{P}_n, P_{\beta,n})$ with $\beta > 1$, we have as $n \to \infty$*

$$h_n^{(\beta+3)/2}\left(\log\frac{d_\lambda}{(n!)^{1/(\beta+2)}} - b_n\right) \xrightarrow{d} N\left(0, \sigma_{\beta,d}^2\right),$$

where the normalizing constant b_n and limiting variance $\sigma_{\beta,d}^2$ are given by

$$b_n = \frac{\beta+1}{\beta+2}\sum_{k=1}^{\infty}\mu_{n,k}\log n - \sum_{k=1}^{\infty}\mu_{n,k}\log\mu_{n,k} + \sum_{k=1}^{\infty}\mu_{n,k} - \frac{\beta+1}{\beta+2}n,$$

$$-\frac{\beta+1}{\beta+2}\left(\log\Gamma(\beta+2)\zeta(\beta+2)\right)\left(\sum_{k=1}^{\infty}\mu_{n,k} - n\right)$$

and

$$\sigma_{\beta,d}^2 = \int_0^\infty\int_0^\infty Cov\left(X_\beta(s), X_\beta(t)\right)\log\Psi_\beta(s)\log\Psi_\beta(t)dsdt.$$

To conclude this chapter, we want to mention another interesting example of multiplicative measure which is given by the exponential generating function. Let $a = (a_k, k \geq 1)$ be a parameter function determined by $g(x) = \exp(\sum_{k \geq 1} a_k x^k)$. Define a probability $P_{a,n}$ on $\mathcal{P}_n$ by

$$P_{a,n}(\lambda) = \frac{1}{Z_{a,n}} \prod_{k=1} \frac{a_k^{r_k}}{r_k!}, \quad \lambda \in \mathcal{P}_n,$$

where $Z_{a,n}$ is the partition function.

In terms of the form of parameter function, the measure $P_{a,n}$ substantially differ from either $P_{u,n}$ or $P_{\beta,n}$. The reader is referred to Erlihson and Granovsky (2008) and the reference therein for the limit shape and functional central limit theorem for the fluctuation.

Chapter 5

Random Plancherel Partitions

5.1 Introduction

In this chapter we shall consider another probability measure, namely Plancherel measure, in $\mathcal{P}_n$ and study its asymptotic properties as $n \to \infty$. As in Chapter 4, our main concerns are again the fluctuations of a typical Plancherel partition around its limit shape.

To start, let us recall the classic elegant Burnside identity

$$\sum_{\lambda \in \mathcal{P}_n} d_\lambda^2 = n!,$$

where d_λ is the number of standard Young tableaux with a shape λ, see (2.13). This naturally induces a probability measure

$$P_{p,n}(\lambda) = \frac{d_\lambda^2}{n!}, \quad \lambda \in \mathcal{P}_n$$

where p in the subscript stands for Plancherel. $P_{p,n}$ is often referred to as Plancherel measure because the Fourier transform

$$L^2(\mathcal{S}_n, \mu_{s,n}) \stackrel{\text{Fourier}}{\longrightarrow} L^2(\widehat{\mathcal{S}}_n, P_{p,n}),$$

is an isometry just like in the classical Plancherel theorem, where $\mu_{s,n}$ is the uniform measure on $\mathcal{S}_n$ and $\widehat{\mathcal{S}}_n$ is the set of irreducible representations of $\mathcal{S}_n$.

Plancherel measure naturally arises in many representation-theoretic, combinatorial and probabilistic problems. To illustrate, we consider the length of longest increasing subsequences in $\mathcal{S}_n$. For a given $\pi = (\pi_1, \pi_2, \cdots, \pi_n) \in \mathcal{S}_n$ and $i_1 < i_2 < \cdots < i_k$, we say π_{i_1}, π_{i_2}, $\cdots$, π_{i_k} is an increasing subsequence if $\pi_{i_1} < \pi_{i_2} < \cdots < \pi_{i_k}$. Let $\ell_n(\pi)$ be the length of longest increasing subsequences of π. For example, let $n = 10$,

$\pi = (7, 2, 8, 1, 3, 4, 10, 6, 9, 5)$. Then $\ell_n(\pi) = 5$, and the longest increasing subsequences are $1, 3, 4, 6, 9$ and $2, 3, 4, 6, 9$.

The study of $\ell_n(\pi)$ dates back to Erdös and Szekeres in the 1930s. A celebrated theorem states every π of $\mathcal{S}_n$ contains an increasing and/or decreasing subsequence of length at least $\sqrt{n}$ (see Steele (1995, 1997)). This can be proved by an elementary pigeon-hole principle. But it also follows from an algorithm developed by Robinson, Schensted and Knuth (see Sagan (2000)) to obtain Young tableaux with the help of permutations. Let $\mathcal{T}_n$ be the set of standard Young tableaux with n squares. According to this algorithm, for any $n \geq 1$ there is a bijection, the so-called RSK correspondence, between $\mathcal{S}_n$ and pairs of $T, T' \in \mathcal{T}_n$ with the same shape:

$$\mathcal{S}_n \ni \pi \overset{\mathrm{RSK}}{\longleftrightarrow} \big(T(\pi), T'(\pi)\big) \in \mathcal{T}_n \times \mathcal{T}_n.$$

The RSK correspondence is very intricate and has no obvious algebraic meaning at all, but it is very deep and allows us to understand many things. In particular, it gives an explicit proof of the Burnside identity (2.13). More interestingly, $\ell_n(\pi)$ is exactly the number of squares in the first row of $T(\pi)$ or $T'(\pi)$, namely $\ell_n(\pi) = \lambda_1(T(\pi))$. Consequently,

$$
\begin{aligned}
\mu_{s,n}\big(\pi \in \mathcal{S}_n : \ell_n(\pi) = k\big) &= \frac{|\{\pi \in \mathcal{S}_n : \ell_n(\pi) = k\}|}{n!} \\
&= \sum_{\lambda \in \mathcal{P}_n : \lambda_1 = k} \frac{d_\lambda^2}{n!} \\
&= P_{p,n}\big(\lambda \in \mathcal{P}_n : \lambda_1 = k\big).
\end{aligned}
$$

In words, the Plancherel measure $P_{p,n}$ on $\mathcal{P}_n$ is the push-forward of the uniform measure $\mu_{s,n}$ on $\mathcal{S}_n$. Thus the analysis of $\ell_n(\pi)$ is equivalent to a statistical problem in the geometry of the Young diagram. See an excellent survey Deift (2000) for more information.

A remarkable feature is that there also exists a limit shape for random Plancherel partitions. Define for $\lambda \in \mathcal{P}$

$$\psi_\lambda(0) = \lambda_1, \quad \psi_\lambda(x) = \lambda_{\lceil x \rceil}, \quad x > 0.$$

Note that $\psi_\lambda(x), x \geq 0$ is a nonincreasing step function such that $\int_0^\infty \psi_\lambda(x)dx = |\lambda|$. Also, $\psi_{\lambda'}(x) = \varphi_\lambda(x)$ where λ' is a dual partition of λ and φ_λ was defined by (4.10).

The so-called limit shape is a function $y = \omega(x)$ defined as follows:

$$x = \frac{2}{\pi}(\sin\theta - \theta\cos\theta), \quad y = x + 2\cos\theta$$

where $0 \leq \theta \leq 2\pi$ is a parameter, see Figure 5.1 below.

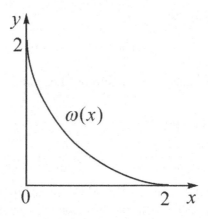

Fig. 5.1 ω curve

Logan and Shepp (1977) used the variational argument to prove

Theorem 5.1. *Under* $(\mathcal{P}_n, P_{p,n})$, *the rescaled function* $\psi_\lambda(\sqrt{n}x)/\sqrt{n}$ *converges to the function* $\omega(x)$ *in a sense of weak convergence in a certain metric d. Here the metric d is defined by (1.15) of Logan and Shepp (1977).*

We remark that one *cannot* derive from Theorem 5.1

$$\frac{1}{\sqrt{n}}\psi_\lambda(\sqrt{n}\,x) \xrightarrow{P} \omega(x)$$

for every $x \geq 0$.

Independently, Vershik and Kerov (1977, 1985) developed a slightly different strategy to establish a uniform convergence for $\psi_\lambda(\sqrt{n}x)/\sqrt{n}$. To state their results, it is more convenient to use the rotated coordinate system

$$u = x - y, \quad v = x + y.$$

Then in the (u, v)-plane, the step function $\psi_\lambda(x)$ transforms into a piecewise linear function $\Psi_\lambda(u)$. Note $\Psi'_\lambda(u) = \pm 1$, $\Psi_\lambda(u) \geq |u|$ and $\Psi_\lambda(u) = |u|$ for sufficiently large u. Likewise, $\omega(x)$ transforms into $\Omega(u)$ (see (1.34) and Figure 1.4):

$$\Omega(u) = \begin{cases} \frac{2}{\pi}\left(u\arcsin\frac{u}{2} + \sqrt{4 - u^2}\right), & |u| \leq 2 \\ |u|, & |u| \geq 2. \end{cases} \tag{5.1}$$

Define

$$\Psi_{n,\lambda}(u) = \frac{1}{\sqrt{n}}\Psi_\lambda(\sqrt{n}\,u), \quad \Delta_{n,\lambda}(u) = \Psi_{n,\lambda}(u) - \Omega(u). \tag{5.2}$$

Theorem 5.2. *Under* $(\mathcal{P}_n, P_{p,n})$,

$$\sup_{-\infty < u < \infty} \left| \frac{1}{\sqrt{n}} \Psi_\lambda(\sqrt{n}\, u) - \Omega(u) \right| \xrightarrow{P} 0, \quad n \to \infty. \tag{5.3}$$

As an immediate corollary, we can improve Logan-Shepp's result to a uniform convergence.

Corollary 5.1. *Under* $(\mathcal{P}_n, P_{p,n})$,

$$\sup_{0 \le x < \infty} \left| \frac{1}{\sqrt{n}} \psi_\lambda(\sqrt{n}\, x) - \omega(x) \right| \xrightarrow{P} 0, \quad n \to \infty. \tag{5.4}$$

We remark that $\omega(0) = 2$ and $\omega(2) = 0$. Compared to $\Psi(x)$ of (4.12), $\omega(x)$ looks more balanced. This can be seen from the definition of Plancherel measure. Roughly speaking, the more balanced a Young diagram is, the more likely it appears. For instance, fix $n = 10$, and consider two partitions $\lambda^{(1)} = (1^{10})$ and $\lambda^{(2)} = (1, 2, 3, 4)$. Then

$$P_{p,10}(\lambda^{(1)}) = \frac{1}{10!}, \quad P_{p,10}(\lambda^{(2)}) = \frac{256}{1575} \approx \frac{1}{6}.$$

Corollary 5.2. *Under* $(\mathcal{P}_n, P_{p,n})$,

$$\frac{\lambda_1}{\sqrt{n}} \xrightarrow{P} 2, \quad \frac{\lambda_1'}{\sqrt{n}} \xrightarrow{P} 2, \quad n \to \infty.$$

Consequently,

$$\frac{\ell_n(\pi)}{\sqrt{n}} \xrightarrow{P} 2, \quad n \to \infty. \tag{5.5}$$

(5.5) provides a satisfactory solution to Ulam's problem.

The rest of this section shall be devoted to a rigorous proof of Theorem 5.2 due to and Kerov (1977). It will consist of a series of lemmas. A key technical ingredient is to prove a certain quadratic integral attains its minimum at Ω. Start with a rough upper bound.

Lemma 5.1.

$$P_{p,n}\big(\max\{\lambda_1, \lambda_1'\} \ge 2e\sqrt{n} \big) \le e^{-2e\sqrt{n}}. \tag{5.6}$$

Proof. We need an equivalent representation of $\ell_n(\pi)$. Let $X_1, X_2, \cdots, X_n$ be a sequence of i.i.d. uniform random variables on $[0, 1]$. Let $\ell_n(X)$ be the length of the longest increasing subsequences of $X_1, X_2, \cdots, X_n$. Trivially,

$$\ell_n(\pi) \overset{d}{=} \ell_n(X),$$

from which we in turn derive

$$\lambda_1 \overset{d}{=} \ell_n(\pi) \overset{d}{=} \ell_n(X).$$

Then it follows

$$P_{p,n}(\lambda_1 \geq k) = P(\ell_n(X) \geq k)$$

$$\leq \binom{n}{k} \frac{1}{k!}.$$

In particular,

$$P_{p,n}(\lambda_1 \geq 2e\sqrt{n}) \leq e^{-2e\sqrt{n}}.$$

We conclude the proof. $\qquad\square$

Next let us take a look at $d_\lambda^2/n!$.

Lemma 5.2. *As* $n \to \infty$

$$P_{p,n}\left(-\log \frac{d_\lambda^2}{n!} > 2c\sqrt{n}\right) \to 0, \tag{5.7}$$

where $c = \pi/\sqrt{6}$ *as in Chapter 4.*

Proof. Denote by A_n the event in (5.7). Then by (4.3)

$$P_{p,n}(A_n) = \sum_{\lambda \in A_n} \frac{d_\lambda^2}{n!}$$

$$\leq \sum_{\lambda \in A_n} e^{-2c\sqrt{n}}$$

$$\leq p(n)e^{-2c\sqrt{n}} \to 0,$$

as desired. $\qquad\square$

Observe that it follows from the hook formula (4.70)

$$\frac{d_\lambda^2}{n!} = \frac{n!}{H_\lambda^2}.$$

So we have

$$-\log \frac{d_\lambda^2}{n!} = -\log \frac{n!}{H_\lambda^2}$$

$$= -\log n! + 2\log H_\lambda$$

$$= -\log n! + 2 \sum_{(i,j) \in \lambda} \log \left(\lambda_i - i + \lambda_j' - j + 1\right)$$

$$= -\log n! + n\log n + 2 \sum_{(i,j) \in \lambda} \log \frac{1}{\sqrt{n}}(\lambda_i - i + \lambda_j' - j + 1).$$

For simplicity of notations, write $\psi_{n,\lambda}(x)$ for $\psi_\lambda(\sqrt{n}x)/\sqrt{n}$. Then

$$\log \frac{1}{\sqrt{n}} (\lambda_i - i + \lambda'_j - j + 1)$$

$$= \log \int \int_\square \frac{1}{\sqrt{n}} (\psi_\lambda(x) - x + \psi_\lambda^{-1}(y) - y) dx dy$$

$$\geq \int \int_\square \log \frac{1}{\sqrt{n}} (\psi_\lambda(x) - x + \psi_\lambda^{-1}(y) - y) dx dy,$$

where $\square$ stands for the (i,j)th unit square, $\psi_{n,\lambda}^{-1}$ denotes the inverse of $\psi_{n,\lambda}$ and the last inequality follows from the concavity property of logarithmic function.

Hence we obtain

$$-\log \frac{d_\lambda^2}{n!} \geq -\log n! + n \log n$$

$$+ 2n \int \int_{0 \leq y < \psi_{n,\lambda}(x)} \log (\psi_{n,\lambda}(x) - x + \psi_{n,\lambda}^{-1}(y) - y) dx dy$$

$$=: nI(\psi_{n,\lambda}) + \epsilon_n,$$

where $\epsilon_n = O(\log n)$ and

$$I(\psi_{n,\lambda}) = 1 + 2 \int \int_{0 \leq y < \psi_{n,\lambda}(x)} \log (\psi_{n,\lambda}(x) - x + \psi_{n,\lambda}^{-1}(y) - y) dx dy.$$

As a direct consequence of Lemma 5.2, it follows for any $\varepsilon > 0$

$$P_{p,n}(I(\psi_{n,\lambda}) > \varepsilon) \to 0. \tag{5.8}$$

Making a change of variables, we have

$$I(\psi_{n,\lambda}) = 1 + \frac{1}{2} \int \int_{v < u} \log(u - v)(1 - \Psi'_{n,\lambda}(u))(1 + \Psi'_{n,\lambda}(v)) du dv$$

$$=: J(\Psi_{n,\lambda}).$$

In terms of $J(\Psi_{n,\lambda})$, (5.8) can be written as

$$P_{p,n}(J(\Psi_{n,\lambda}) > \varepsilon) \to 0. \tag{5.9}$$

Similarly, define

$$J(\Omega) = 1 + \frac{1}{2} \int \int_{v < u} \log(u - v)(1 - \Omega'(u))(1 + \Omega'(v)) du dv.$$

A remarkable contribution due to Vershik and Kerov (1977, 1985) is the following

Lemma 5.3. *With notations above, we have*

(i)

$$J(\Omega) = 0; \tag{5.10}$$

(ii)

$$J(\Psi_{n,\lambda}) = -\frac{1}{4} \int_{-\infty}^{\infty} \int_{-\infty}^{\infty} \log|u - v| \Delta_{n,\lambda}'(u) \Delta_{n,\lambda}'(v) du dv$$

$$+ \int_{|u|>2} (\Psi_{n,\lambda}(u) - |u|) \operatorname{arccosh} \frac{|u|}{2} du. \tag{5.11}$$

Consequently,

$$J(\Psi_{n,\lambda}) \geq -\frac{1}{4} \int_{-\infty}^{\infty} \int_{-\infty}^{\infty} \log|u - v| \Delta_{n,\lambda}'(u) \Delta_{n,\lambda}'(v) du dv. \tag{5.12}$$

Proof. Start with the proof of (5.10). Let

$$\varrho_0(x) = -\log|x|, \quad \varrho_1(x) = \int_0^x \varrho_0(y) dy, \quad \varrho_2(x) = \int_0^x \varrho_1(y) dy.$$

A simple calculus shows

$$\varrho_1(x) = x - x \log|x|, \quad \varrho_2(x) = \frac{3x^2}{4} - \frac{x^2}{2} \log|x|$$

and

$$\varrho_1(-x) = -\varrho_1(x), \quad \varrho_2(-x) = \varrho_2(x).$$

Note $\Omega'(u) = 1$ for $u \geq 2$ and $\Omega'(u) = -1$ for $u \leq -2$. Then

$$J(\Omega) = 1 - \frac{1}{4} \int_{-2}^{2} \int_{-2}^{2} \varrho_0(u - v) du dv + \frac{1}{2} \int_{-2}^{2} \left(\int_{-2}^{u} \varrho_0(u - v) dv \right) \Omega'(u) du$$

$$- 2 \int_{-2}^{2} \left(\int_{v}^{2} \varrho_0(u - v) du \right) \Omega'(v) dv$$

$$+ \frac{1}{4} \int_{-2}^{2} \int_{-2}^{2} \varrho_0(u - v) \Omega'(u) \Omega'(v) du dv. \tag{5.13}$$

First, it is easy to see

$$\int_{-2}^{2} \int_{-2}^{2} \varrho_0(u - v) du dv = 2\varrho_2(4).$$

Also, it follows

$$\int_{-2}^{u} \varrho_0(u - v) dv = \varrho_1(2 + u), \quad \int_{v}^{2} \varrho_0(u - v) du = \varrho_1(2 - v).$$

To calculate the last three double integral in the right hand side of (5.13), set

$$H_2(u) = \int_{-2}^{2} \varrho_1(u-v)\Omega''(v)dv, \quad -\infty < u < \infty.$$

Then

$$H_2''(u) = \frac{2}{\pi}\int_{-2}^{2}\frac{1}{(u-v)\sqrt{4-v^2}}dv = 0, \quad -2 \le u \le 2$$

and so for each $-2 \le u \le 2$

$$H_2'(u) = H_2'(0) = \frac{2}{\pi}\int_{-2}^{2}\frac{\log|v|}{\sqrt{4-v^2}}dv = 0.$$

This in turn implies

$$H_2(u) = H_2(0) = -\int_{-2}^{2}\varrho_1(v)\Omega''(v)dv = 0 \qquad (5.14)$$

since $\varrho_1(v)$ is odd.

It follows by integration by parts

$$\int_{-2}^{2}\left(\int_{v}^{2}\varrho_0(u-v)du\right)\Omega'(v)dv = \int_{-2}^{2}\varrho_1(2-v)\Omega'(v)dv$$

$$= -\varrho_2(4) + \int_{-2}^{2}\varrho_2(2-v)\Omega''(v)dv$$

$$= -\varrho_2(4) + \int_{-2}^{2}\varrho_2(v)\Omega''(v)dv$$

$$= 2 - \varrho_2(4),$$

where we used (5.14) and the fact

$$\int_{-2}^{2}\varrho_2(v)\Omega''(v)du = 2.$$

Similarly,

$$\int_{-2}^{2}\left(\int_{-2}^{u}\varrho_0(u-v)dv\right)\Omega'(u)du = \int_{-2}^{2}\varrho_1(2+u)\Omega'(u)du$$

$$= \varrho_2(4) - 2.$$

Again, by (5.14)

$$\int_{-2}^{2}\varrho_0(u-v)\Omega'(v)dv = \varrho_1(2-u) - \varrho_1(2+u) + \int_{-2}^{2}\varrho_1(u-v)\Omega''(v)dv$$

$$= \varrho_1(2-u) - \varrho_1(2+u).$$

Hence we have

$$\int_{-2}^{2}\int_{-2}^{2}\varrho_0(u-v)\Omega'(u)\Omega'(v)dudv = \int_{-2}^{2}\left(\varrho_1(2-u)-\varrho_1(2+u)\right)\Omega'(u)du$$
$$= 4 - 2\varrho_2(4).$$

In combination, we have proven (5.10).

Turn to the proof of (5.11). First, observe there is an $a = a_n$ (may depend on λ) such that $[-a, a]$ contains the support of $\Delta_{n,\lambda}(u)$. Hence we have

$$J(\Psi_{n,\lambda}) = 1 - \frac{1}{4}\int_{-a}^{a}\int_{-a}^{a}\varrho_0(u-v)dudv$$
$$+ \frac{1}{2}\int_{-a}^{a}\left(\int_{-a}^{u}\varrho_0(u-v)dv\right)\Psi_{n,\lambda}'(u)du$$
$$- \frac{1}{2}\int_{-a}^{a}\left(\int_{v}^{a}\varrho_0(u-v)du\right)\Psi_{n,\lambda}'(v)dv$$
$$+ \frac{1}{4}\int_{-a}^{a}\int_{-a}^{a}\varrho_0(u-v)\Psi_{n,\lambda}'(u)\Psi_{n,\lambda}'(v)dudv.$$

A simple calculus shows

$$\int_{-a}^{a}\int_{-a}^{a}\varrho_0(u-v)dudv = 2\varrho_2(2a),$$

$$\int_{-a}^{a}\left(\int_{-a}^{u}\varrho_0(u-v)dv\right)\Psi_{n,\lambda}'(u)du = \int_{-a}^{a}\varrho_1(a+u)\Psi_{n,\lambda}'(u)du$$
$$= \rho_2(2a),$$

$$\int_{-a}^{a}\left(\int_{v}^{a}\varrho_0(u-v)du\right)\Psi_{n,\lambda}'(v)dv = \int_{-a}^{a}\varrho_1(a-v)\Psi_{n,\lambda}'(v)dv$$
$$= -\rho_2(2a).$$

On the other hand, it follows

$$-\frac{1}{4}\int_{-\infty}^{\infty}\int_{-\infty}^{\infty}\log|u-v|\Delta_{n,\lambda}'(u)\Delta_{n,\lambda}'(v)dudv$$
$$= \frac{1}{4}\int_{-a}^{a}\int_{-a}^{a}\varrho_0(u-v)\Omega'(u)\Omega'(v)dudv$$
$$- \frac{1}{2}\int_{-a}^{a}\int_{-a}^{a}\varrho_0(u-v)\Omega'(u)\Psi_{n,\lambda}'(v)dudv$$
$$+ \frac{1}{4}\int_{-a}^{a}\int_{-a}^{a}\varrho_0(u-v)\Psi_{n,\lambda}'(u)\Psi_{n,\lambda}'(v)dudv.$$

Note $\Omega''(u) = 0$ for $|u| > 2$ and so

$$\int_{-a}^{a} \varrho_0(u-v)\Omega'(u)du = \varrho_1(a-v) - \varrho_1(a+v) + H_2(v). \qquad (5.15)$$

Since $H_2(v) = 0$ for $|v| \leq 2$, we have

$$\int_{-a}^{a}\int_{-a}^{a} \varrho_0(u-v)\Omega'(u)\Omega'(v)dudv$$

$$= \int_{-a}^{a} \varrho_1(a-v)\Omega'(v)dv - \int_{-a}^{a} \varrho_1(a+v)\Omega'(v)dv$$

$$- \int_{-a}^{-2} H_2(u)du + \int_{2}^{a} H_2(u)du.$$

Also, it is easy to see

$$\int_{-a}^{a} \varrho_1(a-v)\Omega'(v)dv = -\varrho_2(2a) + \int_{-2}^{2} \varrho_2(a-v)\Omega''(v)dv$$

$$= 2 - \varrho_2(2a) + \int_{-2}^{2} \left(\varrho_2(a-v) - \varrho_2(2-v)\right)\Omega''(v)dv$$

$$= 2 - \varrho_2(2a) + \int_{2}^{a}\int_{-2}^{2} \varrho_1(u-v)\Omega''(v)dvdu$$

$$= 2 - \varrho_2(2a) + \int_{2}^{a} H_2(u)du,$$

and similarly

$$\int_{-a}^{a} \varrho_1(a+v)\Omega'(v)dv = \varrho_2(2a) - 2 + \int_{-a}^{-2} H_2(u)du.$$

By (5.15),

$$\int_{-a}^{a} \left(\int_{-a}^{a} \varrho_0(u-v)\Omega'(u)du\right)\Psi'_{n,\lambda}(v)dv$$

$$= -2\rho_2(2a) + \int_{2}^{a} H_2(v)\Psi'_{n,\lambda}(v)dv + \int_{-a}^{-2} H_2(v)\Psi'_{n,\lambda}(v)dv.$$

In combination, we get

$$J(\Psi_{n,\lambda})$$
$$= -\frac{1}{4}\int_{-\infty}^{\infty}\int_{-\infty}^{\infty} \log|u-v|\Delta'_{n,\lambda}(u)\Delta'_{n,\lambda}(v)dudv$$

$$+ \frac{1}{2}\int_{2}^{a} H_2(u)\left(\Psi_{n,\lambda}(u) - u\right)'du + \frac{1}{2}\int_{-a}^{-2} H_2(u)(\Psi_{n,\lambda}(u) + u)'dv.$$

To proceed, note for $u > 2$

$$H_2'(u) = \int_{-2}^{2} \rho_0(u-v)\Omega''(v)dv$$

$$= -\int_{-2}^{2} \log(u-v)\frac{2}{\pi\sqrt{4-v^2}}dv$$

$$= -2\arccos\frac{u}{2}.$$

Thus by integration by parts and using the fact $\Psi_{n,\lambda}(a) = a$ and $H_2(2) = 0$,

$$\frac{1}{2}\int_{2}^{a} H_2(u)\big(\Psi_{n,\lambda}(u) - u\big)'du = -\frac{1}{2}\int_{2}^{a} H_2'(u)\big(\Psi_{n,\lambda}(u) - u\big)du$$

$$= \int_{2}^{a} \big(\Psi_{n,\lambda}(u) - u\big)\arccos\frac{u}{2}du.$$

Similarly, it follows

$$\frac{1}{2}\int_{-a}^{-2} H_2(u)\big(\Psi_{n,\lambda}(u) + u\big)'du = -\frac{1}{2}\int_{-a}^{-2} H_2'(u)\big(\Psi_{n,\lambda}(u) + u\big)du$$

$$= \int_{-a}^{-2} \big(\Psi_{n,\lambda}(u) + u\big)\arccos\frac{|u|}{2}du.$$

In combination, we now conclude the proof of (5.11).

Finally, (5.12) holds true since $\Psi_{n,\lambda}(u) \geq |u|$ for all $u \in \mathbb{R}$. $\qquad\square$

The following lemma is interesting and useful since it introduces the Sobolev norm into the study of random partitions. Define

$$\|f\|_s^2 = \int_{-\infty}^{\infty}\int_{-\infty}^{\infty} \left(\frac{f(u) - f(v)}{u - v}\right)^2 dudv,$$

where s in the subscript stands for Sobolev.

Lemma 5.4.

$$-\int_{-\infty}^{\infty}\int_{-\infty}^{\infty} \log|u-v|f'(u)f'(v)dudv = \frac{1}{2}\|f\|_s^2. \tag{5.16}$$

Proof. Denote by $H(f)$ the Hilbert transform, namely

$$H(f)(v) = \int_{-\infty}^{\infty} \frac{f(u)}{v - u}du.$$

Then it is easy to see

$$\widehat{H(f)}(\omega) = \int_{-\infty}^{\infty} e^{i2\pi\omega v} H(f)(v)dv$$

$$= \mathrm{i}\,\mathrm{sgn}\omega\,\widehat{f}(\omega),$$

where $\hat{f}$ is the Fourier transform of f. Then by integration formula by parts and the Pasval-Plancherel identity

$$\text{LHS of (5.16)} = \int_{-\infty}^{\infty} H(f)(v)f'(v)dv$$

$$= \int_{-\infty}^{\infty} \widehat{H(f)}(\omega)\overline{\hat{f'}(\omega)}d\omega$$

$$= \int_{-\infty}^{\infty} \mathrm{i\,sgn}\omega\,\hat{f}(\omega)\overline{\mathrm{i}\omega\hat{f}(\omega)}d\omega$$

$$= \int_{-\infty}^{\infty} |\omega||\hat{f}(\omega)|^2 d\omega.$$

To finish the proof, we need a key observation due to Vershik and Kerov (1985)

$$\int_{-\infty}^{\infty} |\omega||\hat{f}(\omega)|^2 d\omega = \frac{1}{2}\|f\|_s^2.$$

The proof is complete. □

Combining (5.12) and (5.16) yields

$$J(\Psi_{n,\lambda}) \geq \frac{1}{8}\|\Delta_{n,\lambda}\|_s^2. \tag{5.17}$$

Now we are ready to give

Proof of Theorem 5.2. In view of Lemma 5.5, we can and do consider only the case in which the support of $\Delta_{n,\lambda}$ is contained in a finite interval, say, $[-a, a]$. We will divide the double integral into two parts:

$$\|\Delta_{n,\lambda}\|_s^2 = \int_{-a}^{a}\int_{-a}^{a}\left(\frac{\Delta_{n,\lambda}(u) - \Delta_{n,\lambda}(v)}{u - v}\right)^2 dudv + 4a\int_{-a}^{a}\frac{\Delta_{n,\lambda}^2(u)}{a^2 - u^2}du,$$

which together with (5.17) implies

$$\int_{-a}^{a} \Delta_{n,\lambda}^2(u)du \leq a^2\int_{-a}^{a}\frac{\Delta_{n,\lambda}^2(u)}{a^2 - u^2}du$$

$$\leq \frac{a}{4}\|\Delta_{n,\lambda}\|_s^2$$

$$\leq 2aJ(\Psi_{n,\lambda}).$$

Also, since $|\Delta_{n,\lambda}'(u)| \leq 2$, then

$$\sup_{-a \leq u \leq a} |\Delta_{n,\lambda}| \leq 6^{1/3}\left(\int_{-a}^{a}\Delta_{n,\lambda}^2(u)du\right)^{1/3}$$

$$\leq (12a)^{1/3}J(\Psi_{n,\lambda})^{1/3}.$$

By virtue of (5.9), it follows

$$\sup_{-a \leq u \leq a} |\Delta_{n,\lambda}(u)| \xrightarrow{P} 0.$$

The proof is complete. □

We have so far proven that the limit shape exists and found its explicit form. To proceed, it is natural to look at the fluctuations of a typical Plancherel partition around the limit curve. This question was first raised by Logan and Shepp in 1977. The following words are excerpted from page 211 of Logan and Shepp (1977):

It is of course natural to expect that for appropriate normalizing constants $c_n \to \infty$ (perhaps $c_n = n^{1/4}$ would do) the stochastic processes

$$\bar{\lambda}_n(t) = c_n(\lambda_n(t) - f_0(t)), \quad t \geq 0$$

would tend weakly to a nonzero limiting process $W(t), t \geq 0$, as $c_n \to \infty$. It would be of interest to know what the process W is. It is clear only that W integrates pathwise to zero and that $W(t) \geq 0$ for $t \geq 2$. Perhaps $W(t) = 0$ for $t \geq 2$ and is the Wiener process in $0 \leq t \leq 2$ conditioned to integrate to zero over $[0, 2]$ and to vanish at 0 and 2, but this is just a guess.

This turns out to be an interesting and challenging problem. To see the fluctuation at a fixed point, we need to consider two cases separately: at the edge and in the bulk. At $x = 0$, $\psi_\lambda(0)$ is equal to λ_1, the largest part of a partition. Around 2000, several important articles, say Baik, Deift and Johansson (1999), Johansson (2001), Okounkov (2000), were devoted to studying the asymptotic distribution of the λ_1 after appropriately normalized. It was proved that

$$\frac{\lambda_1 - 2\sqrt{n}}{n^{1/6}} \xrightarrow{d} F_2, \quad n \to \infty$$

where F_2 is the Tracy-Widom law, which was first discovered by Tracy and Widom in the study of random matrices, see Tracy and Widom (1994, 2002). The analogs were proven to hold for each λ_k, $k \geq 2$. By symmetry, one can also discuss the limiting distribution at $x = 2$. The graph of F_2 is shown in Figure 5.2 below. The picture looks completely different in the bulk. It will be proved in Section 5.3 that for each $0 < x < 2$,

$$\frac{1}{\frac{1}{2\pi}\sqrt{\log n}} \left(\psi_\lambda(\sqrt{n}x) - \sqrt{n}\omega(x) \right) \xrightarrow{d} \xi(x), \quad n \to \infty$$

where $\xi(x)$ is a centered normal random variable. Note that the normalizing constant $\sqrt{\log n}$ is much smaller than $n^{1/6}$. In addition, we will also see that $\xi(x)$, $0 < x < 2$, constitutes a white noise, namely $Cov\big(\xi(x_1), \xi(x_2)\big) = 0$

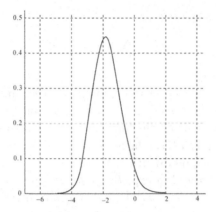

Fig. 5.2   Tracy-Widom law

for two distinct points x_1 and x_2. Thus, one cannot expect a kind of weak convergence of the stochastic process. However, we will in Section 5.2 establish a functional central limit theorem, namely Kerov's integrated central limit theorem for $\psi_\lambda(\sqrt{n}x) - \sqrt{n}\omega(x)$.

5.2   Global fluctuations

In this section we shall establish an integrated central limit theorem, which is used to described the global fluctuation of $\Psi_{n,\lambda}(\sqrt{n}\,u)$ around the limit shape $\Omega(u)$. Let $u_k(u), k \geq 0$ be a sequence of modified Chebyshev polynomials, i.e.,

$$u_k(u) = \sum_{j=0}^{[k/2]} (-1)^j \binom{k-j}{j} u^{k-2j}. \qquad (5.18)$$

Note

$$u_k(2\cos\theta) = \frac{\sin(k+1)\theta}{\sin\theta}$$

and

$$\int_{-2}^{2} u_k(u)u_l(u)\rho_{sc(u)}du = \delta_{k,l}.$$

Theorem 5.3. *Define*

$$\mathcal{X}_{n,k}(\lambda) = \int_{-\infty}^{\infty} u_k(u)\big(\Psi_\lambda(\sqrt{n}u) - \sqrt{n}\Omega(u)\big)du, \quad \lambda \in \mathcal{P}_n.$$

Then under $(\mathcal{P}_n, P_{p,n})$ as $n \to \infty$

$$(\mathcal{X}_{n,k}, \quad k \geq 1) \xrightarrow{d} \left(\frac{2}{\sqrt{k+1}}\xi_k, \quad k \geq 1\right).$$

Here $\xi_k, k \geq 1$ is a sequence of standard normal random variables, the convergence holds in terms of finite dimensional distribution.

This theorem is referred to the Kerov CLT since it was Kerov who first presented it and outlined the main ideas of the proof in Kerov (1993). A complete and rigorous proof was not given by Ivanov and Olshanski (2002) until 2002. The proof uses essentially the moment method and involves a lot of combinatorial and algebraic techniques, though the theorem is stated in standard probability terminologies. We need to introduce some basic notations and lemmas.

Begin with Frobenius coordinates. Let $\lambda = (\lambda_1, \lambda_2, \cdots, \lambda_l)$ be a partition from $\mathcal{P}$. Define

$$\bar{a}_i = \lambda_i - i, \quad \bar{b}_i = \lambda_i' - i, \quad i = 1, 2, \cdots, \ell, \tag{5.19}$$

and

$$a_i = \lambda_i - i + \frac{1}{2}, \quad b_i = \lambda_i' - i + \frac{1}{2}, \quad i = 1, 2, \cdots, \ell, \tag{5.20}$$

where $\ell := \ell(\lambda)$ is the length of the main diagonal in the Young diagram of λ. The natural numbers $\{\bar{a}_i, \bar{b}_i, i = 1, \cdots, \ell\}$ are called the usual Frobenius coordinates, while the half integer numbers $\{a_i, b_i, i = 1, \cdots, \ell\}$ are called the modified Frobenius coordinates. We sometimes represent $\lambda = (a_1, a_2, \cdots, a_\ell | b_1, b_2, \cdots, b_\ell)$.

Lemma 5.5. *For any $\lambda \in \mathcal{P}$*

$$\Phi(z; \lambda) := \prod_{i=1}^{\infty} \frac{z + i - \frac{1}{2}}{z - \lambda_i + i - \frac{1}{2}}$$

$$= \prod_{i=1}^{\ell} \frac{z + b_i}{z - a_i}. \tag{5.21}$$

Proof. First, observe the infinite series in (5.21) is actually finite because $\lambda_i = 0$ when i is large enough. Second, the second product is an noncontractible function since the numbers $a_1, a_2, \cdots, a_\ell, -b_1, -b_2, \cdots, -b_\ell$ are pairwise distinct. The identity (5.21) follows from a classical Frobenius lemma, see Proposition 1.4 of Ivanov and Olshanski (2002). $\square$

As a direct consequence, it follows

$$\log \Phi(z;\lambda) = \sum_{k=1}^{\infty} \frac{\bar{p}_k(\lambda)}{k} z^{-k}, \tag{5.22}$$

where

$$\bar{p}_k(\lambda) = \sum_{i=1}^{\ell} \left(a_i^k - (-b_i)^k\right). \tag{5.23}$$

Motivated by (5.23), we introduce the algebra $\mathbb{A}$ over $\mathbb{R}$ generated by $\bar{p}_1, \bar{p}_2, \cdots$. By convention, $1 \in \mathbb{A}$.

Lemma 5.6. *The generators $\bar{p}_k \in \mathbb{A}$ are algebraically independent, so that $\mathbb{A}$ is isomorphic to $\mathbb{R}[\bar{p}_1, \bar{p}_2, \cdots]$.*

Proof. See Proposition 1.5 of Ivanov and Olshanski (2002). □

Recall that the algebra of symmetric functions, denoted as $\mathbb{F}$, is the graded algebra defined as the projective limit of Λ_n, where Λ_n is the algebra of symmetric polynomials in n variables defined in Section 2.2.

Set $\mathbb{F} \ni p_k \mapsto \bar{p}_k \in \mathbb{A}$, we get an algebra isomorphism $\mathbb{F} \mapsto \mathbb{A}$. We call it the canonical isomorphism, and call the grading in $\mathbb{A}$ inherited from that of $\mathbb{F}$ the canonical grading of $\mathbb{A}$.

For each $\lambda \in \mathcal{P}$, we define the functions $\tilde{p}_2, \tilde{p}_3, \cdots$ by setting

$$\tilde{p}_k(\lambda) = k(k-1) \int_{-\infty}^{\infty} u^{k-2} \frac{1}{2} \left(\Psi_\lambda(u) - |u|\right) du. \tag{5.24}$$

Similarly, define

$$\tilde{p}_k(\Psi_{n,\lambda}) = k(k-1) \int_{-\infty}^{\infty} u^{k-2} \frac{1}{2} \left(\Psi_{n,\lambda}(u) - |u|\right) du$$

and

$$\tilde{p}_k(\Omega) = k(k-1) \int_{-\infty}^{\infty} u^{k-2} \frac{1}{2} \left(\Omega(u) - |u|\right) du. \tag{5.25}$$

Lemma 5.7. *For each $k \geq 2$ we have*
(i)

$$\tilde{p}_k(\Omega) = \begin{cases} \frac{(2m)!}{(m!)^2}, & k = 2m, \\ 0, & k = 2m+1; \end{cases} \tag{5.26}$$

(ii)

$$\tilde{p}_k(\lambda) = \sum_{i=1}^{q+1} x_i^k - \sum_{i=1}^{q} y_i^k, \quad \lambda \in \mathcal{P} \tag{5.27}$$

where the x_i's are the local minima and the y_j's are the local maxima of the function Ψ_λ and $x_1 < y_1 < x_2 < \cdots < x_q < y_q < x_{q+1}$, see Figure 5.3 below.

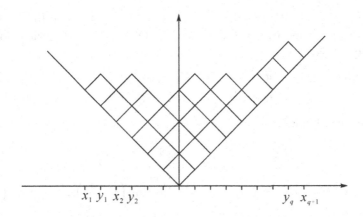

Fig. 5.3 Frobenius representation

Proof.   (i) Note by integration formula by parts

$$\tilde{p}_k(\Omega) = \int_{-\infty}^{\infty} u^k \frac{1}{2} \big(\Omega(u) - |u|\big)'' du$$

$$= \int_{-\infty}^{\infty} u^k \frac{2}{\pi\sqrt{4 - u^2}} du.$$

(5.26) now easily follows.

Turn to (ii). Note

$$\frac{1}{2}\big(\Psi_\lambda(u) - |u|\big)'' = \sum_{i=1}^{q+1} \delta_{x_i} - \sum_{j=1}^{q} \delta_{y_j} - \delta_0.$$

Then we have

$$\tilde{p}_k(\Psi_\lambda) = \int_{-\infty}^{\infty} u^k \frac{1}{2}\big(\Psi_\lambda(u) - |u|\big)'' du$$

$$= \int_{-\infty}^{\infty} u^k \Big(\sum_{i=1}^{q+1} \delta_{x_i} - \sum_{j=1}^{q} \delta_{y_j} - \delta_0\Big) du$$

$$= \sum_{i=1}^{q+1} x_i^k - \sum_{i=1}^{q} y_i^k,$$

as desired.   □

Lemma 5.8. *The functions $\tilde{p}_2, \tilde{p}_3, \cdots$ belong to the algebra $\mathbb{A}$. In particular, we have for any $\lambda \in \mathcal{P}$,*

$$\frac{\tilde{p}_{k+1}(\lambda)}{k+1} = \sum_{j=0}^{[k/2]} \frac{1}{2^{2j}(2j+1)} \binom{k}{2j} \bar{p}_{k-2j}(\lambda), \quad k \geq 1. \tag{5.28}$$

Proof.   Fix $\lambda \in \mathcal{P}$ and let $x_1, \cdots, x_{q+1}, y_1, \cdots, y_q$ be as in (5.27). Then according to Proposition 2.6 in Ivanov and Olshanski (2002), the following identity holds:

$$\frac{\Phi\left(z - \frac{1}{2}; \lambda\right)}{\Phi\left(z + \frac{1}{2}; \lambda\right)} = \frac{\prod_{j=1}^{q}\left(1 - \frac{y_j}{z}\right)}{\prod_{i=1}^{q+1}\left(1 - \frac{x_i}{z}\right)}.$$

This in turn implies

$$\log \Phi\left(z - \frac{1}{2}; \lambda\right) - \log \Phi\left(z + \frac{1}{2}; \lambda\right)$$

$$= \sum_{j=1}^{q} \log\left(1 - \frac{y_j}{z}\right) - \sum_{i=1}^{q+1} \log\left(1 - \frac{x_i}{z}\right). \tag{5.29}$$

By (5.22), the left hand side of (5.29) equals

$$\log \Phi\left(z - \frac{1}{2}; \lambda\right) - \log \Phi\left(z + \frac{1}{2}; \lambda\right)$$

$$= \sum_{l=1}^{\infty} \frac{\bar{p}_l}{l}\left((z - \frac{1}{2})^{-l} - (z + \frac{1}{2})^{-l}\right). \tag{5.30}$$

Also, by Lemma 5.7, the right hand side of (5.29) equals

$$\sum_{k=1}^{\infty} \frac{\tilde{p}_k}{k} z^{-k}. \tag{5.31}$$

By comparing coefficients of z^{-k} in both (5.30) and (5.31), we easily get (5.28). $\qquad\square$

In a simpler way, (5.28) can be interpreted as

$$\frac{\tilde{p}_{k+1}(\lambda)}{k+1} = \bar{p}_k + \text{a linear combination of } \bar{p}_1, \bar{p}_2, \cdots, \bar{p}_{k-1}, \quad k \geq 1.$$

Conversely,

$$\bar{p}_k(\lambda) = \frac{\tilde{p}_{k+1}(\lambda)}{k+1} + \text{a linear combination of } \tilde{p}_2, \tilde{p}_2, \cdots, \tilde{p}_k, \quad k \geq 1.$$

Hence the functions $\tilde{p}_2(\lambda), \tilde{p}_3(\lambda), \cdots$ are algebraically independent generators of the algebra $\mathbb{A}$:

$$\mathbb{A} = \mathbb{R}[\tilde{p}_2, \tilde{p}_3, \cdots].$$

The weight grading of $\mathbb{A}$ is defined as

$$wt(\tilde{p}_k) = k, \quad k = 2, 3, \cdots.$$

Equivalently, the weight grading is the image of the standard grading of $\mathbb{F}$ under the algebra morphism:

$$\mathbb{R}[p_1, p_2, \cdots] = \mathbb{F} \mapsto \mathbb{A} = \mathbb{R}[\tilde{p}_2, \tilde{p}_3, \cdots]$$

$$p_1 \to 0, \quad p_k \to \tilde{p}_k, \quad k = 2, 3, \cdots.$$

The weight grading induces a filtration in $\mathbb{A}$, which we call the weight filtration and denote by the same symbol $wt(\cdot)$. In particular,

$$wt(\bar{p}_k) = k + 1, \quad k \geq 1$$

since the weight of the top homogeneous component of $\bar{p}_k$ is $\tilde{p}_{k+1}/(k+1)$.

Define for each $k \geq 1$

$$p_k^\sharp(\lambda) = \begin{cases} n^{\downarrow k} \frac{\chi_\lambda(k, 1^{n-k})}{d_\lambda}, & \lambda \in \mathcal{P}_n, \ n \geq k \\ 0, & \lambda \in \mathcal{P}_n, \ n < k \end{cases} \tag{5.32}$$

where $n^{\downarrow k} = n(n-1)\cdots(n-k+1)$.

Lemma 5.9. *Fix $k \geq 1$ and $\lambda \in \mathcal{P}$. Then $p_k^\sharp(\lambda)$ equals the coefficient of z^{-1} in the expansion of the function*

$$-\frac{1}{k}\left(z - \frac{1}{2}\right)^{\downarrow k} \frac{\Phi(z; \lambda)}{\Phi(z - k; \lambda)} \tag{5.33}$$

in descending powers of z about the point $z = \infty$.

Proof. We treat two cases separately. First, assume $\lambda \in \mathcal{P}_n$ where $n < k$. Then by definition, $p_k^\sharp(\lambda) = 0$. Also, by Lemma 5.5, it is easy to see that the function of (5.33) is indeed a polynomial of z. So the claim is true.

Next, consider the case $n \geq k$. Recall the following formula due to Frobenius (see Example 1.7.7 of Macdonald (1995) and Ingram (1950)): $p_k^\sharp(\lambda)$ equals the coefficients of z^{-1} in the expansion of the function

$$F(z) = -\frac{1}{k} z^{\downarrow k} \prod_{i=1}^{n} \frac{z - \lambda_i - n + i - k}{z - \lambda_i - n + i}$$

about $z = \infty$. Namely, $p_k^\sharp(\lambda) = -\text{Res}(F(z), z = \infty)$. A simple transformation yields

$$F(z) = -\frac{1}{k}(z - n)^{\downarrow k} \frac{\Phi(z - n + \frac{1}{2}; \lambda)}{\Phi(z - n + \frac{1}{2} - k; \lambda)}.$$

Note the residue at $z = \infty$ will not change under the shift $z \mapsto z + n - 1/2$. Consequently,

$$p_k^\sharp(\lambda) = -\text{Res}\left(-\frac{1}{k}\left(z - \frac{1}{2}\right)^{\downarrow k} \frac{\Phi(z; \lambda)}{\Phi(z - k; \lambda)}, z = \infty\right).$$

The proof is complete. $\qquad\qquad\square$

We shall employ the following notation. Given a formal series $A(t)$, let

$$[t^k]\{A(t)\} = \text{the coefficient of } t^k \text{ in } A(t).$$

Lemma 5.10. *The functions $p_k^\sharp(\lambda)$ belong to the algebra $\mathbb{A}$. In particular, it can be described through the generators $\bar{p}_1, \bar{p}_2, \cdots$ as follows*

$$p_k^\sharp(\lambda) = [t^{k+1}]\Big\{ -\frac{1}{k} \prod_{j=1}^{k} \big(1 - (j - \frac{1}{2})t\big)$$

$$\cdot \exp\Big(\sum_{j=1}^{\infty} \frac{\bar{p}_j(\lambda)t^j}{j}\big(1 - (1 - kt)^{-j}\big)\Big)\Big\}. \qquad (5.34)$$

Proof. This is a direct consequence of Lemmas 5.5 and 5.9. $\qquad\square$

The expression (5.34) can be written in the form

$$p_k^\sharp(\lambda) = -\frac{1}{k}[t^{k+1}]\Big\{(1 + \varepsilon_0(t)) \exp\Big(-\sum_{j=1}^{\infty} k\bar{p}_j(\lambda)t^{j+1}(1 + \varepsilon_1(t))\Big)\Big\}$$

$$= -\frac{1}{k}[t^{k+1}]\Big\{(1 + \varepsilon_0(t)) \sum_{m=0}^{\infty} \frac{(-1)^m}{m!}\Big(\sum_{j=1}^{\infty} k\bar{p}_j(\lambda)t^{j+1}(1 + \varepsilon_1(t))\Big)^m\Big\}.$$

Here each $\varepsilon_r(t)$ is a power series of the form $c_1 t + c_2 t^2 + \cdots$, where the coefficients $c_1, c_2, \cdots$ do not involve the generators $\bar{p}_1, \bar{p}_2, \cdots$.

We can now readily evaluate the top homogeneous component of $p_k^\sharp(\lambda)$ with respect to both the canonical grading and the weight grading in $\mathbb{A}$. In the canonical grading, the highest term of $p_k^\sharp$ equals $\bar{p}_k$:

$$p_k^\sharp = \bar{p}_k + \text{lower terms};$$

while in the weight grading, the top homogeneous component of $p_k^\sharp$ has weight $k + 1$ and can be written as

$$p_k^\sharp = \frac{\tilde{p}_{k+1}}{k+1} + f(\tilde{p}_2, \cdots, \tilde{p}_k) + \text{lower terms}, \qquad (5.35)$$

where $f(\tilde{p}_2, \cdots, \tilde{p}_k)$ is a homogeneous polynomial in $\tilde{p}_2, \cdots, \tilde{p}_k$ of total weight $k + 1$.

Now we invert (5.35) to get

Lemma 5.11. *For $k = 2, 3, \cdots$*

$$\tilde{p}_k(\lambda) = \sum \frac{k^{\downarrow \sum r_i}}{\prod r_i!} \prod_{i \geq 2} \big(p_{i-1}^\sharp(\lambda)\big)^{r_i} + \text{lower terms}, \qquad (5.36)$$

where the sum is taken over all $r_2, r_3, \cdots$ with $2r_2 + 3r_3 + \cdots = k$, and lower terms means a polynomial in $p_1^\sharp, p_2^\sharp, \cdots, p_{k-2}^\sharp$ of total weight $\leq k - 1$, where $wt(p_i^\sharp) = i + 1$.

The proof is left to the reader. See also Proposition 3.7 of Ivanov and Olshanski (2002), which contains a general inversion formula.

We next extend $p_k^\sharp$ in (5.32) to any partition ρ on $\mathcal{P}$. Let $|\rho| = r$, define

$$p_\rho^\sharp(\lambda) = \begin{cases} n^{\downarrow r} \frac{\chi_\lambda(\rho, 1^{n-r})}{d_\lambda}, & \lambda \in \mathcal{P}_n, \ n \geq r, \\ 0, & \lambda \in \mathcal{P}_n, \ n < r. \end{cases} \tag{5.37}$$

The following lemma lists some basic properties. The reader is referred to its proof and more details in Kerov and Olshanski (1993), Okounkov and Olshanski (1998), Vershik and Kerov (1985).

Lemma 5.12. *(i) For any partition ρ, the function $p_\rho^\sharp$ is an element of $\mathbb{A}$.*
(ii) In the canonical grading,

$$p_\rho^\sharp(\lambda) = \bar{p}_\rho(\lambda) + \text{lower terms},$$

where $\lambda = (1^{r_1}, 2^{r_2}, \cdots)$ and

$$\bar{p}_\rho(\lambda) = \prod_{i=1} \bar{p}_i(\lambda)^{r_i}.$$

(iii) The functions $p_\rho^\sharp$ form a basis in $\mathbb{A}$.
(iv) For any partitions σ and τ, in the canonical grading

$$p_\sigma^\sharp p_\tau^\sharp = p_{\sigma \cup \tau}^\sharp + \text{lower terms}. \tag{5.38}$$

We remark that the basis $p_\rho^\sharp$ is inhomogeneous both in the canonical grading and weight grading. For each $f \in \mathbb{A}$, let $(f)_\rho$ be the structure constants of f in the basis of $p_\rho^\sharp$. Namely,

$$f(\lambda) = \sum_\rho (f)_\rho p_\rho^\sharp(\lambda).$$

Define for any index set $\mathbb{J} \subseteq \mathbb{N}$

$$\|\rho\|_\mathbb{J} = |\rho| + \sum_{j \in \mathbb{J}} r_j(\rho), \quad deg_\mathbb{J}(f) = \max_{\rho : (f)_\rho \neq 0} \|\rho\|_\mathbb{J}.$$

We will be particularly interested in $\mathbb{J} = \emptyset, \{1\}$ and $\mathbb{N}$ below. For simplicity, denote

$$\|\rho\|_0 = \|\rho\|_\emptyset, \quad \|\rho\|_1 = \|\rho\|_1, \quad \|\rho\|_\infty = \|\rho\|_\mathbb{N},$$

and

$$deg_0(f) = deg_\emptyset(f), \quad deg_1(f) = deg_1(f), \quad deg_\infty(f) = deg_\mathbb{N}(f).$$

Lemma 5.13. *For any partition σ,*

$$p_\sigma^\sharp p_1^\sharp = p_{\sigma \cup 1}^\sharp + |\sigma| \cdot p_\sigma^\sharp. \tag{5.39}$$

Proof. This directly follows from the definition (5.37). Indeed, we need only to show for each $\lambda \in \mathcal{P}_n$ with $n \geq |\sigma|$ since otherwise both sides are equal to 0. When $n = |\sigma|$ and $\lambda \in \mathcal{P}_n$,

$$p_\sigma^\sharp(\lambda)p_1^\sharp(\lambda) = n \cdot p_\sigma^\sharp(\lambda) = |\sigma| \cdot p_\sigma^\sharp(\lambda).$$

Next assume $n \geq |\sigma| + 1$. Setting $k = |\sigma|$, then

$$p_\sigma^\sharp(\lambda)p_1^\sharp(\lambda) = n^{\downarrow(k)} \cdot n\frac{\chi_\lambda(\sigma, 1^{n-k})}{d_\lambda}.$$

Hence the claim follows from a simple relation

$$n^{\downarrow k} \cdot n = n^{\downarrow(k+1)} + n^{\downarrow k} \cdot k.$$

$\square$

Lemma 5.14. *For any partitions σ and τ,*

$$p_\sigma^\sharp p_\tau^\sharp = p_{\sigma \cup \tau}^\sharp + \text{ lower terms,} \tag{5.40}$$

where lower terms means a linear combination of $p_\rho^\sharp$ with $\|\rho\|_{\mathbb{N}} < \|\sigma\|_{\mathbb{N}} + \|\tau\|_{\mathbb{N}}$.

Proof. Set

$$p_\sigma^\sharp p_\tau^\sharp = \sum_\rho (p_\sigma^\sharp p_\tau^\sharp)_\rho p_\rho^\sharp.$$

We claim that only partitions ρ with $\|\rho\|_{\mathbb{N}} \leq \|\sigma\|_{\mathbb{N}} + \|\tau\|_{\mathbb{N}}$ can really contribute. Indeed, assume $(p_\sigma^\sharp p_\tau^\sharp)_\rho \neq 0$, and fix a set X of cardinality $|\rho|$ and a permutation $s : X \to X$ whose cycle structure is given by ρ. Then according to Proposition 4.5 of Ivanov and Olshanski (2002) (see also Proposition 6.2 and Theorem 9.1 of Ivanov and Kerov (2001)), there must exist a quadruple $\{X_1, s_1, X_2, s_2\}$ such that

(i) $X_1 \subseteq X$, $X_2 \subseteq X$, $X_1 \cup X_2 = X$;

(ii) $|X_1| = |\sigma|$ and $x_1 : X_1 \mapsto X_1$ is a permutation of cycle structure σ;

(iii) $|X_2| = |\tau|$ and $x_2 : X_2 \mapsto X_2$ is a permutation of cycle structure τ;

(iv) denoting by $\bar{s}_1 : X \to X$ and $\bar{s}_2 : X \to X$ the natural extensions of $s_{1,2}$ from $X_{1,2}$ to the whole X. I.e., $\bar{s}_{1,2}$ is trivial on $X \setminus X_{1,2}$, then $\bar{s}_1 \bar{s}_2 = s$.

Fix any such quadruple and decompose each of the permutations s, s_1, s_2 into cycles. Let $C_{\mathbb{N}}(s_1)$ denote the set of all cycles of s_1, $A_{\mathbb{N}}(s_1)$ the subset of those cycles of s_1 that entirely contained in $X_1 \setminus X_2$, $B_{\mathbb{N}}(s_1)$ the subset of those cycles of s_1 that have a nonempty intersection with $X_1 \cap X_2$. Then

$$C_{\mathbb{N}}(s_1) = A_{\mathbb{N}}(s_1) + B_{\mathbb{N}}(s_1).$$

Define similarly $C_{\mathbb{N}}(s_2)$, $A_{\mathbb{N}}(s_2)$ and $B_{\mathbb{N}}(s_2)$, then we have

$$C_{\mathbb{N}}(s_2) = A_{\mathbb{N}}(s_2) + B_{\mathbb{N}}(s_2).$$

Similarly again, let $C_{\mathbb{N}}(s)$ denote the set of all cycles of s, $B_{\mathbb{N}}(s)$ the subset of those cycles of s that intersect both X_1 and X_2. Then

$$C_{\mathbb{N}}(s) = A_{\mathbb{N}}(s_1) + A_{\mathbb{N}}(s_2) + B_{\mathbb{N}}(s).$$

The claimed inequality $\|\rho\|_{\mathbb{N}} \leq \|\sigma\|_{\mathbb{N}} + \|\tau\|_{\mathbb{N}}$ is equivalent to

$$|X| + |B_{\mathbb{N}}(s)| \leq |X_1| + |B_{\mathbb{N}}(s_1)| + |X_2| + |B_{\mathbb{N}}(s_2)|. \tag{5.41}$$

To prove (5.41), it suffices to establish a stronger inequality:

$$|B_{\mathbb{N}}(s)| \leq |X_1 \cap X_2|. \tag{5.42}$$

To see (5.42), it suffices to show each cycle $\in o \in B_{\mathbb{N}}(s)$ contains a point of $X_1 \cap X_2$. By the definition of $B_{\mathbb{N}}(s)$, cycle o contains both points of X_1 and X_2. Therefore there exist points $x_1 \in X_1 \cap c$ and $x_2 \in X_2 \cap o$ such that $sx_1 = x_2$. By (iv), it follows that either x_1 or x_2 lies in $X_1 \cap X_2$. Thus the claim is true, as desired.

Now assume $(p_\sigma^\sharp p_\tau^\sharp)_\rho \neq 0$ and $\|\rho\|_{\mathbb{N}} = \|\sigma\|_{\mathbb{N}} + \|\tau\|_{\mathbb{N}}$, then both $B_{\mathbb{N}}(s_1)$ and $B_{\mathbb{N}}(s_2)$ are empty, which implies $X_1 \cap X_2 = \emptyset$. Therefore $\rho = \sigma \cup \tau$. Finally, by (5.38),

$$(p_\sigma^\sharp p_\tau^\sharp)_{\sigma \cup \tau} = 1.$$

It concludes the proof. $\qquad\square$

Lemma 5.15. *(i) For any two partitions σ and τ with no common part,*

$$p_\sigma^\sharp p_\tau^\sharp = p_{\sigma \cup \tau}^\sharp + \text{lower terms},$$

where lower terms means terms with $\deg_1(\cdot) < \|\sigma \cup \tau\|_1$.
(ii) For any partition $\sigma \in \mathcal{P}$ and $k \geq 2$, if $r_k(\sigma) \geq 1$, then

$$p_\sigma^\sharp p_k^\sharp = p_{\sigma \cup k}^\sharp + k r_k(\sigma) p_{(\sigma \backslash k) \cup 1^k}^\sharp + \text{lower terms}, \tag{5.43}$$

where lower terms means terms with $\deg_1(\cdot) < \|\sigma\|_1 + k$.

Proof. (i) can be proved in a way similar to that of Lemma 5.14 with minor modification. Turn to (ii). Set

$$p_\sigma^\sharp p_\tau^\sharp = \sum_\rho (p_\sigma^\sharp p_\tau^\sharp)_\rho p_\rho^\sharp.$$

Again, only partitions ρ with $\|\rho\|_1 \leq \|\sigma\|_1 + k$ can really contribute. We need below only search for partitions ρ such that $(p_\sigma^\sharp p_\tau^\sharp)_\rho \neq 0$ and $\|\rho\|_1 = \|\sigma\|_1 + k$. As in Lemma 5.14, we get

$$B_1(s_1) = \emptyset, \quad B_1(s_2) = \emptyset, \quad |B_1(s)| = |X_1 \cap X_2|.$$

This means that $X_1 \cap X_2 = \emptyset$ or $X_1 \cap X_2$ entirely consists of common nontrivial cycles of the permutations s_1 and s_2^{-1}. The first possibility, $X_1 \cap X_2 = \emptyset$, means that $\rho = \sigma \cup (k)$. Furthermore, by (5.38),

$$(p_\sigma^\sharp p_\tau^\sharp)_{\sigma \cup k} = 1.$$

The second possibility means $X_1 \supseteq X_2$ because s_2^{-1} reduces to a single k-cycle which is also a k-cycle of s_1. This in turn implies that $r_k(\sigma) \geq 1$ and $\rho = (\sigma \setminus k) \cup 1^k$.

It remains to evaluate $(p_\sigma^\sharp p_\tau^\sharp)_\rho = kr_k(\sigma)$. Note that the number of ways to choose a k-cycle inside a $k+r_1(\sigma)$-point set equals $(k+r_1(\sigma))!/k(r_1(\sigma))!$. According to Proposition 6.2 and Theorem 9.1 of Ivanov and Kerov (2001), we know

$$(p_\sigma^\sharp p_\tau^\sharp)_\rho = \frac{kz_\sigma}{z_\rho} \frac{(k+r_1(\sigma))!}{k(r_1(\sigma))!}$$
$$= kr_k(\sigma),$$

where z_λ is defined by (2.15) for a partition λ. The proof is complete. $\quad\square$

In the preceding paragraphs we have briefly described the structure of algebra $\mathbb{A}$ and its three families of bases including $\{\bar{p}_k\}$, $\{\tilde{p}_k\}$ and $\{p_\rho^\sharp\}$. Next we need to take average operation with respect to $P_{p,n}$ for elements of $\mathbb{A}$. A basic result is as follows.

Lemma 5.16. *Let* $|\rho| = r$, $n \geq r$. *Then*

$$E_{p,n} p_\rho^\sharp = \begin{cases} n^{\downarrow r}, & \rho = (1^r), \\ 0, & otherwise. \end{cases} \tag{5.44}$$

Proof. By (5.37) and (2.12)

$$E_{p,n} p_\rho^\sharp = E_{p,n} n^{\downarrow r} \frac{\chi_\lambda(\rho, 1^{n-r})}{d_\lambda}$$

$$= \frac{n^{\downarrow r}}{n!} \sum_{\lambda \in \mathcal{P}_n} \chi_\lambda(\rho, 1^{n-r}) d_\lambda$$

$$= \frac{n^{\downarrow r}}{n!} \sum_{\lambda \in \mathcal{P}_n} \chi_\lambda(\rho, 1^{n-r}) \chi_\lambda(\mathbf{e})$$

$$= \begin{cases} n^{\downarrow r}, & \rho = (1^r), \\ 0, & otherwise. \end{cases}$$

The proof is complete. $\square$

To proceed, we shall prove a weak form of limit shape theorem.

Theorem 5.4. *Define*

$$\mathcal{Y}_{n,k}(\lambda) = \int_{-\infty}^{\infty} u^k \big(\Psi_{n,\lambda}(u) - \Omega(u) \big) du, \quad \lambda \in \mathcal{P}_n.$$

Then for each $k \geq 0$

$$\mathcal{Y}_{n,k} \xrightarrow{P} 0, \quad n \to \infty.$$

Proof. Note by (5.24) and (5.25),

$$\mathcal{Y}_{n,k}(\lambda) = \int_{-\infty}^{\infty} u^k \big(\Psi_{n,\lambda}(u) - |u| \big) du - \int_{-\infty}^{\infty} u^k \big(\Omega(u) - |u| \big) du$$

$$= \frac{2}{(k+2)(k+1)} \left(\frac{\tilde{p}_{k+2}(\lambda)}{n^{(k+2)/2}} - \tilde{p}_{k+2}(\Omega) \right).$$

Hence it suffices to prove for each $k \geq 2$

$$\frac{\tilde{p}_k(\lambda)}{n^{k/2}} - \tilde{p}_k(\Omega) \xrightarrow{P} 0, \quad n \to \infty.$$

Equivalently,

$$\frac{\tilde{p}_k(\lambda)}{n^{k/2}} \xrightarrow{P} \tilde{p}_k(\Omega), \quad n \to \infty.$$

In turn, this will be done by checking as $n \to \infty$

$$\frac{E_{p,n} \tilde{p}_k(\lambda)}{n^{k/2}} \longrightarrow \tilde{p}_k(\Omega) \tag{5.45}$$

and

$$\frac{E_{p,n} \tilde{p}_k^2(\lambda)}{n^k} \longrightarrow \tilde{p}_k^2(\Omega). \tag{5.46}$$

Expand $\tilde{p}_k$ in the basis of $p_\rho^\sharp$

$$\tilde{p}_k(\lambda) = \sum_\rho (\tilde{p}_k)_\rho p_\rho^\sharp(\lambda), \tag{5.47}$$

where $(\tilde{p}_k)_\rho$ denotes the structure coefficient. Note by Lemmas 5.11 and 5.14 $deg_\infty(\tilde{p}_k) = k$ so that the summation in (5.47) is over all ρ with $\|\rho\|_\infty \leq k$. Then it follows from (5.44)

$$E_{p,n} \tilde{p}_k = \sum_{\|\rho\|_\infty \leq k} (\tilde{p}_k)_\rho E_{p,n} p_\rho^\sharp$$

$$= \sum_{2r \leq k} (\tilde{p}_k)_{1^r} n^{\downarrow r}.$$

In addition, according to (5.36) and (5.39), if $k = 2r$

$$(\tilde{p}_k)_{1^r} = \frac{k^{\downarrow r}}{r!}. \tag{5.48}$$

Hence for $k = 2m$,

$$\frac{E_{p,n}\tilde{p}_{2m}}{n^m} \to \frac{(2m)!}{(m!)^2},$$

while for $k = 2m + 1$,

$$\frac{E_{p,n}\tilde{p}_k}{n^{k/2}} \to 0.$$

This proves (5.45). Analogously, we can prove (5.46). Indeed, it follows from (5.47)

$$\tilde{p}_k^2(\lambda) = \sum_{\|\rho\|_\infty \le k, \|\sigma\|_\infty \le k} (\tilde{p}_k)_\rho (\tilde{p}_k)_\sigma p_\rho^\sharp(\lambda) p_\sigma^\sharp(\lambda)$$

which in turn implies

$$E_{p,n}\tilde{p}_k^2 = \sum_{\|\rho\|_\infty \le k, \|\sigma\|_\infty \le k} (\tilde{p}_k)_\rho (\tilde{p}_k)_\sigma E_{p,n} p_\rho^\sharp p_\sigma^\sharp.$$

Also, by (5.40),

$$E_{p,n} p_\rho^\sharp p_\sigma^\sharp = E_{p,n} p_{\rho \cup \sigma}^\sharp + E_{p,n} \text{ lower terms}$$

where lower terms means a linear combination of $p_\tau^\sharp$'s with $\|\tau\|_\infty < \|\rho\|_\infty + \|\sigma\|_\infty$.

Again by (5.48),

$$E_{p,n} p_{\rho \cup \sigma}^\sharp = \begin{cases} n^{\downarrow r}, & \rho \cup \sigma = (1^r), \\ 0, & \text{otherwise.} \end{cases}$$

In summary, we have

$$E_{p,n}\tilde{p}_k^2 = \sum_{2r_1 \le k, 2r_2 \le k} (\tilde{p}_k)_{1^{r_1}} (\tilde{p}_k)_{1^{r_2}} n^{\downarrow(r_1+r_2)} + E_{p,n} \text{ lower terms}$$

In particular, we have

$$E_{p,n}\tilde{p}_{2m}^2 = \left(\frac{(2m)^{\downarrow m}}{m!}\right)^2 n^{\downarrow 2m} + O(n^{2m-1})$$

and

$$E_{p,n}\tilde{p}_{2m+1}^2 = O(n^{2m}).$$

Therefore it follows

$$\frac{E_{p,n}\tilde{p}_k^2}{n^k} \to \begin{cases} \left(\frac{(2m)^{\downarrow m}}{m!}\right)^2, & k = 2m, \\ 0, & k = 2m + 1. \end{cases}$$

The proof is now complete. $\qquad\qquad\square$

Remark 5.1. One can derive the strong form of limit shape theorem, i.e., Theorem 5.2, using the equivalence between weak topology and uniform topology. The interested reader is referred to Theorem 5.5 of Ivanov and Olshanski (2002).

Theorem 5.5. *Define*

$$\mathscr{Z}_{n,k}(\lambda) = \frac{p_k^\sharp(\lambda)}{n^{k/2}}, \quad \lambda \in \mathcal{P}_n.$$

Then under $(\mathcal{P}_n, P_{p,n})$ *as* $n \to \infty$

$$\left(\mathscr{Z}_{n,k}, \quad k \geq 2\right) \xrightarrow{d} \left(\sqrt{k}\,\xi_k, \quad k \geq 2\right).$$

Here $\xi_k, k \geq 2$ *is a sequence of standard normal random variables, the convergence holds in terms of finite dimensional distribution.*

Proof. For simplicity of notations, we will mainly focus on the 1-dimensional case. Namely, we shall below prove for each $k \geq 2$

$$\mathscr{Z}_{n,k} \xrightarrow{d} \sqrt{k}\,\xi_k, \quad n \to \infty.$$

Adapt the moment method. Fix $l \geq 1$, we need to check

$$E_{p,n}(\eta_k)^l \to E\xi_k^l, \quad n \to \infty$$

where $\eta_k = \mathscr{Z}_{n,k}/\sqrt{k}$. This is equivalent to proving

$$E_{p,n}h_l(\eta_k) \longrightarrow Eh_l(\xi_k), \quad n \to \infty$$

where h_l is a classical Hermite orthogonal polynomial of order l with respect to the weight function $e^{-x^2/2}/\sqrt{2\pi}$, see (3.3). Trivially, by the orthogonality property, $Eh_l(\xi_k) = 0$. So we shall only prove

$$E_{p,n}h_l(\eta_k) \to 0. \tag{5.49}$$

Note by (5.43),

$$p_{k^l}^\sharp p_k^\sharp = p_{k^{l+1}}^\sharp + klp_{k^{l-1}\cup 1^k}^\sharp + \text{ lower terms}, \tag{5.50}$$

where lower terms means a term with $deg_1(\cdot) < k(l+1)$.

Also, by definition

$$p_{k^{l-1}\cup 1^k}^\sharp(\lambda) = \frac{n^{\downarrow kl}\chi_\lambda(k^{l-1}\cup 1^k, 1^{n-kl})}{d_\lambda}$$

$$= \frac{n^{\downarrow kl}}{n^{\downarrow k(l-1)}}p_{k^{l-1}}^\sharp(\lambda), \quad \lambda \in \mathcal{P}_n. \tag{5.51}$$

Inserting (5.51) back into (5.50) and dividing by $(\sqrt{k}n^{k/2})^{l+1}$, we get

$$\eta_{k^l}\eta_k = \eta_{k^{l+1}} + l\frac{n^{\downarrow kl}}{n^{\downarrow k(l-1)}n^k}\eta_{k^{l-1}} + \frac{1}{(\sqrt{k}n^{k/2})^{l+1}} \text{ lower terms,} \quad (5.52)$$

Recall that $h_l(x)$ is characterized by the recurrence relation

$$xh_l(x) = h_{l+1}(x) + lh_{l-1}(x)$$

together with the initial data $h_0 = 1$ and $h_1(x) = x$. Hence we make repeatedly use of (5.52) to yield

$$\eta_{k^l} = h_l(\eta_k) + \frac{1}{(\sqrt{k}n^{k/2})^l} \text{ lower terms}$$

where lower terms means a term with $deg_1(\cdot) < kl$.

In particular, it follows

$$h_l(\eta_k) = \eta_{k^l} + \frac{1}{(\sqrt{k}n^{k/2})^l} \text{ lower terms.}$$

Thus by (5.44)

$$E_{p,n}h_l(\eta_k) = E_{p,n}\eta_{k^l} + \frac{1}{(\sqrt{k}n^{k/2})^l}E_{p,n} \text{ lower terms}$$

$$= O(n^{-1/2}),$$

which proves (5.49) as desired.

To treat m-dimensional case, we need to prove for any positive integers $l_2, \cdots, l_m$

$$E_{p,n}\prod_{k=2}^{m}\eta_k^{l_k} \to \prod_{k=2}^{m}E\xi_k^{l_k} = 0, \quad n \to \infty.$$

Equivalently, for $h_{l_2}, \cdots, h_{l_m}$

$$E_{p,n}\prod_{k=2}^{m}h_{l_k}(\eta_k) \to \prod_{k=2}^{m}Eh_{l_k}(\xi_k), \quad n \to \infty.$$

The details are left to the reader. $\qquad\qquad\square$

Now we are ready to prove Theorem 5.3. Define $q_1 = 0$ and for any $k \geq 2$

$$q_k(\lambda) = \frac{1}{(k+1)n^{k/2}}\left(\tilde{p}_{k+1}(\lambda) - n^{(k+1)/2}\tilde{p}_{k+1}(\Omega)\right), \quad \lambda \in \mathcal{P}_n. \quad (5.53)$$

Lemma 5.17. *For any* $k \geq 2$,

$$q_k(\lambda) = \sum_{j=0}^{[(k-1)/2]}\binom{k}{j}\frac{p_{k-2j}^{\sharp}(\lambda)}{n^{(k-2j)/2}}$$

$$+ \frac{1}{n^{k/2}} \text{ lower terms with } deg_1(\cdot) \leq k - 1. \quad (5.54)$$

Proof. Using Lemma 5.11, one can express $\tilde{p}_{k+1}$ as a polynomial $p_1^\sharp, p_2^\sharp, \cdots$ up to terms of lower weight. In particular, we have

$$\tilde{p}_{k+1}(\lambda) = (k+1) \sum_{j=0}^{[(k-1)/2]} n^j \binom{k}{j} p_{k-2j}^\sharp(\lambda) + n^{(k+1)/2} \tilde{p}_{k+1}(\Omega)$$

$$+\text{lower terms with } deg_1(\cdot) \leq k-1. \tag{5.55}$$

A nontrvial point in (5.55) is the switch between two weight filtrations. Its proof is left to the reader. See also Proposition 7.3 of Ivanov and Olshanski (2002) for details. □

Inverting (5.54) easily gives

Lemma 5.18. *For any* $k \geq 2$,

$$\frac{p_k^\sharp(\lambda)}{n^{k/2}} = \sum_{j=0}^{[(k-1)/2]} (-1)^j \frac{k}{k-j} \binom{k-j}{j} q_{k-2j}(\lambda),$$

$$+\frac{1}{n^{k/2}} \text{ lower terms with } deg_1(\cdot) \leq k-1. \tag{5.56}$$

Proof. Recall the following combinatorial inversion formula due to Riordan (1968): assume that $\alpha_0, \alpha_1, \cdots; \beta_0, \beta_1, \cdots$ are two families of formal variables, then

$$\alpha_k = \sum_{j=0}^{[k/2]} \binom{k}{j} \beta_{k-2j}, \quad k = 0, 1, \cdots$$

$$\Updownarrow$$

$$\beta_k = \sum_{j=0}^{[k/2]} (-1)^j \frac{k}{k-j} \binom{k-j}{j} \alpha_{k-2j}, \quad k = 0, 1, \cdots.$$

Set $\alpha_0 = \alpha_1 = 0$, $\alpha_k = q_k(\lambda)$, $k \geq 2$; $\beta_0 = \beta_1 = 0$, $\beta_k = p_k^\sharp(\lambda)/n^{k/2}$, $k \geq 2$. If we neglect the lower terms in (5.54), then it obviously follows

$$\frac{p_k^\sharp(\lambda)}{n^{k/2}} = \sum_{j=0}^{[(k-1)/2]} (-1)^j \frac{k}{k-j} \binom{k-j}{j} q_{k-2j}.$$

The appearance of remainder terms affect only similar remainder terms in the reverse relations. □

Proof of Theorem 5.3. By (5.18) and integrating term by term, we obtain

$$\mathcal{X}_{n,k}(\lambda) = \sum_{j=0}^{[k/2]} (-1)^j \binom{k-j}{j} \int_{-\infty}^{\infty} u^{k-2j} \big(\Psi_\lambda(\sqrt{n}u) - \sqrt{n}\Omega(u)\big) du$$

$$= \sum_{j=0}^{[k/2]} (-1)^j \binom{k-j}{j} \Big[\int_{-\infty}^{\infty} u^{k-2j} \big(\Psi_\lambda(\sqrt{n}u) - \sqrt{n}|u|\big) du$$

$$- \int_{-\infty}^{\infty} u^{k-2j} \sqrt{n}\big((\Omega(u) - |u|)\big) du \Big]$$

$$= \sum_{j=0}^{[k/2]} (-1)^j \binom{k-j}{j} \frac{2}{(k+2-2j)(k+1-2j)}$$

$$\cdot \Big(\frac{\tilde{p}_{k+2-2j}(\lambda)}{n^{(k+1-2j)/2}} - \sqrt{n}\tilde{p}_{k+2-2j}(\Omega) \Big).$$

By the definition of (5.53) and noting

$$\frac{1}{k+2-2j} \binom{k-j}{j} = \frac{1}{k+1-j} \binom{k+1-j}{j},$$

we further get

$$\mathcal{X}_{n,k}(\lambda) = \sum_{j=0}^{[k/2]} (-1)^j \frac{2}{k+1-j} \binom{k+1-j}{j} q_{k+1-2j}(\lambda).$$

By (5.56),

$$\mathcal{X}_{n,k}(\lambda) = \frac{2p_{k+1}^{\sharp}(\lambda)}{(k+1)n^{(k+1)/2}}$$

$$+ \frac{1}{n^{(k+1)/2}} \text{ lower terms with } deg_1(\cdot) \leq k.$$

Since the remainder terms of negative degree do not affect the asymptotics, then we can use Theorem 5.5 to conclude the proof. □

To conclude this section, we remark that Theorem 5.5 for character ratios is of independent interest. Another elegant approach was suggested by Hora (1998), in which a central limit theorem was established for adjacency operators on the infinite symmetric group. Still, Fulman (2005, 2006) developed the Stein method and martingale approach to prove asymptotic normality for character ratios.

5.3   Fluctuations in the bulk

In this section we shall turn to the study of fluctuations of a typical Plancherel Young diagrams around their limit shape in the bulk of the partition *spectrum*. Here we use the term spectrum informally by analogy with the GUE, to refer to the variety of partition's terms $\lambda_i \in \lambda$. Recall $\psi_\lambda(x)$ and $\omega(x)$ introduced in Section 5.1 and define the random process

$$\Xi_n(x) = \psi_\lambda(\sqrt{n}\,x) - \sqrt{n}\,\omega(x), \quad x \geq 0, \quad \lambda \in \mathcal{P}_n.$$

According to the Corollary 5.1, it follows under $(\mathcal{P}_n, P_{p,n})$

$$\frac{1}{\sqrt{n}} \sup_{x \geq 0} |\Xi_n(x)| \xrightarrow{P} 0, \quad n \to \infty.$$

This is a kind of weak law of large numbers. The following theorem describes the second order fluctuation of Ξ_n at each fixed $0 < x < 2$.

Theorem 5.6. *Under* $(\mathcal{P}_n, P_{p,n})$

$$\frac{\Xi_n(x)}{\frac{1}{2\pi}\sqrt{\log n}} \xrightarrow{d} N\big(0, \varrho^2(x)\big), \quad n \to \infty \tag{5.57}$$

for each $0 < x < 2$, *where* $\varrho^{-2}(x) = \frac{1}{\pi} \arccos \frac{|\omega(x)-x|}{2}$.

The asymptotics of finite dimensional distributions of the random process $\Xi_n(x)$ reads as follows.

Theorem 5.7. *Assume* $0 < x_1 < \cdots < x_m < 2$, *then under* $(\mathcal{P}_n, P_{p,n})$,

$$\frac{1}{\frac{1}{2\pi}\sqrt{\log n}}\big(\Xi_n(x_i),\, 1 \leq i \leq m\big) \xrightarrow{d} (\xi_i,\, 1 \leq i \leq m), \quad n \to \infty$$

where $\xi_i, 1 \leq i \leq m$ *are independent normal random variables.*

Remark 5.2. (i) The work of this section, in particular Theorems 5.6 and 5.7, are motivated by Gustavsson (2005), in which he investigated the Gaussian fluctuation of eigenvalues in the GUE. There is a surprising similarity between Plancherel random partitions and GUE from the viewpoint of asymptotics, though no direct link exists between two finite models.

(ii) Compared with the uniform random partitions, the normalizing constant $\sqrt{\log n}$ is much smaller than $n^{1/4}$, see Theorem 4.5. This means that Plancherel Young diagrams concentrated more stably around their limit shape.

(iii) The random process $\Xi_n(x)$ weakly converges to a Gaussian white noise in the finite dimensional sense. Thus one cannot expect a usual process convergence for Ξ_n in the space of continuous functions on $[0, 2]$.

(iv) As we will see, if $x_n \to x$ where $0 < x < 2$, then (5.57) still holds for $\Xi_n(x_n)$, namely

$$\frac{\Xi_n(x_n)}{\frac{1}{2\pi}\sqrt{\log n}} \xrightarrow{d} N\big(0, \varrho^2(x)\big), \quad n \to \infty. \tag{5.58}$$

(v) Since $\varrho(x) \to \infty$ as $x \to 0$ or 2, the normal fluctuation is no longer true at either 0 or 2. In fact, it was proved

$$\frac{\psi_\lambda(0) - 2\sqrt{n}}{n^{1/6}} \xrightarrow{d} F_2, \quad n \to \infty$$

where F_2 is Tracy-Widom law.

It is instructive to reformulate Theorems 5.6 and 5.7 in the *rotated* coordinates u and v. Recall $\Psi_\lambda(u)$ and $\Omega(u)$ are rotated versions of $\psi_\lambda(x)$ and $\omega(x)$. Define

$$\Upsilon_n(u) = \Psi_\lambda(\sqrt{n}\,u) - \sqrt{n}\,\Omega(u), \quad -\infty < u < \infty. \tag{5.59}$$

We can restate Theorem 5.6 in the following elegant version, whereby— quite surprisingly—the normalization does not depend on the location in the spectrum.

Theorem 5.8. *Under* $(\mathcal{P}_n, P_{p,n})$

$$\frac{\Upsilon_n(u)}{\frac{1}{\pi}\sqrt{\log n}} \xrightarrow{d} N(0,1), \quad n \to \infty$$

for $-2 < u < 2$.

Proof. Fix $-2 < u < 2$ and assume $\lambda \in \mathcal{P}_n$. A key step is to express the error $\Psi_\lambda(\sqrt{n}\,u) - \sqrt{n}\Omega(u)$ in terms of ψ_λ and ω. Let local extrema consist of two interlacing sequences of points

$$\check{u}_1 < \hat{u}_1 < \check{u}_2 < \hat{u}_2 < \cdots < \check{u}_m < \hat{u}_m < \check{u}_{m+1},$$

where $\check{u}_i$'s are the local minima and $\hat{u}_i$'s are the local maxima of the function $\Psi_\lambda(\sqrt{n}\,\cdot)$. Without loss of generality, we may and will assume that u is between $\hat{u}_k$ and $\check{u}_{k+1}$ for some $1 \le k \le m$. Denote by $\sqrt{n}(x_n, x_n)$ and $\sqrt{n}(x_n^*, x_n^*)$ the projections of $(\sqrt{n}\,u, \Psi_\lambda(\sqrt{n}\,u))$ and $\sqrt{n}(u, \Omega(u))$ in the line $u = v$, respectively. Then we obviously have

$$\Psi_\lambda(\sqrt{n}\,u) - \sqrt{n}\,\Omega(u) = 2\sqrt{n}(x_n - x_n^*).$$

According to Theorem 5.2, it follows

$$x_n - x_n^* \xrightarrow{P} 0, \quad n \to \infty$$

and so

$$x_n \xrightarrow{P} x := \frac{1}{2}(\Omega(u) + u).$$

On the other hand, if we let $\sqrt{2}h_n$ be the distance between $\sqrt{n}(x_n, x_n)$ and $(\sqrt{n}\,u, \Psi_\lambda(\sqrt{n}\,u))$, then we have

$$\sqrt{n}(x_n - x_n^*) = h_n - \sqrt{n}\,\omega(x_n^*)$$
$$= h_n - \sqrt{n}\,\omega(x_n) + \sqrt{n}(\omega(x_n) - \omega(x_n^*)). \qquad (5.60)$$

Now using the Taylor expansion for the function ω at x_n and solving equation (5.60), we obtain

$$\sqrt{n}(x_n - x_n^*) = \frac{h_n - \sqrt{n}\,\omega(x_n)}{1 - \omega'(\tilde{x}_n)},$$

where $\tilde{x}_n$ is between x_n and x_n^*.

Since $x_n, x_n^* \xrightarrow{P} x \in (0, 2)$, then it holds

$$\frac{1}{1 - \omega'(\tilde{x}_n)} \xrightarrow{P} \frac{1}{1 - \omega'(x)} = \frac{1}{\pi}\arccos\frac{\omega(x) - x}{2}.$$

Hence it suffices to prove $h_n - \sqrt{n}\,\omega(x_n)$ after properly scaled converges in distribution. Observe that

$$\psi_\lambda(\sqrt{n}\,x_n + 1) \le h_n \le \psi_\lambda(\sqrt{n}\,x_n)$$

since u is between $\hat{u}_k$ and $\check{u}_{k+1}$.

Note $x_n \xrightarrow{P} x \in (0, 2)$. Then for each subsequence $\{n'\}$ of integers there exists a further subsequence $\{n''\} \subseteq \{n'\}$ such that $x_{n''} \to x$ a.e. Thus by (5.58) it holds

$$\frac{\psi_\lambda(\sqrt{n''}\,x_{n''}) - \sqrt{n''}\omega(x_{n''})}{\frac{1}{2\pi}\sqrt{\log n''}} \xrightarrow{d} N\big(0, \varrho^2(x)\big).$$

By a standard subsequence argument, it holds

$$\frac{\psi_\lambda(\sqrt{n}\,x_n) - \sqrt{n}\,\omega(x_n)}{\frac{1}{2\pi}\sqrt{\log n}} \xrightarrow{d} N\big(0, \varrho^2(x)\big).$$

Similarly, since $\sqrt{n}\big(\omega(x_n + \frac{1}{\sqrt{n}}) - \omega(x_n)\big) = O_p(1)$,

$$\frac{\psi_\lambda(\sqrt{n}\,x_n + 1) - \sqrt{n}\,\omega(x_n)}{\frac{1}{2\pi}\sqrt{\log n}} \xrightarrow{d} N\big(0, \varrho^2(x)\big).$$

In combination, we have

$$\frac{h_n - \sqrt{n}\,\omega(x_n)}{\frac{1}{2\pi}\sqrt{\log n}} \xrightarrow{d} N\big(0, \varrho^2(x)\big).$$

We conclude the proof. $\qquad\qquad\qquad\qquad\qquad\qquad\qquad\qquad\square$

Let us now turn to the proof of Theorems 5.6 and 5.7. A basic strategy is to adapt the conditioning argument—the Poissonization and de-Poissonization techniques. Define the Poissonized Plancherel measure $Q_{p,\theta}$ on $\mathcal{P}$ as follows:

$$Q_{p,\theta}(\lambda) = e^{-\theta}\theta^{|\lambda|}\left(\frac{d_\lambda}{|\lambda|!}\right)^2$$

$$= \sum_{n=0}^{\infty} e^{-\theta}\frac{\theta^n}{n!}P_{p,n}(\lambda)1_{(\lambda\in\mathcal{P}_n)}, \quad \lambda \in \mathcal{P} \tag{5.61}$$

where $\theta > 0$ is a model parameter. This is a mixture of a Poisson random variable with mean θ and the Plancherel measures. Let $\Xi_\theta(x)$ be given by (5.57) with n replaced by θ, namely

$$\Xi_\theta(x) = \psi_\lambda(\sqrt{\theta}\,x) - \sqrt{\theta}\,\omega(x), \quad x \geq 0, \quad \lambda \in \mathcal{P}. \tag{5.62}$$

Theorem 5.9. *Under* $(\mathcal{P}, Q_{p,\theta})$,

$$\frac{\Xi_\theta(x)}{\frac{1}{2\pi}\sqrt{\log\theta}} \xrightarrow{d} N(0, \varrho^2(x)), \quad \theta \to \infty$$

for each $0 < x < 2$.

Theorem 5.10. *Assume* $0 < x_1 < \cdots < x_m < 2$. *Under* $(\mathcal{P}, Q_{p,\theta})$,

$$\frac{1}{\frac{1}{2\pi}\sqrt{\log\theta}}(\Xi_\theta(x_i), 1 \leq i \leq m) \xrightarrow{d} (\xi_i, 1 \leq i \leq m), \quad \theta \to \infty$$

where $\xi_i, 1 \leq i \leq m$ *are independent normal random variables.*

Before giving the proof of Theorems 5.9 and 5.10, let us prove Theorems 5.6 and 5.7 with the help of the de-Poissonization technique.

Lemma 5.19. *For* $0 < \alpha < 1$, *define*

$$\theta_n^{\pm} = n \pm \sqrt{n(\log n)^\alpha}.$$

Then uniformly in $x \geq 0$ *and* $z \in \mathbb{R}$,

$$Q_{p,\theta_n^+}(\lambda \in \mathcal{P} : \psi_\lambda(x) \leq z) - \varepsilon_n \leq P_{p,n}(\lambda \in \mathcal{P}_n : \psi_\lambda(x) \leq z)$$

$$\leq Q_{p,\theta_n^-}(\lambda \in \mathcal{P} : \psi_\lambda(x) \leq z) + \varepsilon_n \tag{5.63}$$

where $\varepsilon_n \to 0$ *as* $n \to \infty$.

Proof. We need only give the proof of the lower bound, since the upper bound is similar. Let X be a Poisson random variable with mean θ_n^+. Then we have $EX = \theta_n^+$, $VarX = \theta_n^+$ and the following tail estimate

$$\varepsilon_n := P(|X - \theta_n^+| > \sqrt{n(\log n)^\alpha})$$

$$= O(\log^{-\alpha} n).$$

It follows by (5.61)

$$Q_{p,X}(\lambda) = EP_{p,X}(\lambda), \quad \lambda \in \mathcal{P}$$

and so for any event $A \subseteq \mathcal{P}$

$$Q_{p,X}(A) = EP_{p,X}(A), \quad A \subseteq \mathcal{P}.$$

Note

$$EP_{p,X}(A) = EP_{p,X}(A)\mathbf{1}_{(X<n)} + EP_{p,X}(A)\mathbf{1}_{(X\geq n)}.$$

Trivially,

$$EP_{p,X}(A)\mathbf{1}_{(X<n)} \leq P(X < n) \leq \varepsilon_n. \tag{5.64}$$

In addition, set $A = \{\lambda \in \mathcal{P} : \psi_\lambda(x) \leq z\}$. Then using a similar argument to that of Lemma 2.4 of Johansson (1998b), A is a monotonoic event under $(P_{p,n}, n \geq 1)$, namely $P_{p,n+1}(A) \leq P_{p,n}(A)$. Hence it follows

$$EP_{p,X}(A)\mathbf{1}_{(X\geq n)} \leq P_{p,n}(A). \tag{5.65}$$

Combining (5.64) and (5.65) together implies the lower bound. $\qquad\square$

Proof of Theorem 5.6. Set

$$x_n^+ = \frac{\sqrt{n}\,x}{\sqrt{\theta_n^+}}, \quad x_n^- = \frac{\sqrt{n}\,x}{\sqrt{\theta_n^-}}$$

where $\theta_n^\pm$ are as in Lemma 5.19. Trivially, $x_n^+, x_n^- \to x$. Also, since $0 < \alpha < 1$, then it follows

$$\frac{\sqrt{\theta_n^\pm} - \sqrt{n}}{\sqrt{\log n}} \to 0, \quad n \to \infty.$$

Note

$$\frac{\psi_\lambda(\sqrt{n}\,x) - \sqrt{n}\,\omega(x)}{\sqrt{\log n}} = \frac{\psi_\lambda(\sqrt{\theta_n^\pm}\,x_n^\pm) - \sqrt{\theta_n^\pm}\,\omega(x_n^\pm)}{\sqrt{\log \theta_n^\pm}} \cdot \frac{\sqrt{\log \theta_n^\pm}}{\sqrt{\log n}}$$

$$+ \frac{\sqrt{\theta_n^\pm}\,\omega(x_n^\pm) - \sqrt{n}\,\omega(x)}{\sqrt{\log n}} \tag{5.66}$$

and as $n \to \infty$

$$\frac{\sqrt{\theta_n^\pm}\,\omega(x_n^\pm) - \sqrt{n}\,\omega(x)}{\sqrt{\log n}} \to 0, \quad \frac{\sqrt{\log \theta_n^\pm}}{\sqrt{\log n}} \to 1. \tag{5.67}$$

On the other hand, by Theorem 5.9, under $(\mathcal{P}, Q_{p,\theta_n^\pm})$

$$\frac{\psi_\lambda(\sqrt{\theta_n^\pm}\,x_n^\pm) - \sqrt{\theta_n^\pm}\,\omega(x_n^\pm)}{\frac{1}{2\pi}\sqrt{\log \theta_n^\pm}} \xrightarrow{d} N\big(0, \varrho^2(x)\big).$$

Hence it follows from (5.66) and (5.67) that under $\left(\mathcal{P}, Q_{p,\theta_n^{\pm}}\right)$

$$\frac{\psi_\lambda(\sqrt{n}\,x) - \sqrt{n}\,\omega(x)}{\frac{1}{2\pi}\sqrt{\log n}} \xrightarrow{\,d\,} N\big(0, \varrho^2(x)\big).$$

Taking Lemma 5.19 into account, we now conclude the proof. $\qquad\square$

To prove Theorem 5.9, we will again apply the Costin-Lebowitz-Soshnikov theorem for determinantal point processes, see Theorem 3.7. To do this, we need the following lemma due to Borodin, Okounkov and Olshanski (2000), in which they proved the Tracy-Widom law for the largest parts. Set for $\lambda = (\lambda_1, \lambda_2, \cdots, \lambda_l) \in \mathcal{P}$

$$\mathcal{X}(\lambda) = \{\lambda_i - i,\, 1 \le i \le l\}. \tag{5.68}$$

For $k = 1, 2, \cdots$, the k-point correlation function ρ_k is defined by

$$\rho_k(x_1, x_2, \cdots, x_k) = Q_{p,\theta}\big(\lambda \in \mathcal{P} : x_1, x_2, \cdots, x_k \in \mathcal{X}(\lambda)\big),$$

where $x_1, x_2, \cdots, x_k$ are distinct integers.

Lemma 5.20. ρ_k *has a determinantal structure as follows:*

$$\rho_k(x_1, x_2, \cdots, x_k) = \det\big(K_\theta(x_i, x_j)\big)_{1 \le i,j \le k}, \tag{5.69}$$

with the kernel K_θ of the form

$$K_\theta(x, y) = \begin{cases} \sqrt{\theta}\,\dfrac{J_x J_{y+1} - J_{x+1} J_y}{x - y}, & x \ne y \\[2mm] \sqrt{\theta}\big(J_x' J_{x+1} - J_{x+1}' J_x\big), & x = y \end{cases} \tag{5.70}$$

where $J_m \equiv J_m\big(2\sqrt{\theta}\big)$ is the Bessel function of integral order m.

We will postpone the proof to Section 5.5. Now we are ready to give
Proof of Theorem 5.9. Fix $0 < x < 2$. It suffices to show that for any $z \in \mathbb{R}$

$$Q_{p,\theta}\left(\Xi_\theta(x) \le \frac{\varrho(x)}{2\pi}\sqrt{\log\theta}\,z\right) \to \Phi(z), \quad \theta \to \infty, \tag{5.71}$$

where Φ denotes the standard normal distribution function. Equivalently, it suffices to show that for any $z \in \mathbb{R}$

$$Q_{p,\theta}\big(\psi_\lambda(\sqrt{\theta}\,x) - \lceil\sqrt{\theta}\,x\rceil \le a_\theta\big) \to \Phi(z),$$

where

$$a_\theta := a_\theta(x, z) = \sqrt{\theta}(\omega(x) - x) + \frac{\varrho(x)}{2\pi}\sqrt{\log\theta}\,z. \tag{5.72}$$

Consider the semi-infinite interval $I_\theta := [a_\theta, \infty)$ and let N_θ be the number of points of $\lambda_i - i \in \mathcal{X}(\lambda)$ contained in I_θ. Using that the sequence $\lambda_i - i$ is strictly decreasing, it is easy to see that relation (5.71) is reduced to

$$Q_{p,\theta}\big(N_\theta \leq \lceil \sqrt{\theta}\, x \rceil\big) \to \Phi(z). \tag{5.73}$$

In this situation, one can apply the Costin-Lebowitz-Soshnikov theorem as in Section 3.4. Since the $\mathcal{X}(\lambda)$ is by Lemma 5.20 of determinantal, then

$$\frac{N_\theta - E_{p,\theta} N_\theta}{\sqrt{Var_{p,\theta}(N_\theta)}} \xrightarrow{d} N(0,1) \tag{5.74}$$

provided that $Var_{p,\theta}(N_\theta) \to \infty$ as $\theta \to \infty$.

In order to derive (5.73) from (5.74), we need some basic asymptotic estimation of the first two moments of the random variable N_θ. This will be explicitly given in Lemma 5.21 below. Module Lemmas 5.20 and 5.21, the proof is complete. $\qquad\square$

Lemma 5.21. *Fix* $0 < x < 2$, $z \in \mathbb{R}$ *and let* $I_\theta = [a_\theta, \infty)$ *be as in (5.72).* *Then as* $\theta \to \infty$

$$E_{p,\theta} N_\theta = \sqrt{\theta}\, x - \frac{z}{2\pi}\sqrt{\log \theta} + O(1)$$

and

$$Var_{p,\theta}(N_\theta) = \frac{\log \theta}{4\pi^2}(1 + o(1)).$$

The proof of Lemma 5.21 essentially involves the computation of moments of the number of points lying in an interval for a discrete determinantal point process. Let $k \in \mathbb{Z}$ be a integer (possibly depending on the model parameter θ), and let N_k be the number of points of $\mathcal{X}(\lambda)$ lying in $[k, \infty)$. Then we have by Lemma 5.20

$$E_{p,\theta} N_k = \sum_{j=k}^{\infty} P_{p,\theta}\big(\lambda \in \mathcal{P} : j \in \mathcal{X}(\lambda)\big)$$

$$= \sum_{j=k}^{\infty} K_\theta(j,j)$$

$$= \sum_{j=k}^{\infty} \sum_{s=1}^{\infty} J_{k+s}\big(2\sqrt{\theta}\big)^2$$

$$= \sum_{m=k}^{\infty} (m-k) J_m\big(2\sqrt{\theta}\big)^2$$

and

$$Var_{p,\theta}(N_k) = \sum_{i=k}^{\infty} K_\theta(i,i) - \sum_{i,j=k}^{\infty} K_\theta(i,j)^2$$

$$= \sum_{i=k}^{\infty} \sum_{j=-\infty}^{k-1} K_\theta(i,j)^2$$

$$= \sum_{i=k}^{\infty} \sum_{j=-\infty}^{k-1} \frac{\sqrt{\theta}}{i-j} \left(J_i(2\sqrt{\theta})J_{j+1}(2\sqrt{\theta}) - J_{i+1}(2\sqrt{\theta})J_j(2\sqrt{\theta}) \right)^2.$$

To figure out these infinite sums, we need some precise asymptotic behaviours of Bessel functions in the whole real line. They behave rather differently in three regions so that we must take more care in treating the sums over two critical values. The lengthy computation will appear in the forthcoming paper, see Bogachev and Su (2015).

5.4   Berry-Esseen bounds for character ratios

This section is particularly devoted to the study of convergence rate of random character ratios. Define for $n \geq 2$

$$W_n(\lambda) = \frac{(n-1)\chi_\lambda(1^{n-2},2)}{\sqrt{2}d_\lambda}, \quad \lambda \in \mathcal{P}_n.$$

Note by (5.32) $W_n(\lambda) = p_2^\sharp(\lambda)/\sqrt{2n}$. It was proved in Section 5.2

$$W_n \xrightarrow{d} N(0,1), \quad n \to \infty$$

using the moment method. Namely,

$$\sup_{-\infty < x < \infty} \left| P_{p,n}\left(W_n(\lambda) \leq x\right) - \Phi(x) \right| \to 0, \quad n \to \infty. \qquad (5.75)$$

Having (5.75), it is natural to ask how fast it converges. This was first studied by Fulman. In fact, in a series of papers, he developed a Stein method and martingale approach to the study of the Plancherel measure. In particular, Fulman (2005, 2006) obtained a speed of n^{-s} for any $0 < s < 1/2$ and conjectured the correct speed is $n^{-1/2}$. Following this, Shao and Su (2006) confirmed the conjecture to get the optimal rate. The main result reads as follows.

Theorem 5.11.

$$\sup_{-\infty < x < \infty} \left| P_{p,n}\left(\lambda \in \mathcal{P}_n : W_n(\lambda) \leq x\right) - \Phi(x) \right| = O(n^{-1/2}).$$

The basic strategy of the proof is to construct a $W_n'(\lambda)$ such that $(W_n(\lambda), W_n'(\lambda))$ is an exchangeable pair and to apply the Stein method to $(W_n(\lambda), W_n'(\lambda))$. Let us begin with a Bratelli graph, namely an oriented graded graph $\mathcal{G} = (\mathcal{V}, \mathcal{E})$. Here the vertex set $\mathcal{V} = \mathcal{P} = \cup_{n=0}^{\infty} \mathcal{P}_n$ and there is an oriented edge from $\lambda \in \mathcal{P}_n$ to $\Lambda \in \mathcal{P}_{n+1}$ if Λ can be obtained from λ by adding one square, denoted by $\lambda \nearrow \Lambda$, see Figure 5.4 below.

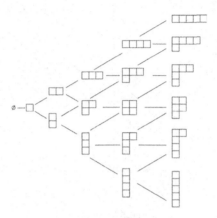

Fig. 5.4   Bratelli graph

Lemma 5.22. *The Plancherel measure is coherent in $\mathcal{G}$, namely*

$$P_{p,n}(\lambda) = \sum_{\Lambda : \lambda \nearrow \Lambda} \frac{d_\lambda}{d_\Lambda} P_{p,n+1}(\Lambda).$$

Proof.   According to the hook formula (4.70), it suffices to prove

$$\sum_{\Lambda : \lambda \nearrow \Lambda} \frac{H_\lambda}{H_\Lambda} = 1.$$

Let us compute the quotient H_λ / H_Λ. Assume that the new square is located in the rth row and sth column of the diagram Λ. Since the squares outside the rth row or sth column have equal hook lengths in the diagrams λ and Λ, we have by the hook formula,

$$\frac{H_\lambda}{H_\Lambda} = \prod_{i=1}^{s-1} \frac{h_{ri}(\lambda)}{h_{ri}(\lambda) + 1} \prod_{j=1}^{r-1} \frac{h_{js}(\lambda)}{h_{js}(\lambda) + 1},$$

where $h_\square(\lambda)$ denotes the hook length of $\square$ in λ.

Next we want to express the quotient in terms of local extrema. Let λ have interlacing local extrema

$$x_1 < y_1 < x_2 < y_2 < \cdots < x_q < y_q < x_{q+1}$$

and suppose the square that distinguishes Λ from λ is attached to the minimum x_k of λ, see Figure 5.5.

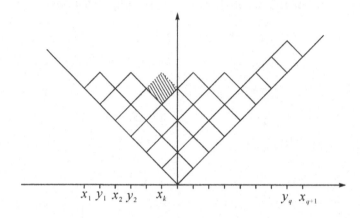

Fig. 5.5 $\lambda \nearrow \Lambda$

Then it follows

$$\prod_{i=1}^{s-1} \frac{h_{ri}(\lambda)}{h_{ri}(\lambda)+1} = \prod_{m=1}^{k-1} \frac{x_k - y_m}{x_k - x_m}$$

and

$$\prod_{j=1}^{r-1} \frac{h_{js}(\lambda)}{h_{js}(\lambda)+1} = \prod_{m=k+1}^{q+1} \frac{x_k - y_{m-1}}{x_k - x_m}.$$

Thus we rewrite

$$\frac{H_\lambda}{H_\Lambda} = \prod_{m=1}^{k-1} \frac{x_k - y_m}{x_k - x_m} \prod_{m=k+1}^{q+1} \frac{x_k - y_{m-1}}{x_k - x_m} =: a_k.$$

It remains to check

$$\sum_{k=1}^{q+1} a_k = 1. \tag{5.76}$$

To do this, note that these numbers coincides with the coefficients of the partial fraction expansion

$$\frac{\prod_{i=1}^{q}(u - y_i)}{\prod_{i=1}^{q+1}(u - x_i)} = \sum_{k=1}^{q+1} \frac{a_k}{u - x_k}.$$

Multiplying both sides by u and letting $u \to \infty$ yields (5.76). $\qquad\square$

To construct an exchange pair, we introduce a Markov chain $X_0^{(n)}$, $X_1^{(n)}$, $\cdots$, $X_k^{(n)}$, $\cdots$ with state space $\mathcal{P}_n$ and transition probability

$$p_n(\lambda, \mu) := P\big(X_1^{(n)} = \mu \big| X_0^{(n)} = \lambda\big)$$

$$= \frac{d_\mu}{(n+1)d_\lambda} \sharp P(\lambda, \mu), \tag{5.77}$$

where $P(\lambda, \mu) = \{\tau \in \mathcal{P}_{n-1}, \tau \nearrow \lambda, \tau \nearrow \mu\}$.

Lemma 5.23. *(i)* $p_n(\lambda, \mu)$ *is a well defined transition probability:*

$$\sum_{\mu \in \mathcal{P}_n} p_n(\lambda, \mu) = 1. \tag{5.78}$$

(ii) $P_{p,n}$ *is a stationary distribution of the Markov chain* $X^{(n)}$, *namely*

$$P_{p,n}(\mu) = \sum_{\lambda \in \mathcal{P}_n} P_{p,n}(\lambda) p_n(\lambda, \mu).$$

(iii) The Markov chain $X^{(n)}$ *is , namely for any* $\lambda, \mu \in \mathcal{P}_n$

$$P_{p,n}(\lambda) p_n(\lambda, \mu) = P_{p,n}(\mu) p_n(\mu, \lambda).$$

Proof. Note the following formula (see the note following the proof of Lemma 3.6 in Fulman (2005))

$$\sharp P(\lambda, \mu) = \frac{1}{n!} \sum_{\pi \in \mathcal{S}_n} \chi_\mu(\pi) \chi_\lambda(\pi) \big(r_1(\pi) + 1\big), \tag{5.79}$$

where $r_1(\pi)$ is the number of fixed points in π. Hence it follows from (5.77)

$$\sum_{\mu \in \mathcal{P}_n} p_n(\lambda, \mu)$$

$$= \sum_{\mu \in \mathcal{P}_n} \frac{d_\mu}{(n+1)d_\lambda} \frac{1}{n!} \sum_{\pi \in \mathcal{S}_n} \chi_\mu(\pi) \chi_\lambda(\pi) \big(r_1(\pi) + 1\big)$$

$$= \frac{1}{(n+1)d_\lambda} \sum_{\pi \in \mathcal{S}_n} \Big(\frac{1}{n!} \sum_{\mu \in \mathcal{P}_n} d_\mu \chi_\mu(\pi)\Big) \chi_\lambda(\pi) \big(r_1(\pi) + 1\big). \tag{5.80}$$

By (2.11),

$$\frac{1}{n!} \sum_{\mu \in \mathcal{P}_n} d_\mu \chi_\mu(\pi) = \begin{cases} 1, & \pi = 1^n, \\ 0, & \pi \neq 1^n. \end{cases} \tag{5.81}$$

Inserting into (5.80) easily yields (5.78), as desired.

(ii) is a direct consequence of Lemma 5.22, while (iii) follows from (5.77) and the Frobenius formula. □

Lemma 5.24. *Given a* $\lambda \in \mathcal{P}_n$,

$$EW'_n(\lambda) = \left(1 - \frac{2}{n+1}\right)W_n(\lambda). \tag{5.82}$$

Consequently,

$$E_{p,n}W_n = 0. \tag{5.83}$$

Proof. By definition,

$$
\begin{aligned}
EW'_n(\lambda) &= EW_n\big(X_1^{(n)}\big|X_0^{(n)} = \lambda\big) \\
&= \sum_{\mu \in \mathcal{P}_n} W_n(\mu)p_n(\lambda,\mu) \\
&= \frac{n-1}{(n+1)\sqrt{2}d_\lambda} \sum_{\mu \in \mathcal{P}_n} \chi_\mu\big(1^{n-2},2\big)\sharp P(\lambda,\mu). \tag{5.84}
\end{aligned}
$$

Substituting (5.79) and noting (2.12), then (5.84) becomes

$$
\begin{aligned}
&\frac{n-1}{(n+1)\sqrt{2}d_\lambda} \sum_{\pi \in \mathcal{S}_n} \left(\frac{1}{n!}\sum_{\mu \in \mathcal{P}_n} \chi_\mu\big(1^{n-2},2\big)\chi_\mu(\pi)\right)\chi_\lambda(\pi)\big(r_1(\pi)+1\big) \\
&= \left(1 - \frac{2}{n+1}\right)W_n(\lambda).
\end{aligned}
$$

This completes the proof of (5.82).

To see (5.83), note

$$
\begin{aligned}
E_{p,n}W_n &= E_{p,n}W'_n = E_{p,n}\big(EW'_n(\lambda)\big) \\
&= \left(1 - \frac{2}{n+1}\right)E_{p,n}W_n.
\end{aligned}
$$

The conclusion holds. □

Lemma 5.25.

$$
\begin{aligned}
E\big(W'_n(\lambda)\big)^2 = 1 - \frac{1}{n} &+ \frac{2(n-1)(n-2)^2}{n(n+1)} \cdot \frac{\chi_\lambda\big(1^{n-3},3\big)}{d_\lambda} \\
&+ \frac{(n-1)(n-2)(n-3)^2}{2n(n+1)} \cdot \frac{\chi_\lambda\big(1^{n-4},2^2\big)}{d_\lambda}. \tag{5.85}
\end{aligned}
$$

Proof. Similarly to (5.84), it follows

$$
\begin{aligned}
E(W'_n(\lambda))^2 &= \sum_{\mu \in \mathcal{P}_n} W_n^2(\mu)p_n(\lambda,\mu) \\
&= \frac{(n-1)^2}{2(n+1)d_\lambda} \sum_{\mu \in \mathcal{P}_n} \frac{\chi_\mu^2\big(1^{n-2},2\big)}{d_\mu}\sharp P(\lambda,\mu). \tag{5.86}
\end{aligned}
$$

Substituting (5.79), (5.86) becomes

$$\frac{(n-1)^2}{2(n+1)d_\lambda} \sum_{\pi \in \mathcal{S}_n} \left(\frac{1}{n!} \sum_{\mu \in \mathcal{P}_n} \frac{\chi_\mu^2(1^{n-2},2)}{d_\mu} \chi_\mu(\pi) \right) \chi_\lambda(\pi) \big(r_1(\pi)+1\big).$$

To proceed, we need the following equation (see Exercise 6.76 of Stanley (1999))

$$\frac{1}{n!} \sum_{\mu \in \mathcal{P}_n} \frac{\chi_\mu^2(1^{n-2},2)}{d_\mu} \chi_\mu(\pi) = \binom{n}{2}^{-2} \sharp(\pi), \qquad (5.87)$$

where $\sharp(\pi)$ is the number of pairs of (σ,τ) such that σ and τ come from the same conjugacy class $(1^{n-2},2)$ and $\sigma \circ \tau = \pi$. See also Lemma 3.4 of Fulman (2005).

Note $\sigma \circ \tau = \pi$ can assume only values in three distinct conjugacy classes: (1^n), $(1^{n-3},3)$, $(1^{n-4},2^2)$, and $\sharp(\pi)$, $\chi_\lambda(\pi)$ and $r_1(\pi)$ are all class functions. It is easy to see

$$\sharp(1^n) = \binom{n}{2}$$

$$\sharp(1^{n-3},3) = 2(n-2)\binom{n}{2}$$

$$\sharp(1^{n-4},2^2) = \binom{n}{2}\binom{n-2}{2}.$$

In combination, we easily get the desired conclusion (5.85). $\qquad\square$

As a direct consequence, we obtain the following

Corollary 5.3.

$$Var_{p,n}(W_n) = Var_{p,n}(\dot{W}'_n) = 1 - \frac{1}{n}.$$

The last lemma we need is to control the difference between $W_n(\lambda)$ and $W'_n(\lambda)$.

Lemma 5.26. *Let $\Delta_n(\lambda) = W_n(\lambda) - W'_n(\lambda)$, then*

$$P_{p,n}\left(|\Delta_n(\lambda)| \geq \frac{4e\sqrt{2}}{\sqrt{n}}\right) \leq 2e^{-2e\sqrt{n}}.$$

Consequently,

$$E_{p,n}|\Delta_n(\lambda)|^2 \mathbf{1}_{(|\Delta_n(\lambda)| \geq 4e\sqrt{2}/\sqrt{n})} = O(n^{-1/2}). \qquad (5.88)$$

Proof.   Recall the Frobenius formula

$$W_n(\lambda) = \frac{\sqrt{2}}{n} \sum_i \left(\binom{\lambda_i}{2} - \binom{\lambda_i'}{2} \right).$$

Given $X_0^{(n)} = \lambda$, $X_1^{(n)} = \mu$ only when μ is obtained from λ by moving a box from row i and column j to row s and column t. Then

$$\Delta_n(\lambda) = W_n(\mu) - W_n(\lambda)$$
$$= \frac{\sqrt{2}}{n}(\lambda_i + \lambda_t' - \lambda_s - \lambda_j').$$

Hence we have

$$|\Delta_n(\lambda)| \le \frac{2\sqrt{2}}{n} \max\{\lambda_1, \lambda_1'\}.$$

According to Lemma 5.1, we have (5.88), as desired. The proof is now complete.   □

Having the preceding preparation, we are now ready to prove Theorem 5.11. The proof is based on the following refinement of Stein's result for exchangeable pairs.

Theorem 5.12. *Let (W, W') be an exchangeable pair of real-valued random variables such that*

$$E(W'|W) = (1 - \tau)W,$$

with $0 < \tau < 1$. Assume $E(W^2) \le 1$. Then for any $a > 0$,

$$\sup_{-\infty < x < \infty} |P(W \le x) - \Phi(x)| \le \sqrt{E\left(1 - \frac{1}{2\tau}E(\Delta^2|W)\right)^2} + \frac{a^3}{\tau}$$
$$+2a + E\Delta^2 1_{(|\Delta|>a)}.$$

Proof.   See Theorem 2.1 of Shao and Su (2006).   □

Proof of Theorem 5.11.   This is a direct application of Theorem 5.12 to exchangeable pairs (W_n, W_n'). Set $\tau_n = 2/(n+1)$, $a_n = 4e\sqrt{2}/\sqrt{n}$ and $\Delta_n = W_n - W_n'$. In view of (5.88), we need only prove

$$E_{p,n}\left(1 - \frac{1}{2\tau_n}E(\Delta_n^2|W_n)\right)^2 = O(n^{-1}).$$

In fact, a simple algebra yields

$$E_{p,n}\left(1 - \frac{1}{2\tau_n}E(\Delta_n^2|W_n)\right)^2 = \frac{3n^2 - 5n + 6}{4n^3}. \tag{5.89}$$

To see this, note

$$E_{p,n}\left(1 - \frac{1}{2\tau_n}E\left(\Delta_n^2\big|W_n\right)\right)^2 = 1 - \frac{1}{\tau_n}E_{p,n}E\left(\Delta_n^2\big|W_n\right)$$
$$+ \frac{1}{4\tau_n^2}E_{p,n}\left(E\left(\Delta_n^2\big|W_n\right)\right)^2.$$

By Lemma 5.24, we have

$$E\left(\Delta_n^2\big|W_n\right) = E\left(W_n^2 + (W_n')^2 - 2W_nW_n'\big|W_n\right)$$
$$= \left(\frac{4}{n+1} - 1\right)W_n^2 + E\left((W_n')^2\big|W_n\right),$$

and so by Corollary 5.3,

$$E_{p,n}E\left(\Delta_n^2\big|W_n\right) = \frac{4}{n+1}\left(1 - \frac{1}{n}\right).$$

Again, by Lemma 5.25,

$$E\left(\Delta_n^2\big|W_n\right) = A + B + C + D$$

where

$$A = 1 - \frac{1}{n},$$
$$B = \frac{2(n-1)(n-2)^2}{n(n+1)} \cdot \frac{\chi_\lambda(1^{n-3}, 3)}{d_\lambda},$$
$$C = \frac{(n-1)(n-2)(n-3)^2}{2n(n+1)} \cdot \frac{\chi_\lambda(1^{n-4}, 2^2)}{d_\lambda},$$
$$D = \left(\frac{4}{n+1} - 1\right)\frac{(n-1)^2}{2} \cdot \frac{\chi_\lambda^2(1^{n-2}, 2)}{d_\lambda^2}.$$

What we next need is to compute explicitly $E_{p,n}(A + B + C + D)^2$. We record some data as follows.

$$E_{p,n}AB = E_{p,n}AC = 0;$$

$$E_{p,n}AD = \frac{(n-1)^2}{n^2}\left(\frac{4}{n+1} - 1\right);$$

$$E_{p,n}BC = \frac{(n-1)^2(n-2)^3(n-3)^2}{n^2(n+1)^2}E_{p,n}\frac{\chi_\lambda(1^{n-3}, 3)}{d_\lambda} \cdot \frac{\chi_\lambda(1^{n-4}, 2^2)}{d_\lambda}$$
$$= \frac{(n-1)^2(n-2)^3(n-3)^2}{n^2(n+1)^2}\frac{1}{n!}\sum_{\lambda \in \mathcal{P}_n}\chi_\lambda(1^{n-3}, 3)\chi_\lambda(1^{n-4}, 2^2)$$
$$= 0;$$

$$E_{p,n}BD = -\frac{(n-1)^3(n-2)^2(n-3)}{n(n+1)^2}E_{p,n}\frac{\chi_\lambda\left(1^{n-3},3\right)}{d_\lambda}\cdot\frac{\chi_\lambda^2\left(1^{n-2},2\right)}{d_\lambda^2}$$

$$= -\frac{(n-1)^3(n-2)^2(n-3)}{n(n+1)^2}\frac{1}{n!}\sum_{\lambda\in\mathcal{P}_n}\frac{\chi_\lambda\left(1^{n-3},3\right)\chi_\lambda^2\left(1^{n-2},2\right)}{d_\lambda}$$

$$= -\frac{12(n-1)(n-2)^2(n-3)}{n^3(n+1)^2};$$

$$E_{p,n}CD = -\frac{(n-1)^3(n-2)(n-3)^3}{4n(n+1)^2}E_{p,n}\frac{\chi_\lambda\left(1^{n-4},2^2\right)}{d_\lambda}\cdot\frac{\chi_\lambda^2\left(1^{n-2},2\right)}{d_\lambda^2}$$

$$= -\frac{(n-1)^3(n-2)(n-3)^3}{4n(n+1)^2}\frac{1}{n!}\sum_{\lambda\in\mathcal{P}_n}\frac{\chi_\lambda\left(1^{n-4},2^2\right)\chi_\lambda^2\left(1^{n-2},2\right)}{d_\lambda}$$

$$= -\frac{2(n-1)(n-2)(n-3)^3}{n^3(n+1)^2};$$

$$E_{p,n}B^2 = \frac{4(n-1)^2(n-2)^4}{n^2(n+1)^2}E_{p,n}\frac{\chi_\lambda^2(1^{n-3},3)}{d_\lambda^2}$$

$$= \frac{4(n-1)^2(n-2)^4}{n^2(n+1)^2}\frac{1}{n!}\sum_{\lambda\in\mathcal{P}_n}\chi_\lambda^2(1^{n-3},3)$$

$$= \frac{12(n-1)(n-2)^3}{n^3(n+1)^2};$$

$$E_{p,n}C^2 = \frac{(n-1)^2(n-2)^2(n-3)^4}{4n^2(n+1)^2}E_{p,n}\frac{\chi_\lambda^2(1^{n-4},2^2)}{d_\lambda^2}$$

$$= \frac{(n-1)^2(n-2)^2(n-3)^4}{4n^2(n+1)^2}\frac{1}{n!}\sum_{\lambda\in\mathcal{P}_n}\chi_\lambda^2(1^{n-4},2^2)$$

$$= \frac{2(n-1)(n-2)(n-3)^3}{n^3(n+1)^2};$$

$$E_{p,n}D^2 = \frac{(n-1)^4(n-3)^2}{4(n+1)^2}E_{p,n}\frac{\chi_\lambda^4\left(1^{n-2},2\right)}{d_\lambda^4}$$

$$= \frac{(n-1)^4(n-3)^2}{4(n+1)^2}\frac{1}{n!}\sum_{\lambda\in\mathcal{P}_n}\frac{\chi_\lambda^4\left(1^{n-2},2\right)}{d_\lambda^2}$$

$$= \frac{2(n-1)(n-3)^2}{n^3(n+1)^2}\left(3\binom{n}{2}+4(n-3)\right).$$

In combination, we obtain (5.89). The proof is now complete. □

5.5 Determinantal structure

The goal of this section is to provide a self-contained proof of Lemma 5.20 following the line of Borodin, Okounkov and Olshanski (2000). Let us first prove an equivalent form of Lemma 5.20. Let $\lambda = (\lambda_1, \lambda_2, \cdots, \lambda_l) \in \mathcal{P}$. Recall the ordinary Fronebius coordinates are

$$\bar{a}_i = \lambda_i - i, \quad \bar{b}_i = \lambda_i' - i, \quad 1 \le i \le \ell$$

and the modified Frobenius coordinates are

$$a_i = \bar{a}_i + \frac{1}{2}, \quad b_i = \bar{b}_i + \frac{1}{2}, \quad 1 \le i \le \ell,$$

where ℓ is the length of the main diagonal in the Young diagram of λ. Define

$$\mathcal{F}(\lambda) = \{a_i, -b_i, \quad 1 \le i \le \ell\}. \tag{5.90}$$

This is a finite set of half integers, $\mathcal{F}(\lambda) \subset \mathbb{Z} + \frac{1}{2}$. Note $\mathcal{F}(\lambda)$ consists of equally many positive half integers and negative half integers. Interestingly, $\mathcal{F}(\lambda)$ have a nice determinantal structure under the Poissonized Plancherel measure. In particular, denote the k-point correlation function

$$\varrho_k(x_1, \cdots, x_k) = Q_{p,\theta}(\lambda \in \mathcal{P} : x_1, \cdots, x_k \in \mathcal{F}(\lambda)),$$

where $x_i \in \mathbb{Z} + \frac{1}{2}, 1 \le i \le k$. Then we have

Lemma 5.27.

$$\varrho_k(x_1, \cdots, x_k) = \det \big(M(x_i, x_j)\big)_{k \times k},$$

where the kernel function

$$M(x, y) = \begin{cases} \sqrt{\theta} \dfrac{K_+(|x|, |y|)}{|x| - |y|}, & xy > 0, \\ \sqrt{\theta} \dfrac{K_-(|x|, |y|)}{|x| - |y|}, & xy < 0. \end{cases} \tag{5.91}$$

Here

$$K_+(x, y) = J_{x - \frac{1}{2}} J_{y - \frac{1}{2}} - J_{x + \frac{1}{2}} J_{y + \frac{1}{2}},$$
$$K_-(x, y) = J_{x - \frac{1}{2}} J_{y - \frac{1}{2}} - J_{x + \frac{1}{2}} J_{y + \frac{1}{2}}.$$

Its proof is based on the following three lemmas. The first one shows that the Poissonizaed Plancherel measure can be expressed by a determinant. Set

$$L(x, y) = \begin{cases} 0, & xy > 0, \\ \dfrac{1}{x - y} \cdot \dfrac{\theta^{(|x| + |y|)/2}}{\Gamma(|x| + \frac{1}{2})\Gamma(|y| + \frac{1}{2})}, & xy < 0. \end{cases}$$

Lemma 5.28.

$$Q_{p,\theta}(\lambda) = \frac{\det \big(L(x_i, x_j)\big)_{2\ell \times 2\ell}}{\det(1 + L)}, \tag{5.92}$$

where $x_i = a_i, \quad x_{i+\ell} = -b_i, \quad 1 \le i \le \ell$.

We remark that the numerator in (5.92) is a determinant of $2\ell \times 2\ell$ matrix, while the denumerator $\det(1 + L)$ is interpreted as the

$$\det(1 + L) = \sum_{X \subseteq \mathbb{Z}+\frac{1}{2}} \det\left(L(X)\right),$$

where $\det(L(X)) = 0$ unless X consists of equally many positive and negative half integers.

Proof. Recall a classic determinant formula for d_λ:

$$\frac{d_\lambda}{|\lambda|!} = \det\left(\frac{1}{(\bar{a}_i + \bar{b}_j + 1)\bar{a}_i!\bar{b}_j!}\right)_{\ell \times \ell}.$$

Thus letting

$$A = \begin{pmatrix} L(x_1, x_{\ell+1}) & L(x_1, x_{\ell+2}) & \cdots & L(x_1, x_{2\ell}) \\ L(x_2, x_{\ell+1}) & L(x_2, x_{\ell+2}) & \cdots & L(x_2, x_{2\ell}) \\ \vdots & \vdots & \vdots & \vdots \\ L(x_\ell, x_{\ell+1}) & L(x_\ell, x_{\ell+2}) & \cdots & L(x_\ell, x_{2\ell}) \end{pmatrix},$$

we have

$$\left(\frac{d_\lambda}{|\lambda|!}\right)^2 = \det\begin{pmatrix} \mathbf{0} & A \\ -A' & \mathbf{0} \end{pmatrix}.$$

It follows by definition of L

$$Q_{p,\theta}(\lambda) = Q_{p,\theta}\left(\{x_1, \cdots, x_{2\ell}\}\right)$$
$$= e^{-\theta} \det\left(L(x_i, x_j)\right)_{2\ell \times 2\ell}.$$

As a direct consequence,

$$e^\theta = e^\theta \sum_{\lambda \in \mathcal{P}} Q_{p,\theta}(\lambda)$$
$$= \sum_X \det\left(L(X)\right) = \det(1 + L).$$

We now conclude the proof. $\qquad\qquad\qquad\qquad\qquad\qquad\qquad\square$

We shall below prove that the point process $\mathcal{F}(\lambda)$ is of determinantal. Let

$$M_L = \frac{L}{1 + L}.$$

Then we have

Lemma 5.29. *Given* $x_1, \cdots, x_k \in \mathbb{Z} + \frac{1}{2}$,

$$\varrho_k(x_1, \cdots, x_k) = \det\left(M_L(x_i, x_j)\right)_{k \times k}.$$

Proof. Assume that $g : \mathbb{Z} + \frac{1}{2} \to \mathbb{Z} + \frac{1}{2}$ takes 0 except at finitely many points. According to Lemma 5.28, it follows

$$E_{p,\theta} \prod_{x \in \mathcal{F}(\lambda)} (1 + g(x)) = \sum_{X \subseteq \mathbb{Z} + \frac{1}{2}} \prod_{x \in X} (1 + g(x)) Q_{p,\theta}(\mathcal{F}(\lambda) = X)$$

$$= \sum_{X \subseteq \mathbb{Z} + \frac{1}{2}} \prod_{x \in X} (1 + g(x)) \frac{\det(L(X))}{\det(1 + L)}$$

$$= \sum_{X \subseteq \mathbb{Z} + \frac{1}{2}} \frac{\det\big((1 + g)L(X)\big)}{\det(1 + L)}$$

$$= \frac{\det\big(1 + L + gL\big)}{\det(1 + L)} = \det\big(1 + gM_L\big)$$

$$= \sum_{X \subseteq \mathbb{Z} + \frac{1}{2}} \prod_{x \in X} g(x) \det\big(M_L(X)\big), \qquad (5.93)$$

where in the last two equations we used the properties of Fredholm determinants. On the other hand,

$$E_{p,\theta} \prod_{x \in \mathcal{F}(\lambda)} (1 + g(x)) = \sum_{X \subseteq \mathbb{Z} + \frac{1}{2}} \prod_{x \in X} (1 + g(x)) Q_{p,\theta}(\mathcal{F}(\lambda) = X)$$

$$= \sum_{X \subseteq \mathbb{Z} + \frac{1}{2}} \sum_{Y : Y \subseteq X} \prod_{x \in Y} g(x) Q_{p,\theta}(\mathcal{F}(\lambda) = X)$$

$$= \sum_{Y \subseteq \mathbb{Z} + \frac{1}{2}} \prod_{x \in Y} g(x) \sum_{X : Y \subseteq X} Q_{p,\theta}(\mathcal{F}(\lambda) = X)$$

$$= \sum_{Y \subseteq \mathbb{Z} + \frac{1}{2}} \prod_{x \in Y} g(x) \varrho(Y). \qquad (5.94)$$

Thus comparing (5.93) with (5.94) yields

$$\varrho(X) = \det\big(M_L(X)\big)$$

since g is arbitrary. $\qquad \square$

To conclude the proof of Lemma 5.27, we need only to determine M_L above is exactly equal to M of (5.91).

Lemma 5.30.

$$M_L(x, y) = M(x, y).$$

Proof. Fix $x, y \in \mathbb{Z} + \frac{1}{2}$ and set $z = \sqrt{\theta}$. Note M and L are a function of z. We need to prove for all $z \geq 0$,

$$M + ML - L = 0. \qquad (5.95)$$

Obviously, (5.95) is true at $z = 0$. We shall below prove

$$\dot{M} + \dot{M}L + M\dot{L} - \dot{L} = 0, \tag{5.96}$$

where $\dot{M} = \frac{\partial M}{\partial z}$ and $\dot{L} = \frac{\partial L}{\partial z}$.

It easily follows by definition

$$\dot{L} = \begin{cases} 0, & xy > 0, \\ \text{sgn}(x)\dfrac{z^{|x|+|y|-1}}{\Gamma(|x|+\frac{1}{2})\Gamma(|y|+\frac{1}{2})}, & xy < 0. \end{cases}$$

To compute $\dot{M}$, we use the following formulas:

$$\frac{\partial}{\partial z} J_x(2z) = -2J_{x+1}(2z) + \frac{x}{z}J_x(2z)$$

$$= 2J_{x-1}(2z) - \frac{x}{z}J_x(2z).$$

Then

$$\dot{M} = \begin{cases} J_{|x|-\frac{1}{2}}J_{|y|+\frac{1}{2}} + J_{|x|+\frac{1}{2}}J_{|y|-\frac{1}{2}}, & xy > 0, \\ \text{sgn}(x)(J_{|x|-\frac{1}{2}}J_{|y|-\frac{1}{2}} - J_{|x|+\frac{1}{2}}J_{|y|+\frac{1}{2}}), & xy < 0. \end{cases}$$

It remains to verify (5.96). To do this, recall the following identities: for any $\nu \neq 0, -1, -2, \cdots$ and any $z \neq 0$ we have

$$\Gamma(\nu)J_\nu(2z) = z^\nu \sum_{m=0}^{\infty} \frac{1}{m+\nu} \frac{z^m}{m!} J_m(2z),$$

$$\Gamma(\nu)J_{\nu-1}(2z) = z^{\nu-1} - z^\nu \sum_{m=0}^{\infty} \frac{1}{m+\nu} \frac{z^m}{m!} J_{m+1}(2z).$$

Now the verification of (5.96) becomes a straightforward application, except for the occurrence of the singularity at negative integers ν. This singularity is resolved using the following identity due to Lommel

$$J_\nu(2z)J_{1-\nu}(2z) + J_{-\nu}(2z)J_{\nu-1}(2z) = \frac{\sin \pi\nu}{\pi z}.$$

This concludes the proof of Lemma 5.30, and so Lemma 5.27. □

Turn to the proof of Lemma 5.20. A key observation is the following link between $\mathcal{X}(\lambda)$ and $\mathcal{F}(\lambda)$ due to Frobenius (see (5.68) and (5.90)):

$$\mathcal{F}(\lambda) = \left(\mathcal{X}(\lambda) + \frac{1}{2}\right)\Delta\left(\mathbb{Z}_{\leq 0} - \frac{1}{2}\right). \tag{5.97}$$

Given $x_1, \cdots, x_k \in \mathbb{Z}$, denote $X = \left\{x_1 + \frac{1}{2}, \cdots, x_k + \frac{1}{2}\right\}$. Divide X into positive half integers and negative half integers: $X_+ = X \cap \left(\mathbb{Z}_{\geq 0} + \frac{1}{2}\right)$ and

$X_- = X \cap \left(\mathbb{Z}_{\leq 0} - \frac{1}{2} \right)$. If $X \subseteq \mathcal{X}(\lambda) + \frac{1}{2}$, then by (5.97), $X_+ \subseteq \mathcal{F}(\lambda)$ and there exists a finite subset $S \subseteq \left(\mathbb{Z}_{\leq 0} - \frac{1}{2} \right) \setminus X_-$ such that $S \subseteq \mathcal{F}(\lambda)$. This implies

$$Q_{p,\theta}\left(X \subseteq \mathcal{X}(\lambda) + \frac{1}{2} \right)$$
$$= Q_{p,\theta}\left(\exists S \subseteq \left(\mathbb{Z}_{\leq 0} - \frac{1}{2} \right) \setminus X_- : X_+ \cup S \subseteq \mathcal{F}(\lambda) \right). \tag{5.98}$$

By exclusion-inclusion principle, the right hand side of (5.98) becomes

$$\sum_{S \subseteq \left(\mathbb{Z}_{\leq 0} - \frac{1}{2} \right) \setminus X_-} (-1)^{|S|} Q_{p,\theta}\left(X_+ \cup S \subseteq \mathcal{F}(\lambda) \right)$$
$$= \sum_{S \subseteq \left(\mathbb{Z}_{\leq 0} - \frac{1}{2} \right) \setminus X_-} (-1)^{|S|} \varrho(X_+ \cup S)$$
$$= \sum_{S \subseteq \left(\mathbb{Z}_{\leq 0} - \frac{1}{2} \right) \setminus X_-} (-1)^{|S|} \det \left(M(X_+ \cup S) \right), \tag{5.99}$$

where in the last equation we used Lemma 5.27.

Define a new as follows:

$$M^\triangle(x,y) = \begin{cases} M(x,y), & x \in \mathbb{Z}_{\geq 0} + \frac{1}{2}, \\ -M(x,y), & x \in \mathbb{Z}_{\geq 0} + \frac{1}{2}, y \in \mathbb{Z}_{\leq 0} - \frac{1}{2}, \\ -M(x,y), & x,y \in \mathbb{Z}_{\leq 0} - \frac{1}{2}, x \neq y, \\ 1 - M(x,x), & x,y \in \mathbb{Z}_{\leq 0} - \frac{1}{2}, x = y. \end{cases}$$

Lemma 5.31. *Given* $x,y \in \mathbb{Z}$,

$$M^\triangle\left(x + \frac{1}{2}, y + \frac{1}{2} \right) = \epsilon(x)\epsilon(y)K_\theta(x,y),$$

where $\epsilon(x) = \left(sgn(x) \right)^{x+1}$.

Proof. It suffices to show

$$M\left(x + \frac{1}{2}, y + \frac{1}{2} \right) = \begin{cases} sgn(x)\epsilon(x)\epsilon(y)K_\theta(x,y), & x \neq y, \\ K_\theta(x,y), & x = y > 0, \\ 1 - K_\theta(x,y), & x = y < 0. \end{cases}$$

Using the relation

$$J_{-n}\left(2\sqrt{\theta} \right) = (-1)^n J_n\left(2\sqrt{\theta} \right)$$

and the definition of M, one can easily verify the case $x \neq y$. Also, the claim remains valid for $x = y > 0$. It remains to consider the case $x = y < 0$. In this case, we have to show that

$$1 - M\left(x + \frac{1}{2}, x + \frac{1}{2} \right) = J(x,x), \quad x \in \mathbb{Z}_{\leq 0}.$$

Equivalently,

$$1 - J(k, k) = J(-k - 1, -k - 1), \quad k \in \mathbb{Z}_{\geq 0}. \tag{5.100}$$

Note for any $k \in \mathbb{Z}$,

$$J(k, k) = \sum_{m=0}^{\infty} (-1) \frac{(2k + m + 2)^{\uparrow m}}{((k + m + 1)!)^2} \frac{\theta^{k+m+1}}{m!},$$

where we use the symbol $(x)^{\uparrow m} = x(x+1) \cdots (x+m-1)$. We need to show

$$1 - \sum_{m=0}^{\infty} (-1)^m \frac{(2k + m + 2)^{\uparrow m}}{((k + m + 1)!)^2} \frac{\theta^{k+m+1}}{m!}$$

$$= \sum_{nml=0}^{\infty} (-1)^l \frac{(-2k + l + 2)^{\uparrow l}}{(\Gamma(-k + l + 1))^2} \frac{\theta^{-k+l}}{l!}. \tag{5.101}$$

Examine the right hand side of (5.101). The terms with $l = 0, 1, \cdots, k - 1$ vanish because then $1/\Gamma(-k+l+1) = 0$. The term with $l = k$ is equal to 1. Next the terms with $l = k + 1, \cdots, 2k$ vanish because for these values of l, the expression $(-2k+l)^{\uparrow l}$ vanishes. Finally, for $l \geq 2k+1$, say $l = 2k+1+m$,

$$(-1)^l \frac{(-2k + l)^{\uparrow l}}{(\Gamma(-k + l + 1))^2} \frac{\theta^{-k+l}}{l!} = (-1)^{m+1} \frac{(m + 1)^{\uparrow l}}{((k + m + 1)!)^2} \frac{\theta^{k+m+1}}{(2k + 1 + m)!}$$

$$= (-1)^{m+1} \frac{(2k + m + 2)^{\uparrow m}}{((k + m + 1)!)^2} \frac{\theta^{k+m+1}}{m!}.$$

Thus we have proved (5.100). $\qquad \square$

Proof of Lemma 5.20 Fix $X = (x_1, x_2, \cdots, x_k) \subseteq \mathbb{Z}$. Then according to (5.27), (5.98) and (5.99), we have

$$\rho_k(x_1, \cdots, x_k) = Q_{p,\theta}\big(\lambda \in \mathcal{P} : x_1, \cdots, x_k \in \mathcal{X}(\lambda)\big)$$

$$= \sum_{S \subseteq (\mathbb{Z}_{\leq 0} - \frac{1}{2}) \setminus X_-} (-1)^{|S|} \det \big(M(X_+ \cup S)\big). \tag{5.102}$$

To compute the alternating sum in (5.102), write

$$M(X_+ \cup S) = \begin{pmatrix} (X_+, X_+) & (X_+, S) \\ (S, X_+) & (S, S) \end{pmatrix},$$

where (X_+, X_+) stands for the matrix $\big(M(x_i + \frac{1}{2}, x_j + \frac{1}{2})\big)$ with $x_i + 1/2$, $x_j + 1/2 \in X_+$, the others are similar. Then by definition

$$M^{\triangle}(X_+ \cup S) = \begin{pmatrix} (X_+, X_+) & (X_+, S) \\ -(S, X_+) & 1 - (S, S) \end{pmatrix}.$$

A simple matrix determinant manipulation shows

$$\sum_{S \subseteq (\mathbb{Z}_{\leq 0} - \frac{1}{2}) \backslash X_-} (-1)^{|S|} \det \left(M(X_+ \cup S) \right) = \det \left(M^{\triangle}(X_+ \cup S) \right).$$

It follows in turn from Lemma 5.31

$$\det \left(M^{\triangle}(x_i + \frac{1}{2}, x_j + \frac{1}{2}) \right)_{k \times k} = \det \left(K_{\theta}(x_i, x_j) \right)_{k \times k}.$$

This concludes the proof. $\qquad\qquad\qquad\qquad\qquad\qquad\qquad\qquad\square$

Bibliography

Anderson, G.W., Guionnet, A. and Zeitouni, O. (2010). *An Introduction to Random Matrices*, Cambridge University Press.

Andrews, G. E. (1976). *The theory of partitions*, Encyclopedia of mathematics and its applications, **2**, Addison-Wesley, Reading, MA Press.

Bai, Z. D. and Silverstein, J. (2010). *Spectral Analysis of Large Dimensional Random Matrices*, Springer Series in Statistics, Science Press.

Baik, J., Deift, P. and Johansson, K. (1999). *On the distribution of the length of the longest increasing subsequence in a radnom permutation*, J. Amer. Math. Soc. **12**, 1119-1178.

Bao, Z. G. and Su, Z. G. (2010). *Local semicircle law and Gaussian fluctuation for Hermite β ensemble*, arXiv:11043431 [math.PR]

Billingsley, P. (1999a). *Convergence of Probability Measures*, 2nd edition, Wiley-Interscience.

Billingsley, P. (1999b). *Probability and Measure*, 3rd edition, Wiley-Interscience.

Bogachev, L. V. and Su, Z.G. (2015). *Gaussian fluctuations for random Plancherel partitions*, in preparation.

Borodin, A., Okounkov, A. and Olshanski, G. (1999). *Asymptotics of Plancherel measures for symmetric groups*, J. Amer. Math. Soc. **13**, 481–515.

Bourgade, P., Hughes, C.P., Nikeghbali, A. and Yor, M. (2008). *The characteristic polynomial os a random unitary matrix: a probabilistic approach*, Duke Math. J. **145**, 45-69.

Brézin, E. and Hikami, S. (2000). *Characteristic polynomials of random matrices*, Comm. Math. Phys. **214**, 111-135.

Brown, B. M. (1971). *Martingale central limit theorems*, Ann. Math. Stat. **42**, 59-66.

Cantero, M. J., Moral, L. and Velázquez, L. (2003). *Five-diagonal matrices and zeros of orthogonal polynomials on the unit circle*, Linear Algebra Appl. **362**, 29-56.

Cavagna, A., Garrahan, J. P. and Giardina, I. (2000). *Index distribution of random matrices with an application to disordered system*, Phys. Rev. B **61**, 3960-3970.

Chen, L. H. Y., Goldstein, L. and Shao, Q. M. (2010). *Normal Approximation by Stein's Method.* Probability and its Applications, Springer-Verlag, New York.

Chow, Y. S. (2003). *Probability Theory: Independence, Interchangeability, Martingales*, Springer Texts in Statistics, 3rd edition, Springer.

Chung, K. L. (2000). *A Course in Probability Theory*, 3rd edition, Academic Press.

Conrey, B. (2005). *Notes on eigenvalue distributions for the classical compact groups.* Recent Perspectives in Random Matrix Theory and Number Theory, LMS Lecture Note Series (No. 322), 111-146, Cambridge University Press.

Costin, O. and Lebowitz, J. (1995). *Gaussian fluctuations in random matrices*, Phys. Rev. Lett. **75**, 69-72.

Deift, P. (2000). *Integrable systems and combinatorial theory*, Notices of Amer. Math. Soc. **47**, 631-640.

Deift, P. and Gioev, D. (2009). *Random Matrix Theory: Invariant Ensembles and Universality*, Amer. Math. Soc. Providence, RI.

Deift, P., Its. A. and Krasovsky, I. (2012). *Toeplitz matrices and Toeplitz determinants under the impetus of the Ising model: Some history and some recent results*, Comm. Pure Appl. Math. **66**, 1360-1438.

Diaconis, P. (2003). *Patterns in eigenvalues: The 70th Josiah Willard Gibbs Lecture*, Bull. Amer. Math. Soc. **40**, 155-178.

Diaconis, P. and Evans, S. N. (2001). *Linear functionals of eigenvalues of random matrices*, Trans. Amer. Math. Soc. **353**, 2615-2633.

Diaconis, P., Shahshahani, M. (1981). *Generating a random permutation with random transpositions*, Z. Wahrsch Verw. Gebiete. **57**, 159-179.

Diaconis, P. and Shahshahani, M. (1994). *On the eignvalues of random matrices*, J. Appl. Probab. **31** A, 49-62.

Durrett, R. (2010). *Probability: Theory and Examples*, Cambridge University Press.

Dumitriu, I. and Edelman, A. (2002). *Matrix models for beta ensembles*, J. Math. Phys. **43**, 5830-5847.

Dyson, F. M. (1962). *The threefold way: Algebraic structure of symmetry groups and ensembles in quantum mechanics*, J. Math. Phys. **3**, 1199.

Dyson, F. M. (1962). *Statistical theory of energy levels of complex systems II*, J. Math. Phys. **62**, 157-165.

Erdös, P., Lehner, J. (1941). *The distribution of the number of summands in the partition of a positive integer*, Duke Math. J. **8**, 335-345.

Erlihson, M. M. and Granovsky, B. L. (2008). *Limit shapes of Gibbs distributions on the set of integer partitions: The expansive case*, Ann. Inst. H. Poincare Probab. Statist., **44**, 915-945.

Eskin, A. and Okounkov, A. (2001). *Asymptotics of numbers of branched coverings of a torus and volumes of moduli spaces of holomorphic differentials*, Invent. Math. **145**, 59-103.

Feller, W. (1968). *An Introduction to Probability Theory and Its Applications*, **1**, 3rd edition, John Wiley & Sons, Inc.

Feller, W. (1971). *An Introduction to Probability Theory and Its Applications*, **2**, 2nd edition, John Wiley & Sons, Inc.

Fischer, H. (2011). *A history of the central limit theorm—From classical to modern probability theory*, Springer.

Forrester, P. J. (2010). *Log-Gas and Random Matrices*, Princeton University Press.

Forrester, P. J. and Frankel, N. E. (2004). *Applications and generalizations of Fisher-Hartwig asymptotics*, J. Math. Phys. **45**, 2003-2028.

Forrester, P. J. and Rains, E. M. (2006). *Jacobians and rank 1 perturbations relating to unitary Hessenberg matrices*, Int. Math. Res. Not. **2006**, 1-36.

Frame, J. S., de B. Robinson, G. and Thrall, R. M. (1954). *The hook graphs of the symmetric groups*, Canada J. Math. **6**, 316-324.

Fristedt, B. (1993). *The structure of random partitions of large integers*, Trans. Amer. Math. Soc. **337**, 703-735.

Frobenius, G. (1903). *Über die charactere der symmetrischen gruppe*, Sitzungsber Preuss, Aadk. Berlin, 328-358.

Fulman, J. (2005). *Steins method and Plancherel measure of the symmetric group*, Trans. Amer. Math. Soc. **357**, 555-570.

Fulman, J. (2006). *Martingales and character ratios*, Trans. Amer. Math. Soc. **358**, 4533-4552.

Gikhman, I. I. and Skorohod, A. V. (1996). *Introduction to The theory of Random Processes*, Dover Publications, New York.

Ginibre, J. (1965). *Statistical ensembles of complex, quaterion, and real matrices*, J. Math. Phys. **6**, 440-449.

Girko, V. L. (1979). *The central limit theorem for random determinants (Russian)*, Translation in Theory Probab. Appl. **24**, 729-740.

Girko, V. L. (1990). *Theory of random determinants*, Kluwer Acadmic Publishers Group, Dordrecht.

Girko, V. L. (1998). *A refinement of the central limit theorem for random determinants (Russian)*, Translation in Theory Probab. Appl. **42**, 121-129.

Gustavsson, J. (2005). *Gaussian fluctuations of eigenvalues in the GUE*, Ann. Inst. H. Poincaré Probab. Stat. **41**, 151-178.

Hall, P. and Heyde, C. C. (1980). *Martingale Limit Theory and Its Application*, Academic Press.

Hardy, G. H. and Ramanujan, S. (1918). *Asymptotic formulae in combiantory analysis*, Proc. London Math. Soc. **17**, 75-115.

Hora, A. (1998). *Central limit theorem for the adjacency operators on the inifnite symmetric groups*, Comm. Math. Phys. **195**, 405-416.

Ingram, R. E. (1950). *Some characters of the symmetric group*, Proc. Amer. Math. Soc. **1**, 358-369.

Ivanov, V. and Kerov, S. (2001). *The algebra of conjugacy classes in symmetric groups, and partial permutations*, J. Math. Sci. (New York), **107**, 4212-4230.

Ivanov, V. and Olshanski, G. (2002). *Kerov's central limit theorem for the Plancherel measure on Young diagrams*, Symmetric Functions 2001: Surveys of Developments and Perspectives (S. Fomin, ed.), 93-151, NATO Sci. Ser. II Math. Phys. Chem. **74**, Kluwer Acad. Publ., Dordrecht.

Johansson, K. (1998a). *On fluctuations of eigenvalues of random Hermitian matrices*, Duke Math. J. **91**, 151-204.

Johansson, K. (1998b). *The longest increasing subsequence in a random permutation and a unitary random matrix model,* Math. Res. Lett. **5**, 63-82.

Johansson, K. (2001). *Discrete orthogonal polynomials ensembles and the Plancherel measure,* Ann. Math. **153**, 259-296.

Keating, J. P. and Snaith, N. C. (2000). *Random matrix theory and* $\zeta(1/2 + it)$, Commun. Math. Phys. **214**, 57-89.

Kerov, S. V. (1993). *Gaussian limit for the Plancherel measure of the symmetric group,* C. R. Acad. Sci. Paris, Sér. I Math. **316**, 303-308.

Kerov, S. V. and Olshanski, G. (1994). *Polynomial functions on the set of Young diagrams,* Comptes Rendus Acad. Sci. Paris Sér. I **319**, 121-126.

Killip, R. (2008). *Gaussian fluctuations for* β *ensembles,* Int. Math. Res. Not. **2008**, 1-19.

Killip, R. and Nenciu, I. (2004). *Matrix models for circular ensembles,* Int. Math. Res. Not. **2004**, 2665-2701.

Krasovsky, I. V. (2007). *Correlations of the characteristic polynomials in the Gaussian unitary ensembles or a singular Hankel determinant,* Duke Math. J. **139**, 581-619.

Ledoux, M. and Talagrand, M. (2011). *Probability in Banach Spaces: Isoperimetry and Processes,* Springer Verlag.

Logan, B. F. and Shepp, L. A. (1977). *A variational problem for random Young tableaux,* Adv. Math. **26**, 206-222.

Lytova, A. and Pastur, L. (2009). *Central limit theorem for linear eigenvalue statistics of random matrices with independent entries,* Ann. Prob. **37**, 1778-1840.

Macchi, O. (1975). *The coincidence approach to stochastic point processes,* Adv. Appl. Prob. **7**, 83-122.

Macdonald, I. (1995). *Symmetric Functions and Hall Polynomials,* 2nd edition, Clarendon Press, Oxford.

Majumdar, S. N., Nadal, C., Scardicchio, A. and Vivo, P. (2009). *The index distribution of Gaussian random matrices,* Phys. Rev. Lett. **103**, 220603.

McLeish, D. L. (1974). *Dependent central limit theorems and Invariance principles,* Ann. Probab. **2**, 620-628.

Mehta, L. A. (2004). *Random Matrices,* 3rd edition, Academic Press.

Okounkov, A. (2000). *Random matrices and random permutations,* Int. Math. Res. Not. **2000**, 1043-1095.

Okounkov, A. (2001). *Infinite wedge and random partitions,* Selecta Math. **7**, 57-81.

Okounkov, A. (2003). *The use of random partitions,* XIVth ICMP, 379-403, World Sci. Publ., Hackensack, New Jersey.

Okounkov, A. and Olshanski, G. (1998). *Shifted Schur functions,* St. Petersburg Math. J. **9**, 239-300.

Okounkov, A. and Pandharipande, R. (2005). *Gromov-Witten theory, Hurwitz numbers, and matrix models,* Algebraic geometry—Seattle 2005, Part I, 325-414, Proc. Sympos. Pure Math., 80, Part I, Amer. Math. Soc., Providence, RI, 2009.

Pastur, L. and Shcherbina, M. (2011). *Eigenvalue Distribution of Large Random*

Matrices, Mathematical Surveys and Monographs, **171**, Amer. Math. Soc.

Petrov, V. (1995). *Limit Theorems of Probability Theory, Sequences of Independent Random Variables*, Oxford University Press.

Pittel, B. (1997). *On a likely shape of the random Ferrers diagram*, Adv. Appl. Math. **18**, 432-488.

Pittel, B. (2002). *On the distribution of the number of Young tableaux for a uniformly random diagram*, Adv. Appl. Math. **29**, 184-214.

Postnikov, A. G. (1988). *Introduction to analytic number theory*, Translation of Mathematical Monographs, **68**, Amer. Math. Soc., Providence, RI.

Ramíer, J., Rider, B. and Virág, B. (2011). *Beta ensembles, stochastic Airy spectrum, and a diffusion*, J. Amer. Math. Soc. **24**, 919-944.

Riordan, J. (1968). *Combinatorial identities*, Wiley, New York.

Ross, N. (2011). *Fundamentals of Steins method*. Probab. Surveys **8**, 210-293.

Sagan, B. E. (2000). *The Symmetric Group: Representations, Combiantorial Algorithms, and Symmetric Functions*, 2nd edition, GTM **203**, Springer.

Serfozo, R. (2009). *Basics of Applied Stochastic Processes*, Probability and its Applications, Springer.

Shao, Q. M. and Su, Z. G. (2006). *The Berry-Esseen bound for character ratios*, Proc. Amer. Math. Soc. **134**, 2153-2159.

Simon, B. (2004). *Orthogonal Polynomials on the Unit Circle*, 1, AMS Colloquium Series, Amer. Math. Soc., Providence, RI.

Soshnikov, A. (2000). *Determinantal random point fields*, Ruassian Math. Surveys, **55**, 923-975.

Soshnikov, A. (2002). *Gaussian limit for determinantal random point fields*, Ann. Probab. **30**, 171-180.

Stanley, R. (1999). *Enumerative combinatorics*, **2**, Cambridge Studies in Advanced Mathematics, **62**, Cambridge University Press.

Steele, J. M. (1995). *Variations on monotone subsequence problem of Erdös and Szekeres*, In Discrete Probability and Algorithms (Aldous, Diaconis, and Steele, Eds.) 111-132, Springer Publishers, New York.

Steele, J. M. (1997). *Probability theory and combinatorial optimization*, CBMS-NSF Regional Conference Sereis in Applied Mathematics, **69**, SIAM.

Stein, C. (1970). *A bound for the error in the normal approximation to the distribution of a sum of dependent random variables*, Proc. Sixth Berkeley Symp. Math. Statist. Prob. **2**, University of California Press.

Stein, C. (1986). *Approximate Computation of Expectations*, IMS, Hayward, California.

Su, Z. G. (2014). *Normal convergence for random partitions with multiplicative measures*, Teor. Veroyatnost i. Primenen., **59**, 97-129.

Szegö, G. (1975). *Orthogonal Polynomials*, **23**, Colloquium Publications, 4th edition, Amer. Math. Soc., Providence, RI.

Tao, T. (2012). *Topics in random matrix theory*, Graduate Studies in Mathematics, **132**, Amer. Math. Soc..

Tao, T. and Vu, V. (2012). *A cnetral limit theorem for the determinant of a Wigner matrix*, Adv. Math. **231**, 74-101.

Tao, T. and Vu, V. (2010). *Random matrices: universality of local eigenvalues,*

Acta. Math. **206**, 127-204.

Temperley, H. (1952). *Statistical mechanics and the partition of numbers, II. The form of crystal surfaces.* Proc. Royal Soc. London, A. **48**, 683-697.

Tracy, C. and Widom, H. (1994). *Level spacing dsitributions and the Airy kernel,* Comm. Math. Phys. **159**, 151-174.

Tracy, C. and Widom, H. (2002). *Dsitribution functions for largest eigenvalues and their applications,* Proc. ICM, **1**, 587-596, Higher Education Press, Beijing.

Trotter, H. (1984). *Eigenvalue distributions of large Hermitian matrices: Wigner's semicircle law and a theorem of Kac, Murdock, and Szegő,* Adv. Math. **54**, 67-82.

Valkó, B. and Virág, B. (2009). *Continuum limits of random matrices and the Brownian carousel,* Invent. Math. **177**, 463-508.

Vershik, A. M. (1994). *Asymptotic combinatorics and algebraic analysis,* Proc. ICM, Zürich, **2**, 1384-1394, Birkhäuser, Basel, 1995.

Vershik, A. M. (1996). *Statistical mechanics of combinatorial partitions, and their limit configurations,* Funct. Anal. Appl. **30**, 90–105.

Vershik, A. M. and Kerov, S. V. (1977). *Asymptotics of the Plancherel measure of the symmetric group and the limiting form of Young tables,* Soviet Math. Doklady **18**, 527–531.

Vershik, A. M. and Kerov, S. V. (1985a). *Asymptotic of the largest and the typical dimensions of irreducible representations of a symmetric group,* Func. Anal. Appl. **19**, 21-31.

Vershik, A. M. and Kerov, S. V. (1985b) *Asymptotic theory of characters of the symmetric group,* Func. Anal. Appl. Vol. 15, 246-255.

Vershik, A. M. and Yakubovich, Yu. (2006). *Fluctuations of the maximal particle energy of the quantum ideal gas and random partitions,* Comm. Math. Phys. **261**, 795-769.

Wieand, K.L. (1998). *Eigenvalue distributions of random matrices in the permutation group and compact Lie groups,* Ph.D. thesis, Harvard University.

Index